STUDENT SOLUTIONS MANUAL

to accompany

CHEMISTRY
& Chemical Reactivity

Fourth Edition

KOTZ & TREICHEL

ALTON J. BANKS

North Carolina State University

SAUNDERS GOLDEN SUNBURST SERIES

Saunders College Publishing

Harcourt Brace College Publishers

Fort Worth Philadelphia San Diego New York Orlando Austin
San Antonio Toronto Montreal London Sydney Tokyo

Portions of this work were published in previous editions.

Printed in the United States of America

ISBN 0-03-023796-3

0 1 2 3 4 5 6 7 8 202 10 9 8 7 6 5

To the student:

The skills involved in solving chemistry problems are acquired only by discovering the conceptual paths which connect the available data to the desired piece(s) of information. These paths are discovered in different ways by different people. What is true is that those discoveries frequently require repetition. Working multiple problems is one very good approach to clarifying and solidifying the fundamental concepts. I would suggest that this **Solutions Manual** will provide maximum benefit if you consult it *after* you have attempted to solve a problem.

The selected Study Questions have been chosen by the authors of your text to allow you to discover the range and depth of your understanding of chemical concepts. The importance of mastering the "basics" cannot be overemphasized. You will find that the text, **Chemistry & Chemical Reactivity**, has a wealth of study questions to assist you in your study of the science we call Chemistry.

Many of the questions contained in your book—and this solutions manual—have multiple parts. In many cases, comments have been added to aid you in the process of gathering available data and applicable conversion factors and connecting them via those fundamental concepts which undergird this branch of science. In these multiple-step questions, you may find an answer which differs slightly from those given here. This may be a result of "rounding" intermediate answers. The procedure followed in this manual was to report intermediate answers to the appropriate number of significant figures, and to calculate the "final" answer without any intermediate rounding. In cases involving atoms and molecular masses, those quantities were expressed with one digit more than the number of digits needed for the data provided.

A word of appreciation is due to several people. Thanks go to the authors, especially Dr. John C. Kotz, for the many conversations held during the development of this manual. The many fine folks at Saunders have been very helpful. Beth Rosato, Sarah Fitz-Hugh and Peter McGahey were always helpful and efficient. Additionally I would like to thank my wife, Dr. Catherine Hamrick Banks for her invaluable assistance in typing and proofreading this manuscript. One of my blessings is to have such a person as a patient wife and a chemical colleague.

While we have worked diligently to remove all errors from this text, I am certain that some have escaped the many inspections. I accept responsibility for those errors.

I hope that you will find this **Solutions Manual** a useful addition in your exploration of chemistry.

Alton J. Banks
Department of Chemistry
North Carolina State University
Raleigh, North Carolina 27695

Table of Contents

Chapter 1
Matter and Measurement

Review Questions

4. Liquids: mercury and water Solid: copper

   Note that the liquid and mercury both conform to the shape of the test tube-one indication of the liquid phase.

   Most dense: mercury Least dense: water

6. Determine if the property is physical or chemical property for the following:

   (a) color a physical property

   (b) transformed into rust a chemical property

   (c) explode a chemical property

   (d) density a physical property

   (e) melts a physical property

   (f) green a physical property

   Physical properties are those that can be observed or measured without changing the composition of the substance. Exploding or transforming into rust result in substances which are **different** from the original substances— and represent chemical properties.

8. The number of significant figures in each of the following numbers:

   (a) 9.87 3 significant figures

   (b) 1050 3 significant figures—the absence of a decimal point following the last zero indicates the uncertainty lies in the 5.

   (c) 0.00823 3 significant figures—the leading zeroes are NOT significant

   (d) 1.67×10^{-6} 3 significant figures— written in exponential notation, only digits in front of the "x 10"..... are significant

10. Qualitative observations: colorless, clear

    Quantitative observations: 2.65 g/cm^3 ; 2.5 g ; 4.6 cm

    Extensive observations: the mass, 2.5 g; the length, 4.6 cm.

    Intensive observations: the density, 2.65 g/cm^3; the color & transparency (colorless and clear)

Numerical and Other Questions

Density

12. What mass of ethylene glycol possesses a volume of 500. mL of the liquid?

$$\frac{500.\text{ mL}}{1} \bullet \frac{1\text{ cm}^3}{1\text{ mL}} \bullet \frac{1.1135\text{ g}}{\text{cm}^3} = 557\text{ g}$$

14. Calculate the mass (in g) of 500. mL of water, given D = 0.997 g/cm^3 at 25 °C.

$$\text{Mass} = \text{D} \bullet \text{V} = \frac{0.997\text{ g}}{1\text{ cm}^3} \bullet 500.\text{ mL} = 499\text{ g}$$

Express this mass in kilograms: $499\text{ g} \bullet \frac{1\text{ kg}}{1000\text{ g}} = 0.499\text{ kg}$

16. The metal will displace a volume of water that is equal to the volume of the metal.
Hence the difference in volumes of water (20.7-6.7) corresponds to the volume of metal.
Since 1 mL = 1 cm^3, the density of the metal is then:

$$\frac{\text{Mass}}{\text{Volume}} = \frac{37.5\text{ g}}{14.0\text{ cm}^3} \text{ or } 2.68\frac{\text{g}}{\text{cm}^3}.$$

From the list of metals provided, the metal with a density closest to this is **Aluminum.**

18. Calculate the density of olive oil if 1 cup (237 mL) has a mass of 205 g:

$$\text{Since Density} = \frac{\text{Mass}}{\text{Volume}} \text{ then } \frac{205\text{ g olive oil}}{237\text{ mL olive oil}} = 0.865\text{ g/mL or } 0.865\text{ g/cm}^3$$

Temperature

20. Express 25 °C in kelvins:
K = (25 °C+ 273) or 298 kelvins

22. Make the following temperature conversions:

	°C	K
(a)	16	16 + 273.15 = 289
(b)	370 - 273	370
	$3.7 \times 10^2 - 2.73 \times 10^2 = 1.0 \times 10^2$	

	°C	K
(c)	- 40	- 40 + 273.15 = 230
		(note no decimal point after -40)

24. The accepted value for a normal human temperature is 98.6 °F.
On the Celsius scale this corresponds to:

$$°C = \frac{5}{9}(98.6 - 32) = 37\ °C$$

Since the melting point of gallium is 29.8°C, the gallium should melt in your hand.

Elements and Atoms

26. Names for the following elements:

a. C - carbon	c. Cl - chlorine	e. Mg - magnesium
b. Na - sodium	d. P - phosphorus	f. Ca - calcium

28. Symbols for the following elements:

a. lithium - Li	c. iron - Fe	e. lead - Pb
b. titanium - Ti	d. silicon - Si	f. zinc - Zn

Units and Unit Conversions

30. Convert the distance 1500 m into kilometers; into centimeters:

$$\frac{1500\ \text{m}}{1} \bullet \frac{1\ \text{km}}{1000\ \text{m}} = 1.5\ \text{km} \quad \text{and} \frac{1500\ \text{m}}{1} \bullet \frac{100\ \text{cm}}{1\ \text{m}} = 1.5 \times 10^5\ \text{cm}$$

Note that BOTH answers have only two significant figures, since 1500 (without a decimal point following the right-most zero has only two sf)

32. Express the area of a 2.5 cm x 2.1 cm stamp in cm^2; in m^2:

$$2.5\ \text{cm} \bullet 2.1\ \text{cm} = 5.3\ \text{cm}^2$$

$$5.3\ \text{cm}^2 \bullet \left(\frac{1\ \text{m}}{100\ \text{cm}}\right)^2 = 5.3 \times 10^{-4}\ \text{m}^2$$

34. Express 250. mL in cm^3; in liters(L); in m^3 :

$$\frac{250.\ mL}{1\ beaker} \bullet \frac{1\ cm^3}{1\ mL} = \frac{250.\ cm^3}{1\ beaker}$$

$$\frac{250.\ cm^3}{1\ beaker} \bullet \frac{1\ L}{1000\ cm^3} = \frac{0.250\ L}{1\ beaker}$$

$$\frac{250.\ cm^3}{1\ beaker} \bullet \frac{1\ m^3}{1 \times 10^6\ cm^3} = \frac{2.50 \times 10^{-4}\ m^3}{1\ beaker}$$

36. Convert book's mass of 2.52 kg into grams:

$$\frac{2.52\ kg}{1\ book} \bullet \frac{1 \times 10^3\ g}{1\ kg} = \frac{2.52 \times 10^3\ g}{book}$$

38. Express the dimensions 8 1/2 x 11 inches in centimeters:

$\frac{8.5\ in}{1} \bullet \frac{2.54\ cm}{1\ in} = 21.59\ cm$ or 22 cm (to 2 significant figures)

$\frac{11\ in}{1} \bullet \frac{2.54\ cm}{1\ in} = 27.94\ cm$ or 28 cm (to 2 significant figures)

The area in square centimeters would be:

$21.59\ cm \bullet 27.94\ cm = 603.2\ cm^2$ or $6.0 \times 10^2\ cm^2$ (to 2 significant figures)

Note that multiplying 22 cm x 28 cm will provide an answer of 620 cm^2. One good habit to develop is to **round once** *after* all your calculations are done. This habit will eliminate the cumulative roundoff errors that can occur.

Accuracy, Precision, and Error

40. Using the data provided, the Averages and their deviations are as follows:

Data point	Method A	deviation	Method B	deviation
1	2.2	0.2	2.703	0.777
2	2.3	0.1	2.701	0.779
3	2.7	0.3	2.705	0.775
4	2.4	0.0	5.811	2.331
Averages:	2.4	0.2	3.480	1.166

(a) The average density for method A is 2.4 ± 0.2 grams while the average density for method B is 3.480 ± 1.166 grams—if one includes all the data points. Data point 4 in Method B has a large deviation, and should probably be excluded from the calculation. If one omits data point 4, Method B gives a density of 2.703 ± 0.001 g

(b) The error for each method :

Error = experimental value - accepted value

From Method A error = (2.4 - 2.702) = - 0.3

From Method B error = (2.703 - 2.702) = 0.001 (omitting data point 4)

error = (3.480 - 2.702) = 0.778 (including all data points)

(c) Precision and Accuracy of each method:

If one counts all data points, the deviations **for all data points** of Method A are less than those for **the data points of** Method B, Method A offers better *precision* . On the other hand, omitting data point 4, Method B offers both *better accuracy* (average closer to the accepted value) and *better precision* (since the value is known to a greater number of significant figures).

Significant Figures

42. Calculate the volume (in cubic centimeters) of a backpack whose dimensions are 22.86 cm x 38.0 cm x 76 cm .

Volume = l x w x h = 22.86 cm x 38.0 cm x 76 cm = 6.6×10^4 cm^3 (2 sf are allowed)

44. Express the product of three numbers to the proper number of significant figures:

$$(1.68)(7.847)\left(\frac{1.0000}{55.85}\right) = 0.236 \qquad \text{(3 sf are allowed)}$$

46. Solve for n and report the answer to the correct number of significant figures:

$$\underset{\underset{4\text{ sf}}{\uparrow}}{\frac{\overset{\overset{3\text{ sf}}{\downarrow}}{43.7}}{760.0}} \cdot \underset{\underset{3\text{ sf}}{\uparrow}}{125} = n \cdot \underset{\underset{5\text{ sf}}{\uparrow}}{0.082057} \cdot \underset{\underset{4\text{ sf}}{\uparrow}}{298.2}$$

$$n = \frac{\frac{43.7}{760.0} \times 125}{0.082057 \times 298.2} = 0.294 \text{ to three significant figures.}$$

General Questions

48. Volume of a 1.50 carat diamond:

$$\frac{1.50 \text{ carat}}{1} \bullet \frac{0.200 \text{ g}}{1 \text{ carat}} \bullet \frac{1 \text{ cm}^3}{3.513 \text{ g}} = 0.0854 \text{ cm}^3$$

50. Separation of carbon atoms in (a) meters and (b) angstroms:

(a) $$0.154 \text{ nm} \bullet \frac{1 \text{ m}}{10^9 \text{ nm}} = 1.54 \times 10^{-10} \text{ m}$$

(b) $$1.54 \times 10^{-10} \text{ m} \bullet \frac{1 \text{ Å}}{1 \times 10^{-10} \text{ m}} = 1.54 \text{ Å}$$

52. Mass of a gold coin 2.2 cm in diameter and 3.0 mm thick:

To calculate the mass of the gram we'll need to determine the volume of the coin, and use the density of gold.

$$\text{Volume} = \pi r^2 \bullet \text{thickness} = 3.14159 \bullet (\frac{2.2 \text{ cm}}{2})^2 \bullet (\frac{3.0 \text{ mm}}{1} \bullet \frac{1\text{cm}}{10\text{mm}}) = 1.1 \text{ cm}^3$$

The mass of the coin is then: $V \bullet D = 1.1 \text{ cm}^3 \bullet \frac{19.3 \text{ g}}{\text{cm}^3} = 22 \text{ g}$ (to 2 sf)

Value of the coin:

$$\frac{\$410}{1 \text{ troy ounce}} \bullet \frac{1 \text{ troy ounce}}{31.10 \text{ g}} \bullet 22 \text{ g gold} = \$290$$

54. To calculate the density of the metal, first calculate the volume of the piece of metal:

$$2.35 \text{ cm} \bullet 1.34 \text{ cm} \bullet 1.05 \text{ cm} = 3.31 \text{ cm}^3$$

Then the density can be calculated by dividing the mass (29.454g) by the volume:

$$D = \frac{29.454 \text{ g}}{3.31 \text{ cm}^3} = 8.91 \frac{\text{g}}{\text{cm}^3}$$

Note that this answer is obtained by dividing the mass by the **unrounded** volume (3.30645 cm^3). Given this calculated density, the metal in question has to be **nickel**.

56. Which occupies a larger volume: 600 g of water or 600 g of lead ?

$$600 \text{ g } H_2O \bullet \frac{1 \text{ cm}^3}{0.995 \text{ g}} = 603 \text{ cm}^3 \quad (600 \text{ cm}^3 \text{ to 1 sf})$$

$$600 \text{ g Pb} \bullet \frac{1 \text{ cm}^3}{11.34 \text{ g}} = 52.9 \text{ cm}^3 \quad (50 \text{ cm}^3 \text{ to 1 sf})$$

58. Estimate the density of water at 20 °C :

Using the approximation that density varies linearly with temperature—over a short temperature range—the density at 20°C would be midway between the density at 15 °C and that at 25 °C.

$$D = \frac{0.99913 + 0.99707}{2} = 0.99810 \frac{\text{g}}{\text{cm}^3}$$

The suggested value is **not reasonable.**

60. Calculate the mass of platinum in 1.53 g of a compound that is 65.0% platinum:

$$\frac{1.53 \text{ g compound}}{1} \bullet \frac{65.0 \text{ g Pt}}{100.0 \text{ g compound}} = 0.995 \text{ g Pt}$$

62. Volume of 38.08% solution of sulfuric acid that contains 125 g of sulfuric acid.

Since the solution is ONLY 38.08% sulfuric, first calculate the MASS of the solution that contains 125 g of sulfuric acid:

$$125 \text{ g } H_2SO_4 \bullet \frac{100 \text{ g solution}}{38.08 \text{ g } H_2SO_4} = 328.3 \text{ g solution or 328 g solution (3 sf)}$$

Now we can calculate the **volume** of solution that has this mass:

$$125 \text{ g } H_2SO_4 \bullet \frac{100 \text{ g solution}}{38.08 \text{ g } H_2SO_4} \bullet \frac{1 \text{ cm}^3 \text{ solution}}{1.285 \text{ g solution}} = 255 \text{ cm}^3 \text{ solution}$$

Conceptual Questions

64. Indicate the relative arrangements of the particles in each of the following:

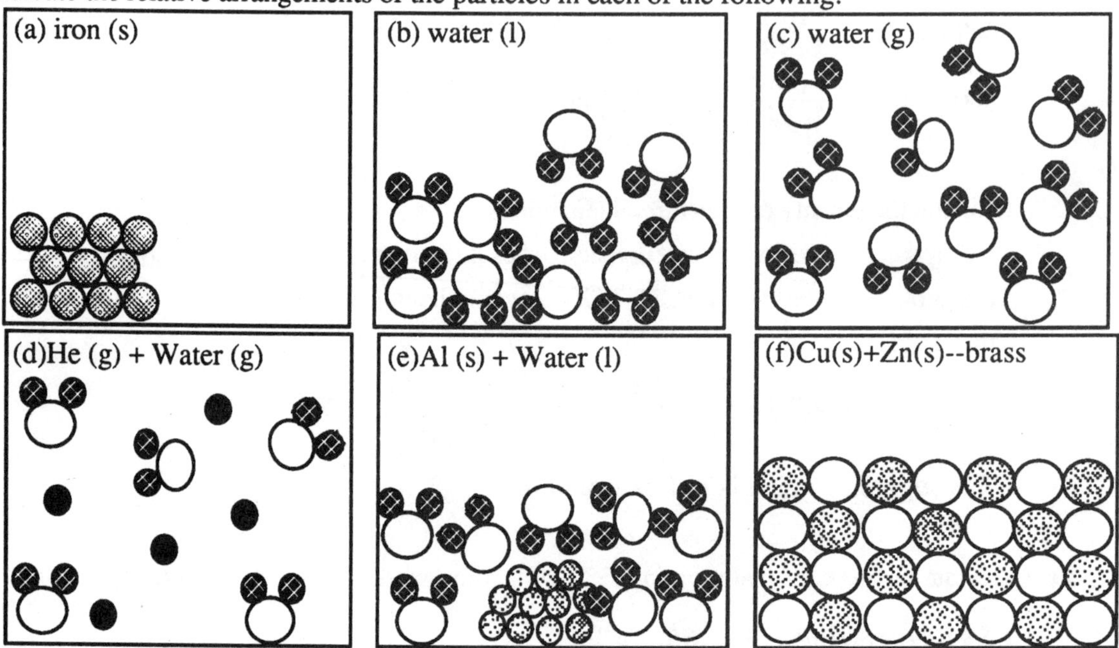

66. Experimental method to determine the identity of a liquid:

    There are several methods. One method would be to weigh an accurately known volume of the liquid. An empty dry weighed 10.0 mL graduated cylinder could be filled to the 10.0 mL mark with the liquid, and reweighed. The mass of liquid divided by the volume would provide the density of the liquid. That density could be compared with published values of the density of water *at that temperature.*

    To determine if the water contains dissolved salts, test the electrical conductivity. Pure water is a very poor conductor. Water containing dissolved salts (and therefore the ions produced when that salt dissolves) would conduct an electric current.

68. Pouring three immiscible liquids into a test tube will result in three discrete layers in which the liquids arrange themselves from the most dense liquid (at bottom) to the least dense liquid (at top).

Least dense liquid: water

Medium density: carbon tetrachloride

Most dense liquid: mercury

70. To determine if a copper-colored metal is pure copper :

One can check some of the properties that identify copper. (1) Melting a sample of the wire and comparing the melting point of the sample to the melting point of pure copper. (2) Carefully determine the density of the wire, and compare that density with the literature value for the density of pure copper. (3) Test the electrical conductivity of the wire and compare that conductivity to copper's electrical conductivity.

Challenging Questions

72. Calculate the mass of 12 ounces of aluminum in grams and from that mass, the volume:

$$\text{Volume} = \frac{\text{Mass}}{\text{Density}} = \frac{12 \text{ oz} \cdot \frac{28.4 \text{ g}}{1 \text{ oz}}}{2.70 \text{ g/cm}^3} = 130 \text{ cm}^3 \quad \text{(to 2 sf)}$$

$$\text{Volume} = \text{Area} \cdot \text{Thickness}$$

Express the area in units of cm^2, then calculate the thickness by dividing the volume by the area:

$$\text{Area} = 75 \text{ ft}^2 \cdot \left(\frac{12 \text{ in}}{1 \text{ ft}}\right)^2 \cdot \left(\frac{2.54 \text{ cm}}{1 \text{ in}}\right)^2 = 7.0 \times 10^4 \text{ cm}^2$$

then $\text{Volume} = \text{Area} \cdot \text{Thickness}$

$$130 \text{ cm}^3 = 7.0 \times 10^4 \text{ cm}^2 \cdot \text{Thickness}$$

and $\text{Thickness} = \frac{130 \text{ cm}^3}{7.0 \times 10^4 \text{ cm}^2} = 1.8 \times 10^{-3} \text{ cm or } 1.8 \times 10^{-2} \text{ mm}$

74. Calculate the volume of the "cylinder" of copper wire with a density of 8.94 g/cm^3 and a mass of (57 kg) 125 lb:

$$125 \text{ lb} \cdot \frac{454 \text{ g}}{1 \text{ lb}} \cdot \frac{1 \text{ cm}^3}{8.94 \text{ g}} = 6350 \text{ cm}^3$$

The volume of a cylinder is equal to $\pi r^2 h$.
We can calculate the "length" of the wire (the height of the cylinder) since we are told that the diameter of the wire is 9.50 mm. Expressing the diameter in centimeters and converting the diameter into a radius (1/2 the diameter) we obtain:

$$\text{Volume} = \pi r^2 h$$

$$6350 \text{ cm}^3 = 3.1415 \cdot (0.475 \text{ cm})^2 \cdot h$$

then $h = \frac{6350 \text{ cm}^3}{3.1415 \cdot (0.475 \text{ cm})^2} = 8.96 \times 10^3 \text{ cm}$

Expressing this number in meters:

$$8.96 \times 10^3 \text{ cm} \cdot \frac{1 \text{ m}}{100 \text{ cm}} = 8.96 \times 10^1 \text{m or } 9.0 \times 10^1 \text{m (2 sf)}$$

76. To determine the thickness of the oil layer, we can think about the oil layer as having a certain volume,($V = l \cdot w \cdot h$), and that our "task" is to determine the "thickness" -- or h in our formula. The volume of oil is 1 teaspoon (5 cm^3). The area covered ($l \cdot w$) is 0.5 acres. So if we divide the volume by the area, we should have the **thickness**.

$$\frac{5 \text{ cm}^3}{1 \text{ teaspoon}} \cdot \frac{1 \text{ teaspoon}}{0.5 \text{ acre}} \cdot \frac{2.47 \text{ acres}}{1.0 \times 10^4 \text{ m}^2} \cdot \left(\frac{1 \text{ m}}{100 \text{ cm}}\right)^2 = 2 \times 10^{-7} \text{ cm}$$

Volume / Area	Acre converted to square meters	Conversion of sq.m to sq. cm.

Assuming that a monolayer of oil is obtained, the thickness (2 x 10^{-7} cm) could correspond to the "length" of the oil molecules.

The Chemical Puzzler

78. When potassium reacts with water, one can write the equation to represent the reaction as:

$$2\ K\ (s)\ +2\ H_2O\ (l) \rightarrow 2\ K^+\ (aq) + 2\ OH^-\ (aq)\ +H_2\ (g)$$

(a) States of matter represented: solid: potassium -- a reactant
liquid: water-- a reactant
gas: hydrogen-- a product
solution: of KOH produced

(b) Types of changes observed: Chemical changes are exhibited—the chemical reactivity of potassium with water. One could also test the resulting solution to find it basic—since KOH is produced. The reaction produces much heat, which usually results in the hydrogen gas burning—a chemical change

Physical changes are also exhibited—the change in the color of elemental potassium metal and potassium ions. The heat would also vaporize liquid water to gaseous water.

(c) Qualitative observations: The color of the flame indicates that potassium is one of the reacting species. Litmus paper could be used to find that the solution is basic (turns red litmus paper blue). Carefully grasping the beaker at the conclusion of the reaction would demonstrate that heat is evolved as potassium reacts with water.

(d) Structure of potassium metal: Potassium atoms are stacked in a repeating array (in the picture, the atoms resemble squares stacked atop one another),with a potassium atom placed in the opening created between the two layers.

Chapter 2
Atoms and Elements

The Composition of Atoms

20. Mass number for

    a. Na (at. no. 11) with 12 neutrons : 23

    b. Ti (at. no. 22) with 26 neutrons : 48

    c. Ge (at. no. 32) with 40 neutrons : 72

    The mass number represents the SUM of the protons + neutrons in the nucleus of an atom. The atomic number represents the # of protons, so

    (atomic no. + # neutrons) = mass number

22. Mass number (A) = no. of protons + no. of neutrons;
    Atomic number (Z) = no. of protons

    a. ${}^{39}_{19}K$ b. ${}^{39}_{18}Ar$ c. ${}^{60}_{27}Co$

24.

substance	protons	neutrons	electrons
a. magnesium-24	12	12	12
b. tin- 119	50	69	50
c. plutonium-244	94	150	94

Note that the number of protons and electrons are equal for any neutral atom. The number of protons is always equal to the atomic number. The mass number equals the sum of the numbers of protons and neutrons.

26.

Symbol	${}^{58}Ni$	${}^{33}S$	^{20}<u>Ne</u>	^{55}<u>Mn</u>
Number of protons	<u>28</u>	<u>16</u>	10	<u>25</u>
Number of neutrons	<u>30</u>	<u>17</u>	10	30
Number of electrons in the neutral atom	<u>28</u>	<u>16</u>	<u>10</u>	25
Name of element	<u>nickel</u>	<u>sulfur</u>	<u>neon</u>	<u>manganese</u>

Isotopes

28. For technetium- 99 (at. no. 43) :

protons : 43
neutrons : (99 - 43) = 56
electrons : 43

30. Isotopes of cobalt (atomic number 27) with 30,31, and 33 neutrons:

would have symbols of $^{57}_{27}Co$, $^{58}_{27}Co$, and $^{60}_{27}Co$ respectively.

Atomic Mass

32. The atomic mass of lithium is:

$$(0.0750)(6.015121) + (0.9250)(7.016003) = 6.94 \text{ amu}$$

Recall that the atomic mass is a weighted average of all isotopes of an element, and is obtained by **adding** the *product* of (relative abundance x mass) for all isotopes.

34. The average atomic weight of gallium is 69.723. If we let **x** represent the abundance of the lighter isotope, and **(1-x)** the abundance of the heavier isotope, the expression to calculate the atomic weight of gallium may be written:

$$(x)(68.9257) + (1 - x)(70.9249) = 69.723$$

[Note that the sum of all the isotopic abundances must add to 100% -- or 1 (in decimal notation.] Simplifying the equation gives:

$$68.9257 \text{ amu } x + 70.9249 \text{ amu} - 70.9249 \text{ amu } x = 69.723 \text{ amu}$$
$$-1.9992 \text{ amu } x = (69.723 \text{ amu} - 70.9249)$$
$$-1.9992 \text{ amu } x = -1.202 \text{ amu}$$
$$x = 0.6012$$

So the relative abundance of isotope 69 is 60.12 % and that of isotope 71 is 39.88 %.

36. Thallium has two stable isotopes ^{203}Tl and ^{205}Tl. The more abundant isotope is:________

The atomic weight of thallium is 204.3833. The fact that this weight is closer to 205 than 203 indicates that the 205 isotope is the more abundant.

The Periodic Table

38. The elements in Group 5A are:
    nitrogen (N - nonmetal), phosphorus (P - nonmetal), arsenic (As - metalloid), antimony (Sb - metalloid), bismuth (Bi - metal)

40. Periods with 8 elements:2 Periods 2 (at.no. 3-10) and 3 (at.no. 11-18)
    Periods with 18 elements:2 Periods 4 (at.no. 19-36) and 5(at.no. 37-54)
    Periods with 32 elements:1 Period 6 (at.no. 55-86)
    [and possibly Period 7—although the full 32 are not presently known]

42. Some of the other Group 2A elements have oxides with formula: MgO, CaO—so the formula for the oxide of barium is expected to be: <u>BaO</u>.

General Questions

44. Given that the average atomic mass for potassium is 39.0983 amu, the lighter isotope, ^{39}K would be the more abundant of the remaining isotopes. Remember that atomic masses are **weighted**, that is the average atomic mass is closer to the most abundant isotope.

46. Crossword puzzle:

    Clues:

    1-2 A metal used in ancient times: tin (Sn)
    3-4 A metal that burns in air and is found in Group 5A: bismuth (Bi)

    1-3 A metalloid: antimony (Sb)
    2-4 A metal used in U.S. coins: nickel (Ni)

    1. A colorful nonmetal: sulfur (S)
    2. A colorless gaseous nonmetal: nitrogen (N)
    3. An element that makes fireworks green: boron (B)
    4. An element that has medicinal uses: iodine (I)

    1-4 An element used in electronics: silicon (Si)

2-3 A metal used with Zr to make wires for superconducting magnets: niobium (Nb)

Using these solutions, the following letters fit in the boxes:

1 S	2 N
3 B	4 I

48. a. Three elements in the series with the greatest density: At. no. 27 (Cobalt), At. no. 28 (Nickel), and At. no. 29 (Copper). The density of all three **metals** is approximately 9 g/cm^3.

b. The element in the second period with the largest density is Boron (atomic number 5) while the element in the third period with the largest density is Aluminum (atomic number 13). Both of these elements belong to group 3A.

c. Elements, from the first 36 elements, with very **low densities** are **gases**. These include Hydrogen, Helium, Nitrogen, Oxygen, Fluorine, Neon, Chlorine, Argon and Krypton.

50. a. An element in Group 2A : beryllium, magnesium, calcium, strontium, barium, radium

b. An element in the third period: sodium, magnesium, aluminum, silicon, phosphorus, sulfur, chlorine, argon

c. An element in the 2nd period in Group 4A: carbon

d. An element in the third period in Group 6A: sulfur

e. A halogen in the fifth period: iodine

f. An alkaline earth element in the third period: magnesium

g. A noble gas element in the fourth period: krypton

h. A nonmetal in Group 6A and the third period: sulfur

i. A metalloid in the fourth period: germanium or arsenic

Conceptual Questions

52. One would predict that calcium's analogues in Group 2A would also form 1 : 1 oxides, MO.

54. In general the relative abundance of elements 1-36 decreases as the atomic number increases.
    One exception to this trend occurs in the "even" elements, with iron (atomic number 26) being a bit more abundant than surrounding elements.
    A difference in the abundances of the "even" and "odd" elements in this range is that the "odd" elements are less abundant than the "even" elements with atomic number ±1.

56.
    a. Sample of a solid element : **5** (regularly distributed particles, closely packed)
    b. Sample of a liquid element : **4** (particles of one kind, conforming to the walls of the container)
    c. Sample of oxygen gas: **2** (one type of particle,widely separated)
    d. Sample of solid bronze—Cu and Zn: **3** (particles of two kinds, intimately mixed)
    e. Sample of a mixture of nitrogen & neon gas: **1** (two types of particles,widely separated)

Structural Chemistry

58. Distance between oxygen atoms in ozone:
    Distances the same or different?
    According to the CAChe viewer, using the model shown to the right, the O-O distances are 1.160 Angstroms and 1.161 Angstroms—the **same distance**

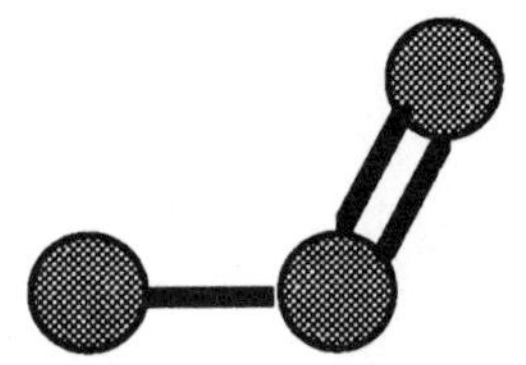

    What is the O-O-O angle ? The angle is 120.8 degrees.

60. Compare structures of diamond and silicon.
Describe similarities and differences.

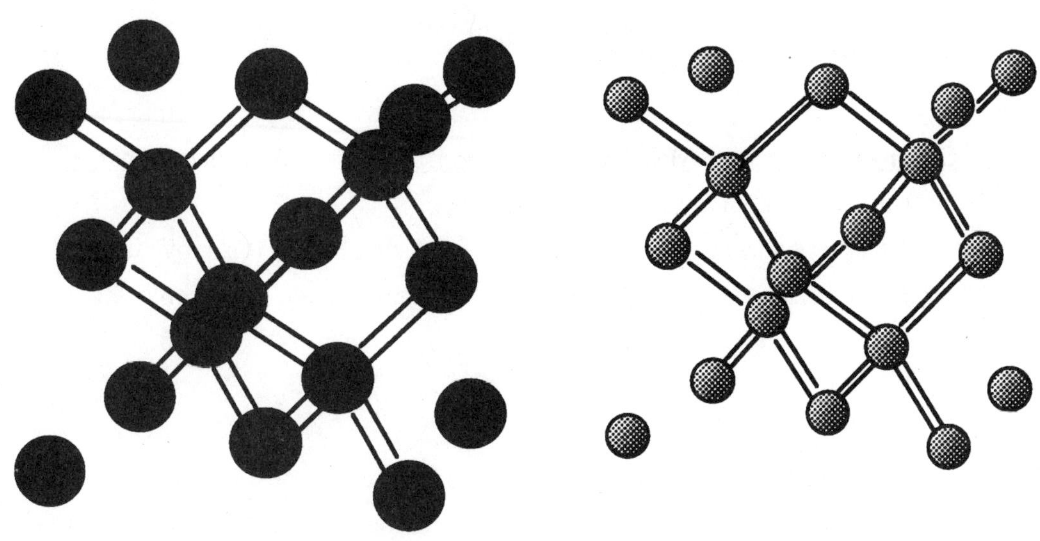

Diamond Silicon

Above are models of diamond and silicon. Both have similar atomic arrangements, and both are network solids. Since **both** carbon and silicon are in succeeding periods and are both from Group 4A, this should not be totally surprising. The Si-Si distance is 3.57 Angstroms. The C-C distance is 2.35 Angstroms. Carbon is referred to as a nonmetal, while Silicon is referred to as a metalloid.

Challenging Questions

62. The ratio of the atomic masses of P/O :
0.744 g of P combined with 0.960 g of O. (1.704 - 0.744)

Given that the compound formed is P_4O_{10}, the masses of phosphorus and oxygen correspond to the ratio of 4 phosphorus atoms to 10 oxygen atoms. Dividing the masses by the appropriate number of atoms

$$\frac{0.744 \text{ g P}}{4} = 0.186 \text{ g P} \qquad \text{and} \qquad \frac{0.960 \text{ g O}}{10} = 0.0960 \text{ g O}$$

So the mass of $\frac{P}{O}$ is $\frac{0.186 \text{ g P atoms}}{0.0960 \text{ g O atoms}} = 1.94$

If the atomic mass of oxygen is 16.000 amu, the atomic mass of phosphorus is

$$16.000 \text{ amu} \cdot \frac{1.94 \text{ g P atoms}}{1.0 \text{ g O atoms}} = 31.0 \text{ amu}$$

64. Data:

Oil Drop	Measured Charge on Drop(C)	Relative charge
1	1.59×10^{-19}	1
2	11.1×10^{-19}	7
3	9.54×10^{-19}	6
4	15.9×10^{-19}	10
5	6.36×10^{-19}	4

Millikan's thesis was that one couldn't be certain to have **only one** electron on an oil droplet, but that the **smallest** number of electrons (and hence the smallest charge) was **one**. Begin by dividing ALL of the data by the *smallest charge* (the relative charge).
a. Hence, the smallest charge found was on **drop 1**, so one could conclude that the charge on the electron is 1.59×10^{-19} Coulombs.

b. The number of electrons on each drop are found in the right-most column above. (1,7,6,10,4)

c. To calculate the average deviation and error, one would normally average the 5 data points. However since the oil droplets have varying numbers of electrons, we should first calculate the average charge for each droplet. Doing so reveals that the **average value** for the charge on the electron is **1.59×10^{-19} Coulombs,** and *to two significant values* the average deviation is zero.
The error is (experimentally determined - accepted value) or (1.59×10^{-19} - 1.60×10^{-19}) or -0.01×10^{-19}, or a percent error = 0.6%.

Chapter 3
Molecules and Compounds

Review Questions

2. The molecular formula for cis-Platin is : $Pt(NH_3)_2Cl_2$ contains

   N atoms in one molecule: 2 nitrogen atoms (1 in each of two ammonia groups)

   H atoms in one molecule: 6 hydrogen atoms (3 in each of two ammonia groups)

   H atoms in one mole of cis-platin:

   One mole of cis-platin contains Avogadro's # of molecules (6.023×10^{23} molecules) so the number of hydrogen atoms is $6 \cdot 6.023 \times 10^{23} = 3.61 \times 10^{24}$.

   The molar mass of $Pt(NH_3)_2Cl_2$ is 300.0 g/mol

4. The **number of electrons** in a strontium atom is **38**, the same as the atomic number for the element. When an atom of strontium forms an ion, it **loses two electrons,** forming an ion having the same electron configuration as the noble gas **krypton.**

6. The molecule vanillin has a molecular formula of $C_8H_8O_3$.

   a. The molar mass of vanillin is 152.14. One can calculate the # of moles of vanillin by dividing the mass (in g) by 152.14.

   $$40 \text{ g} \cdot \frac{1 \text{ mol } C_8H_8O_3}{152.14 \text{ g } C_8H_8O_3} = 0.26 \text{ mol of } C_8H_8O_3$$

   3.0×10^{23} molecules of vanillin would be 1/2 a mole (Since one mole of vanillin would contain Avogadro's number of molecules), and would have a mass of $1/2 \cdot 152.14$ or 76 grams— greater than 40 grams.

   b. Styrene, graphite, and vanillin all share the feature of having 6-member C rings.

8. To determine whether 0.5 mol of sodium or silicon has the **larger mass**, one need only look at the atomic masses of sodium (23 g/mol) and silicon (29 g/mol). One-half mole of sodium would have a mass of about 12.5 g while the same amount of silicon would have a mass of about 14.5 g—so 0.5 mol of silicon has the **larger mass.**

10. An empirical formula indicates the **ratio** of atoms present in a molecule, while the molecular formula represents **both** the **ratio and the number** of atoms of each type present in the molecule.

With C_2H_6 being the molecular formula for ethane, C_1H_3 (or CH_3) is the empirical formula.

Numerical and Other Questions

Molecular Formulas

12. The molecular formula for the compounds:
    a. 4 boron atoms and 10 hydrogen atoms per molecule: B_4H_{10}
    b. 6 carbon atoms, 8 hydrogen atoms, and 6 oxygen atoms per molecule: $C_6H_8O_6$
    c. 14 carbon atoms, 18 hydrogen atoms, 2 nitrogen atoms, and 5 oxygen atoms per molecule: $C_{14}H_{18}N_2O_5$

14. Total atoms of each element in a formula unit:

	element	# atoms
a. CaC_2O_4	Ca	1
	C	2
	O	4
b. $C_6H_5CHCH_2$	C	8
	H	8
c. $Cu_2CO_3(OH)_2$	Cu	2
	C	1
	O	5
	H	2
d. $Co(NH_3)_6Cl_3$	Co	1
	N	6
	H	18
	Cl	3
e. $K_4Fe(CN)_6$	K	4
	Fe	1
	C	6
	N	6

16. Molecular formula for alanine and lactic acids:
    The molecular formula may be written by adding all atoms of a particular element:
    alanine : $C_3H_7NO_2$ lactic acid : $C_3H_6O_3$

Molecular Models

18. The formula for sulfuric acid is H_2SO_4. The molecule is **not flat**. The O atoms are arranged around the sulfur at the corners of a tetrahedron—that is the O-S-O angle would be about 109 degrees. The hydrogen atoms are connected to two of the oxygen atoms also with angles of approximately 109 degrees.

Ions and Ion Charges

20. Most commonly observed ion for:
    a. Magnesium—like all the alkaline earth metals : +2
    b. Zinc : +2
    c. Iron : +3 (although iron—as most transition metals—can form more than one ion—in this case also +2)
    d. Gallium : +3 (an analogue of Aluminum)

22. The symbol and charge for the following ions:

a. strontium ion	Sr^{+2}
b. titanium(IV) ion	Ti^{+4}
c. aluminum ion	Al^{+3}
d. hydrogen carbonate ion	HCO_3^{-1}
e. sulfide ion	S^{-2}
f. perchlorate ion	ClO_4^{-1}
g. cobalt(II) ion	Co^{+2}
h. ammonium ion	NH_4^{+1}

24. When potassium becomes a monatomic ions, potassium—like all alkali metals—**loses 1 electron.** The noble gas atom with the same number of electrons as the potassium ion is **argon**.

Ionic Compounds

26. Barium is in Group 2A, and is expected to form a +2 ion while bromine is in Group 7A and expected to form a -1 ion. Since the compound would have to have an **equal amount** of negative and positive charges, the formula would be $BaBr_2$.

28. Formula, Charge, and Number of ions in:

	cation	# of	anion	# of
a. K_2S	K^{+1}	2	S^{-2}	1
b. $NiSO_4$	Ni^{+2}	1	SO_4^{-2}	1
c. $(NH_4)_3PO_4$	NH_4^{+1}	3	PO_4^{-3}	1
d. $Ca(ClO)_2$	Ca^{+2}	1	ClO^{-1}	2
e. $KMnO_4$	K^{+1}	1	MnO_4^{-1}	1

30. Cobalt oxide

Cobalt(II) oxide	CoO	cobalt ion :	Co^{+2}
Cobalt(III) oxide	Co_2O_3		Co^{+3}

32. Provide correct formulas for compounds:
    a. $AlCl_3$ The tripositive aluminum ion requires three chloride ions.
    b. NaF Sodium is a monopositive cation. Fluoride is a mononegative anion.
    c. and d. are correct formulas

Coulomb's Law

34. Coulomb's Law states that the *force of attraction* between oppositely charged ions *increases as the distance between the ions becomes smaller.* Since the distance between the sodium ion and fluoride ion would be **smaller than** the distance between the sodium ion and iodide ion, the attractive forces would be **greater** between Na^+ and I^- than between Na^+ and F^-.

Naming Ionic Compounds

36. Names for the ionic compounds

a. K_2S	potassium sulfide
b. $NiSO_4$	nickel(II) sulfate
c. $(NH_4)_3PO_4$	ammonium phosphate
d. $Ca(ClO)_2$	calcium hypochlorite

38. Formulas for the ionic compounds

a. ammonium carbonate	$(NH_4)_2CO_3$
b. calcium iodide	CaI_2
c. copper(II) bromide	$CuBr_2$
d. aluminum phosphate	$AlPO_4$
e. silver(I) acetate	$AgCH_3CO_2$

40. Names and formulas for ionic compounds

cation	anion CO_3^{2-}	anion Br^-	anion NO_3^-
K^+	K_2CO_3 potassium carbonate	KBr potassium bromide	KNO_3 potassium nitrate
Ba^{2+}	$BaCO_3$ barium carbonate	$BaBr_2$ barium bromide	$Ba(NO_3)_2$ barium nitrate
NH_4^+	$(NH_4)_2CO_3$ ammonium carbonate	NH_4Br ammonium bromide	NH_4NO_3 ammonium nitrate

Naming Binary, Nonmetal Compounds

42. Names of binary nonionic compounds

a. NF_3	nitrogen trifluoride
b. HI	hydrogen iodide
c. BBr_3	boron tribromide
d. PF_5	phosphorus pentafluoride

44. Formulas for:

a. sulfur dichloride SCl_2
b. dinitrogen pentaoxide N_2O_5
c. silicon tetrachloride $SiCl_4$
d. diboron trioxide B_2O_3

Atoms, Elements, and the Mole

46. Calculate the mass, in grams, of:

a. 2.5 mol B $2.5 \text{ mol} \cdot \frac{10.811 \text{ g B}}{1 \text{ mol B}}$ = 27 g B (2 sf)

b. 1.25×10^{-3} mol Fe $1.25 \times 10^{-3} \text{ mol Fe} \cdot \frac{55.847 \text{ g Fe}}{1 \text{ mol Fe}}$ = 0 .0698 g Fe (3 sf)

c. 0.015 mol O_2 $0.015 \text{ mol } O_2 \cdot \frac{31.9988 \text{ g } O_2}{1 \text{ mol } O_2}$ = 0.48 g O_2 (2 sf)

d. 653 mol He $653 \text{ mol He} \cdot \frac{4.0026 \text{ g He}}{1 \text{ mol He}}$ = 2610 g He (3 sf)

48. Calculate the number of moles represented by :

a. 127.08 g Cu $127.08 \text{ g Cu} \cdot \frac{1 \text{ mol Cu}}{63.546 \text{ g Cu}}$ = 1.9998 mol Cu (5 sf)

b. 0.012 g K $0.012 \text{ g K} \cdot \frac{1 \text{ mol K}}{39.0983 \text{ g K}}$ = 3.1×10^{-4} mol K (2 sf)

c. 5.0 mg Am $5.0 \text{ mg Am} \cdot \frac{1 \text{ g Am}}{1000 \text{ mg Am}} \cdot \frac{1 \text{ mol Am}}{243 \text{ g Am}}$ = 2.1×10^{-5} mol Am (2 sf)

d. 6.75 g Al $6.75 \text{ g Al} \cdot \frac{1 \text{ mol Al}}{26.9815 \text{ g Al}}$ = 0.250 mol Al (3 sf)

50. Number of moles of Kr in 0.00789 g Kr:

$$0.00789 \text{ g Kr} \cdot \frac{1 \text{ mol Kr}}{83.80 \text{ g Kr}} = 9.42 \times 10^{-5} \text{ mol Kr (3 sf)}$$

$$9.42 \times 10^{-5} \text{ mol Kr} \cdot \frac{6.022 \times 10^{23} \text{ atoms Kr}}{1 \text{ mol Kr}} = 5.67 \times 10^{19} \text{ atoms Kr}$$

52. Average mass of one copper atom:

One mole of copper (with a mass of 63.546 g) contains 6.0221×10^{23} atoms. So the average mass of **one** copper atom is:

$$\frac{63.546 \text{ g Cu}}{6.0221 \times 10^{23} \text{ atoms Cu}} = 1.0552 \times 10^{-22} \text{ g/Cu atom}$$

54. The volume of a cube of Na containing 0.125 mol Na:

First we need to know the **mass** of 0.125 mol Na:

$$0.125 \text{ mol Na} \bullet \frac{22.99 \text{ g Na}}{1 \text{ mol Na}} = 2.87 \text{ g Na (3 sf)}$$

Now we can calculate the volume that contains 2.87 g Na: (using the density given)

$$2.87 \text{ g Na} \bullet \frac{1 \text{ cm}^3}{0.968 \text{ g Na}} = 2.97 \text{ cm}^3$$

If the cube is a perfect cube (that is each side is equivalent in length to any other side), what is the length of one edge ?

$2.97 \text{ cm}^3 = 1 \bullet 1 \bullet 1$ so 1.44 cm = length of one edge

Molecules, Compounds, and the Mole

56. Molar mass of the following: (with atomic weights expressed to 4 significant figures)

a. Fe_2O_3 (2)(55.85) + (3)(16.00) = 159.7

b. BCl_3 (1)(10.81) + (3)(35.45) = 117.2

c. $Ni(NO_3)_2 \bullet 6\,H_2O$ (1)(58.69) + (2)(14.01) + (12)(1.008) + (12)(16.00) = 290.8

d. $C_6H_8O_6$ (6)(12.01) + (8)(1.008) + (6)(16.00) = 176.1

58. Moles represented by 1.00 g of the compounds:

Compound	Molar mass	Moles in 1.00 g
a. C_3H_7OH	60.10	0.0166
b. $C_{11}H_{16}O_2$	180.2	0.00555
c. $MgSO_4 \bullet 7\,H_2O$	246.48	0.00406

60. Moles of acetonitrile in 2.50 kg:

1. Molar mass of CH_3CN:

(2)(12.01) + (3)(1.008) + (1)(14.01) = 41.05 g/mol

2. Moles:

$$2.50 \times 10^3 g \cdot \frac{1 \text{ mol } CH_3CN}{41.05 \text{ g}} = 60.9 \text{ mol } CH_3CN$$

62. To begin, express the mass of aspirin, sodium hydrogen carbonate, and citric acid in grams. Convert those masses to moles, using the molar mass of each substance.

a. Moles of aspirin:

$$0.324 \text{ g } C_9H_8O_4 \cdot \frac{1 \text{ mol } C_9H_8O_4}{180.16 \text{ g } C_9H_8O_4} = 1.80 \times 10^{-3} \text{ mol } C_9H_8O_4$$

Moles of sodium hydrogen carbonate:

$$1.904 \text{ g } NaHCO_3 \cdot \frac{1 \text{ mol } NaHCO_3}{84.007 \text{ g } NaHCO_3} = 2.266 \times 10^{-2} \text{ mol } NaHCO_3$$

Moles of citric acid:

$$1.000 \text{ g } C_6H_8O_7 \cdot \frac{1 \text{ mol } C_6H_8O_7}{192.13 \text{ g } C_6H_8O_7} = 5.205 \times 10^{-3} \text{ mol } C_6H_8O_7$$

b. Molecules of aspirin per tablet:

$$1.80 \times 10^{-3} \text{ mol } C_9H_8O_4 \cdot \frac{6.022 \times 10^{23} \text{ molecules aspirin}}{1 \text{ mol aspirin}}$$

$$= 1.08 \times 10^{21} \text{ molecules aspirin}$$

Percent Composition

64. Mass percent for: [4 significant figures]

a. PbS: (1)(207.2) + (1)(32.06) = 239.3 g/mol

$$\%Pb = \frac{207.2 \text{ g Pb}}{239.3 \text{ g PbS}} \cdot 100 = 86.60\ \%$$

$$\%S = 100.00 - 86.60 = 13.40\ \%$$

b. C_3H_8: (3)(12.01)+(8)(1.008) = 44.09 g/mol

$$\%C = \frac{36.03 \text{ g C}}{44.09 \text{ g } C_3H_8} \cdot 100 = 81.71\%$$

$$\%H = 100.00 - 81.71 = 18.29\%$$

c. NH_4NO_3: (2)(14.01) + (4)(1.008) + (3)(16.00) = 80.05 g/mol

$$\%H = \frac{4.032 \text{ g H}}{80.05 \text{ g } NH_4NO_3} \cdot 100 = 5.037 \%H$$

$$\%O = \frac{48.00 \text{ g O}}{80.05 \text{ g } NH_4NO_3} \cdot 100 = 59.96 \%O$$

%N = 100.00 - (5.037 + 59.96) = 35.00 %N

66. For capsaicin, $C_{18}H_{27}NO_3$

a. Molar mass:

(18) (12.01) + (27)(1.008) + (1)(14.01) + (3)(16.00) = 305.4 g/mol

b. Moles of capsaicin corresponding to 55 mg

$$\frac{55 \times 10^{-3} \text{ g capsaicin}}{305.4 \text{ g/mol}} = 1.8 \times 10^{-4} \text{ mol capsaicin.}$$

c. Mass percent of each element:

$$\% C = \frac{216.2 \text{ g C}}{305.4 \text{ g } C_{18}H_{27}NO_3} \cdot 100 = 70.79 \% C$$

$$\% H = \frac{27.22 \text{ g H}}{305.4 \text{ g } C_{18}H_{27}NO_3} \cdot 100 = 8.912 \% H$$

$$\% N = \frac{14.01 \text{ g N}}{305.4 \text{ g } C_{18}H_{27}NO_3} \cdot 100 = 4.587 \% N$$

$$\% O = \frac{48.00 \text{ g O}}{305.4 \text{ g } C_{18}H_{27}NO_3} \cdot 100 = 15.71 \% O$$

d. Mass of C (in mg) in 55 mg of $C_{18}H_{27}NO_3$

$$55 \text{ mg } C_{18}H_{27}NO_3 \cdot \frac{70.79 \text{ g C}}{100.00 \text{ g } C_{18}H_{27}NO_3} = 39 \text{ mg C}$$

Empirical and Molecular Formulas

68. The empirical formula ($C_2H_3O_2$) would have a mass of 59.04 g.

Since the molar mass is 118.1 g/mol we can write

$$\frac{1 \text{ empirical formula}}{59.04 \text{ g succinic acid}} \cdot \frac{118.1 \text{ g succinic acid}}{1 \text{ mol succinic acid}} = \frac{2.0 \text{ empirical formulas}}{1 \text{ mol succinic acid}}$$

So the molecular formula contains 2 empirical formulas (2 • $C_2H_3O_2$) or $C_4H_6O_4$.

70. Calculate the empirical formula of acetylene by calculating the atomic ratios of carbon and hydrogen in 100 g of the compound.

$$92.26 \text{ g C} \bullet \frac{1 \text{ mol C}}{12.011 \text{ g C}} = 7.681 \text{ mol C}$$

$$7.74 \text{ g H} \bullet \frac{1 \text{ mol H}}{1.008 \text{ g H}} = 7.678 \text{ mol H}$$

Calculate the atomic ratio: $\frac{7.68 \text{ mol C}}{7.68 \text{ mol H}} = \frac{1 \text{ mol C}}{1 \text{ mol H}}$

The atomic ratio indicates that there is 1 C atom for 1 H atom (1:1). The empirical formula is then CH. The formula mass is 13.01. Given that the molar mass of the compound is 26.02 g/mol, there are two formula units per molecular unit, hence the molecular formula for acetylene is C_2H_2.

72. Calculate the empirical formula of a nitrogen oxide which is 36.84 % N.

This compound must contain (100.00-36.84) 63.16 % O. With this knowledge we can calculate the molar ratio of atoms in 100 g of the compound.

$$63.16 \text{ g O} \bullet \frac{1 \text{ mol O}}{16.00 \text{ g O}} = 3.948 \text{ mol O}$$

$$36.84 \text{ g N} \bullet \frac{1 \text{ mol N}}{14.01 \text{ g N}} = 2.630 \text{ mol N}$$

Calculating the ratio of atoms:

$$\frac{3.948 \text{ mol O}}{2.630 \text{ mol N}} = \frac{1.5 \text{ mol O}}{1 \text{ mol N}}$$

The empirical formula of this oxide indicates that there are 1.5 O atoms for each N atom. Since we cannot have a fractional part of an atom we write the empirical formula to express this ratio—N_2O_3.

74. Empirical and Molecular formula for Mandelic Acid:

$$63.15 \text{ g C} \bullet \frac{1 \text{ mol C}}{12.0115 \text{ g C}} = 5.258 \text{ mol C}$$

$$5.30 \text{ g H} \bullet \frac{1 \text{ mol}}{1.0079 \text{ g H}} = 5.28 \text{ mol H}$$

$$31.55 \text{ g O} \bullet \frac{1 \text{ mol O}}{15.9994 \text{ g O}} = 1.972 \text{ mol O}$$

Using the smallest number of atoms, we calculate the ratio of atoms:

$$\frac{5.258 \text{ mol C}}{1.972 \text{ mol O}} = \frac{2.666 \text{ mol C}}{1 \text{ mol O}} \quad \text{or} \quad \frac{2\ 2/3 \text{ mol C}}{1 \text{ mol O}} \quad \text{or} \quad \frac{8/3 \text{ mol C}}{1 \text{ mol O}}$$

So 3 mol O combine with 8 mol C and 8 mol H so the empirical formula is $C_8H_8O_3$. The formula mass of $C_8H_8O_3$ is 152.15. Given the data that the molar mass is 152.14 g/mL, the molecular formula for mandelic acid is $C_8H_8O_3$.

76. Empirical and molecular formula for cacodyl :

$$22.88 \text{ g C} \cdot \frac{1 \text{ mol C}}{12.011 \text{ g C}} = 1.905 \text{ mol C}$$

$$5.76 \text{ g H} \cdot \frac{1 \text{ mol H}}{1.008 \text{ g H}} = 5.71 \text{ mol H}$$

$$71.36 \text{ g As} \cdot \frac{1 \text{ mol As}}{74.922 \text{ g As}} = 0.9525 \text{ mol As}$$

Expressing these values as ratios of each element to one element we obtain:

$$\frac{1.905 \text{ mol C}}{0.9525 \text{ mol As}} = \frac{2.000 \text{ mol C}}{1 \text{ mol As}}$$

$$\frac{5.71 \text{ mol H}}{0.9525 \text{ mol As}} = \frac{5.99 \text{ mol H}}{1 \text{ mol As}}$$

From these ratios we know that for **each As atom** in the molecule there are **2 C atoms** and **6 H atoms**, for an empirical formula of AsC_2H_6.
Adding the atomic weights of the empirical formula gives a mass of :

empirical formula = 1(74.922) + 2(12.011) + 6(1.008) = 104.99

$$\frac{210 \text{ g cacodyl}}{1 \text{ mol cacodyl}} \cdot \frac{1 \text{ empirical formula}}{104.99 \text{ g cacodyl}} = \frac{2 \text{ empirical formulas}}{1 \text{ mol cacodyl}}$$

Multiplying each element in the empirical formula by two yields a molecular formula of $As_2C_4H_{12}$.

78. The amount of water present in the 4.74 g sample of alum is: 2.16 g water

$$(4.74 \text{ g sample} - 2.16 \text{ g water}) = 2.58 \text{ g anhydrous } KAl(SO_4)_2$$

$$2.16 \text{ g water} \bullet \frac{1 \text{ mol water}}{18.02 \text{ g water}} = 0.120 \text{ mol water}$$

Calculate the number of moles of the anhydrous salt:

$$2.58 \text{ g } KAl(SO_4)_2 \bullet \frac{1 \text{ mol } KAl(SO_4)_2}{258.2 \text{ g } KAl(SO_4)_2} = 0.00999 \text{ mol } KAl(SO_4)_2$$

Calculating the ratios of water to anhydrous salt gives:

$$\frac{0.120 \text{ mol water}}{0.00999 \text{ mol } KAl(SO_4)_2} = \frac{12 \text{ mol water}}{1 \text{ mol } KAl(SO_4)_2}$$

so there are **12 molecules of water** for each formula unit of $KAl(SO_4)_2$.

80. Given the masses of xenon involved, we can calculate the number of moles of the element:

$$0.526 \text{ g Xe} \bullet \frac{1 \text{ mol Xe}}{131.29 \text{ g Xe}} = 0.00401 \text{ mol Xe}$$

The mass of fluorine present is: 0.678 g compound - 0.526 g Xe = 0.152 g F

$$0.152 \text{ g F} \bullet \frac{1 \text{ mol F}}{19.00 \text{ g F}} = 0.00800 \text{ mol F}$$

Calculating atomic ratios:

$$\frac{0.00800 \text{ mol F}}{0.00401 \text{ mol Xe}} = \frac{2 \text{ mol F}}{1 \text{ mol Xe}}$$ indicating that the empirical formula is XeF_2 .

Using Formulas

82. Galena has the formula PbS. The percentage of galena that is lead is :

$$\% \text{ Pb} = \frac{207.2 \text{ g Pb}}{239.27 \text{ g PbS}} \times 100 = 86.60 \% \text{ Pb}$$

To produce 2.00 kg Pb:

$$2.00 \text{ kg Pb} \bullet \frac{100.0 \text{ g PbS}}{86.60 \text{ g Pb}} = 2.31 \text{ kg PbS}$$

84. $Ca_3(PO_4)_2 + C + SiO_2 \rightarrow P_4$ (not a complete equation)

Calculate the % P in calcium phosphate:

$$\% P = \frac{61.95 \text{ g P}}{310.18 \text{ g } Ca_3(PO_4)_2} \times 100 = 19.97 \% P$$

To produce 15.0 kg of phosphorus:

$$15.0 \text{ kg P} \bullet \frac{100.0 \text{ g } Ca_3(PO_4)_2}{19.97 \text{ g P}} = 75.1 \text{ kg } Ca_3(PO_4)_2$$

General Questions

86. The number of water molecules in 0.05 mL of water:

$$0.05 \text{ mL water} \bullet \frac{1 \text{ cm}^3}{1 \text{ mL}} \bullet \frac{1.00 \text{ g water}}{1 \text{ cm}^3 \text{ water}} \bullet \frac{1 \text{ mol water}}{18.02 \text{ g water}} \bullet \frac{6.02 \times 10^{23} \text{ molecules}}{1 \text{ mol}} =$$

1.67×10^{21} molecules of water (2×10^{21} —to 1 sf)

88. The question resolves into asking which of the two compounds, iron(II) sulfate or iron (II) gluconate, has the **greater percentage of iron** :
For $FeSO_4$ the molar mass is 151.9 g/mol, while that for $Fe(C_6H_{11}O_7)_2$ is 446.1 g/mol.

$$\% \text{ Fe in iron(II) sulfate} = \frac{55.85 \text{ g Fe}}{151.9 \text{ g compound}} \bullet 100 = 36.76 \% \text{ Fe}$$

$$\% \text{ Fe in iron(II) gluconate} = \frac{55.85 \text{ g Fe}}{446.1 \text{ g compound}} \bullet 100 = 12.52 \% \text{ Fe}$$

In 100. mg of each compound there would be 36.8 mg of Fe in iron(II) sulfate and 12.5 mg of Fe in iron(II) gluconate.

90. The empirical formula for iron oxalate:

0.109 g of the compound contain 17.46% iron or (0.1746 x 0.109 = 0.01903) g iron
The law of mass action tells us that the rest is oxalate:
(0.109 g compound - 0.019 g iron) = 0.090 g oxalate ($C_2O_4{}^{2-}$)
Convert these masses into moles of iron and oxalate:

$$\frac{0.019 \text{ g Fe}}{55.847 \text{ g/mol}} = 3.40 \times 10^{-4} \text{ moles of iron}$$

$$\frac{0.090 \text{ g oxalate}}{88.02 \text{ g/mol}} = 1.02 \times 10^{-3} \text{ moles of } C_2O_4{}^{2-}$$

The ratio of oxalate to iron is then: $\frac{1.02 \times 10^{-3} \text{ mol } C_2O_4{}^{2-}}{3.40 \times 10^{-4} \text{ mol iron}} = 3.00$

So the empirical formula for the compound is $Fe(C_2O_4)_3$.

[*Note: Using 66.60 % iron, the formula* FeC_2O_4 is obtained.]

92. β-D-Ribose has the molecular formula : $C_5H_{10}O_5$. Carbohydrates in general are compounds of carbon and water (hydrate). Rewriting the formula gives $C_5(H_2O)_5$. So each **mole of ribose** that decomposes into carbon and water gives **5 moles of water**.

1) Calculate the number of moles of ribose in 10.0 g of ribose,

2) Multiply the number of moles of ribose by 5 (to obtain moles of water) and

3) Convert moles of water into mass of water:

$$10.0 \text{ g} \bullet \frac{1 \text{ mol ribose}}{150.1 \text{ g ribose}} \bullet \frac{5 \text{ mol water}}{1 \text{ mol ribose}} \bullet \frac{18.02 \text{ g water}}{1 \text{ mol water}} = 6.00 \text{ g water}$$

94. Ionic compounds; formulas and names

c. Li_2S	lithium sulfide
d. In_2O_3	indium oxide
g. MgF_2	magnesium fluoride

Pair a & b & f: consist of two non-metals — a covalent compound is anticipated

Pair e : Argon doesn't typically form ionic compounds

96. Formulas for compounds; identify the ionic compounds

a. sodium hypochlorite	$NaClO$	ionic
b. boron triiodide	BI_3	
c. aluminum perchlorate	$Al(ClO_4)_3$	ionic
d. calcium acetate	$Ca(CH_3CO_2)_2$	ionic
e. potassium permanganate	$KMnO_4$	ionic
f. ammonium sulfite	$(NH_4)_2SO_3$	ionic
g. potassium dihydrogen phosphate	KH_2PO_4	ionic
h. disulfur dichloride	S_2Cl_2	
i. chlorine trifluoride	ClF_3	
j. phosphorus trifluoride	PF_3	

98. Empirical and Molecular formula of Azulene:

Given the information that azulene is a hydrocarbon, if it is 93.71 % C, it is also (100.00 - 93.71) or 6.29 % H.

In a 100.00 g sample of azulene there are:

$$93.71 \text{ g C} \cdot \frac{1 \text{ mol C}}{12.011 \text{ g C}} = 7.802 \text{ mol C} \quad \text{and}$$

$$6.29 \text{ g H} \cdot \frac{1 \text{ mol H}}{1.0079 \text{ g H}} = 6.241 \text{ mol H}$$

The ratio of C to H atoms is: 1.25 mol C : 1 mol H or a ratio of 5 mol C: 4 mol H. (C_5H_4) The mass of such an empirical formula is ≈ 64. Given that the molar mass is ~128 g/mol, the molecular formula for azulene is $C_{10}H_8$.

100.

I_2	+	Cl_2	→	I_xCl_y
0.678 g		(1.246 - 0.678)		1.246 g

Calculate the ratio of I : Cl atoms

$$0.678 \text{ g I} \cdot \frac{1 \text{ mol I}}{126.9 \text{ g I}} = 5.34 \times 10^{-3} \text{ mol I atoms}$$

$$0.568 \text{ g Cl} \cdot \frac{1 \text{ mol Cl}}{35.45 \text{ g Cl}} = 1.60 \times 10^{-2} \text{ mol Cl atoms}$$

The ratio of Cl : I is : $\frac{1.60 \times 10^{-2} \text{ mol Cl atoms}}{5.34 \times 10^{-3} \text{ mol I atoms}} = 3.00 \frac{\text{Cl atoms}}{\text{I atoms}}$

The empirical formula is ICl_3 (FW = 233.3)

Given that the molar mass of I_xCl_y was 467 g/mol, we can calculate the number of empirical formulas per mole:

$$\frac{467 \text{ g/mol}}{233.3 \text{ g/empirical formula}} = 2 \frac{\text{empirical formula}}{\text{mol}}$$

for a molecular formula of I_2Cl_6.

102. Mass of Fe in 15.8 kg of FeS_2 :

$$\% \text{ Fe in } FeS_2 = \frac{55.85 \text{ g Fe}}{119.97 \text{ g } FeS_2} \times 100 = 46.55 \text{ \% Fe}$$

and in 15.8 kg FeS_2: $15.8 \text{ kg } FeS_2 \cdot \frac{46.55 \text{ kg Fe}}{100.00 \text{ kg } FeS_2} = 7.35 \text{ kg Fe}$

104. Which of the following is impossible:

a. Ag foil that is 1.2×10^{-4} m thick (or 0.12 mm) - quite doable

b. A sample of K containing 1.784×10^{24} atoms. (2.96 mol of K or about 116 g)- doable

Recall that 1 mol contains 6.022×10^{23} atoms, and that 1 mol would have a mass equal to the atomic mass of the element.

c. A gold coin of mass 1.23×10^{-3} kg (or 1.23 g Au)- quite doable

d. 3.43×10^{-27} mol of S_8:

$$3.43 \times 10^{-27} \text{ mol of } S_8 \cdot \frac{6.022 \times 10^{23} \text{ molecules } S_8}{1 \text{ mol } S_8} = 0.00207 \text{ molecules}$$

not possible— since the smallest particle of S_8 would be 1 molecule.

106. What metal would form a compound with the formula MCl_4 ?

If the compound is 74.75 % Cl, then in 100.00 g of the compound there would be 74.75g Cl— and (100.00-74.75) 25.25 g of M.

Calculate the # of moles of chlorine in the compound, and multiply by 4:

$$74.75 \text{ g Cl} \cdot \frac{1 \text{ mol Cl}}{35.453 \text{ g Cl}} \cdot \frac{1 \text{ mol M}}{4 \text{ mol Cl}} = 0.5271 \text{ mol M}.$$

To calculate the atomic mass (g/mol), divide the mass of metal by the # of mol:

$$\frac{25.25 \text{ g M}}{0.5271 \text{ mol M}} = 47.90 \text{ g/mol M}$$—which most likely makes the metal Ti (47.88 g/mol).

108. To determine the empirical formula for the compound formed between V and S, calculate the # mole of each element, and then the ratio of mol V: mol S :

$$\frac{2.04 \text{ g V}}{50.94 \text{ g/mol V}} = 0.0400 \text{ mol V} \qquad \frac{1.93 \text{ g S}}{32.07 \text{ g/mol S}} = 0.0602 \text{ mol S}$$

The ratio of the two elements is $\frac{0.0602 \text{ mol S}}{0.0400 \text{ mol V}} = 1.50$, which means that for **each mol of V there are 1.5 mol of S.** Since we express formulas with *integral* values, this translates into 2 mol of V with 3 mol of S, giving a formula of V_2S_3.

Conceptual Questions

110. Of the ions Cl^-, K^+, Mg^{2+}, Al^{3+}, which has the strongest attraction for water molecules in the vicinity? We can calculate the proportional charge density by dividing the ionic charges by the ionic radii:

Cl^-	K^+	Mg^{2+}	Al^{3+}
$\frac{1}{167\ pm}$	$\frac{1}{152\ pm}$	$\frac{2}{86\ pm}$	$\frac{3}{68\ pm}$

The charge density of Al^{3+} is greatest, and so this cation would have the greatest attraction to the dipole, H_2O.

112. Using the student data, let's calculate the number of moles of $CaCl_2$ and moles of H_2O:

$$0.739\ g\ CaCl_2 \bullet \frac{1\ mol\ CaCl_2}{111.0\ g\ CaCl_2} = 0.00666\ mol\ CaCl_2$$

$$(0.832\ g - 0.739\ g)\ \text{or}\ 0.093\ g\ H_2O \bullet \frac{1\ mol\ H_2O}{18.02\ g\ H_2O} = 0.0052\ mol\ H_2O$$

The number of moles of water/mol of calcium chloride is then $\frac{0.0052\ mol\ H_2O}{0.00666\ mol\ CaCl_2} = 0.78$

This is a sure sign that they should **heat the crucible again, and then reweigh it.**

Structural Chemistry

114. For the organic acids: formic, acetic, and oxalic :

a. Common features of the molecules:

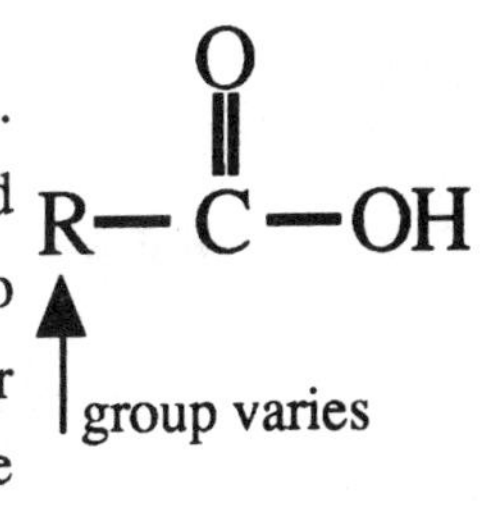

All possess the "carboxylic acid" group common to organic acids. This group possesses a C atom double bonded to an O atom, and singly bonded to the -OH group. The C atom is then bonded to one other group. For formic, the other group is a H atom, for acetic, the other group is CH_3 (a methyl group), for oxalic, the other group is another -COOH group.

b. O-C-O angle in each molecule:

Using the CAChe models from the CD-ROM (disc 1), and measuring the angle we find: In oxalic acid the angle is 120.298 degrees
In formic acid the angle is 114.175 degrees
In acetic acid the angle is 116.358 degrees

Challenging Problems

116. If 0.15 mol of A_2Z_3 has a mass of 15.9 g and 0.15 mol of AZ_2 has a mass of 9.3 g, we can calculate the molar mass of each.

$$A_2Z_3 \quad \frac{15.9 \text{ g } A_2Z_3}{0.15 \text{ mol}} = 106 \text{ g/mol} \qquad AZ_2 \quad \frac{9.3 \text{ g } AZ_2}{0.15 \text{ mol}} = 62 \text{ g/mol}$$

From the compounds compositions we can write:

for A_2Z_3 : (2 mol A atoms)(AW_A) + (3 mol Z atoms)(AW_Z) = 106 g (1)

for AZ_2 : (1 mol A atoms)(AW_A) + (2 mol Z atoms)(AW_Z) = 62 g (2)

[AW_A represents the atomic weight of A].

Solving in the second equation for **mol A atoms** :

(1 mol A atoms)(AW_A) = 62 g - (2 mol Z atoms)(AW_Z)

Substituting into the first equation for the first term (multiply by 2) gives:

(2)(62 g - (2 mol Z atoms)(AW_Z)) + (3 mol Z atoms)(AW_Z) = 106 g

124 g - (4 mol Z atoms)(AW_Z) + (3 mol Z atoms)(AW_Z) = 106 g

or 124 g - (1 mol Z atoms)(AW_Z) = 106 g

-(1 mol Z atoms)(AW_Z) = 106 g - 124 g

(1 mol Z atoms)(AW_Z) = 18 g

$$AW_Z = \frac{18 \text{ g}}{1 \text{ mol Z atoms}} \qquad AW_Z = 18 \text{ g/mol}$$

If the Atomic weight of Z is 18 g/mol we can solve for the atomic weight of A by using <u>either</u> equation (1) or (2)

(1 mol A atoms)(AW_A) + (2 mol Z atoms)(AW_Z) = 62 g (2)

$$(1 \text{ mol A atoms})(AW_A) + (2 \text{ mol Z atoms})\left(\frac{18 \text{ g Z}}{\text{mol Z}}\right) = 62 \text{ g}$$

$$(1 \text{ mol A atoms})(AW_A) = 62 \text{ g} - 36 \text{ g} \text{ or } \qquad AW_A = \frac{26 \text{ g}}{1 \text{ mol A atoms}}$$

The atomic weight of A is then 26 g/mol.

118. Number of atoms of C in 2.0000 g sample of C.

$$2.0000 \text{ g C} \cdot \frac{1 \text{ mol C}}{12.011 \text{ g C}} \cdot \frac{6.0221 \times 10^{23} \text{ atoms C}}{1 \text{ mol C}} = 1.0028 \times 10^{23} \text{ atoms C}$$

If the accuracy of the balance is +/- 0.0001g then the mass could be 2.0001g C, so we would do a similar calculation to obtain 1.0028×10^{23} atoms C (the same number of C atoms).

120. Calculate:

a. moles of nickel—found by density **once** the volume of foil is calculated.

$$V = 1.25 \text{ cm} \times 1.25 \text{ cm} \times 0.0550 \text{ cm} = 8.59 \times 10^{-2} \text{ cm}^3$$

$$\text{Mass} = \frac{8.908 \text{ g}}{1 \text{ cm}^3} \cdot 8.59 \times 10^{-2} \text{ cm}^3 = 0.766 \text{ g Ni}$$

$$0.766 \text{ g Ni} \cdot \frac{1 \text{ mol Ni}}{58.69 \text{ g Ni}} = 1.30 \times 10^{-2} \text{ mol Ni}$$

b. Formula for the fluoride salt:

$$\text{Mass F} = (1.261 \text{ g salt} - 0.766 \text{ g Ni}) = 0.495 \text{ g F}$$

$$\text{Moles F} = 0.495 \text{ g F} \cdot \frac{1 \text{ mol F}}{19.00} = 2.60 \times 10^{-2} \text{ mol F},$$

so 1.30×10^{-2} mol Ni combines with 2.60×10^{-2} mol F, indicating a formula of NiF_2.

c. Name: Nickel(II) fluoride

Chapter 4
Chemical Equations and Stoichiometry

Review Questions

2. A balanced equation for the production of ammonia:

   The formula for ammonia, NH_3, indicates **four** atoms per molecule—one atom of nitrogen and three atoms of hydrogen. Given that this substance is being formed from two diatomic substances—that is substances that have *two atoms per molecule* , we'll definitely need **odd** numbers of molecules of elemental nitrogen and elemental hydrogen to balance the equation. Answer (c) provides such a solution:

   $$N_2 (g) + 3 H_2 (g) \rightarrow 2 NH_3 (g)$$

   One molecule of nitrogen (contributes 2 nitrogen atoms), 3 molecules of hydrogen (contribute 6 hydrogen atoms), and two molecules of ammonia contain a total of 2 nitrogen atoms and 6 hydrogen atoms.

4. Stoichiometric factor needed for the relationship of **ammonia** to **nitrogen**:

   Since ammonia is our *desired substance* and *nitrogen* is our *given substance* , the ratio we need would involve those two—with ammonia in the numerator, viz.

   $\frac{2 \text{ moles } NH_3}{1 \text{ mol } N_2}$, hence if we begin with **3 moles of nitrogen, we can calculate** :

   $$3 \text{ moles } N_2 \bullet \frac{2 \text{ moles } NH_3}{1 \text{ mol } N_2} = 6 \text{ moles } NH_3$$

Balancing Equations

Balancing equations can be a matter of "running in circles" if a reasonable methodology is not employed. While there isn't one "right place" to begin, generally you will suffer fewer complications if you begin the balancing process using a substance that contains the greatest number of elements or the largest subscript values. Noting that you must have at least that many atoms of each element involved, coefficients can be used to increase the "atomic inventory". In the next few questions, you will see one **emboldened** substance in each equation. This emboldened substance is the one that I judge to be a "good" starting place. One last hint -- modify the coefficients of uncombined elements, i.e. those not in compounds, after you modify the coefficients for compounds containing those elements -- not before!

6. a. $4\ Cr\ (s) + 3\ O_2\ (g) \rightarrow$ **$2\ Cr_2O_3$** (s)

1. Note the need for <u>at least</u> 2 Cr and 3 O atoms.
2. Oxygen is diatomic -- we' ll need an <u>even</u> number of oxygen atoms, so try : 2 Cr_2O_3.
3. 3 O_2 would give 6 O atoms on both sides of the equation.
4. 4 Cr would give 4 Cr atoms on both sides of the equation.

b. $Cu_2S\ (s) + O_2\ (g) \rightarrow 2\ Cu(s) +$ **SO_2** (g)

1. A minimum of 2 O in SO_2 is required, and is provided with one molecule of elemental oxygen.
2. 2 Cu atoms (on the right) indicates 2 Cu (on the left).

c. **$C_6H_5CH_3$** $(\ell) + 9\ O_2\ (g) \rightarrow 4\ H_2O\ (g) + 7\ CO_2\ (g)$

1. A minimum of 7 C and 8 H is required.
2. 7 CO_2 furnishes 7 C and 4 H_2O furnishes 8 H atoms.
3. 4 H_2O and 7 CO_2 furnish a total of 18 O atoms, making the coefficient of $O_2 = 9$.

8. Balance and name the products:

a. **Fe_2O_3** $(s) + 3\ Mg(s) \rightarrow 3\ MgO\ (s) + 2\ Fe\ (s)$

1. Note the need for <u>at least</u> 2 Fe and 3 O atoms.
2. 2 Fe atoms would provide the proper iron atom inventory.
3. 3 MgO would give 3 O atoms on both sides of the equation.
4. 3 Mg would give 3 Mg atoms on both sides of the equation.

Products: magnesium oxide and iron

b. **$AlCl_3$** $(s) + 3\ H_2O\ (\ell) \rightarrow Al(OH)_3\ (s) + 3\ HCl\ (aq)$

1. Note the need for <u>at least</u> 1 Al and 3 Cl atoms.
2. 3 HCl molecules would provide the proper Cl atom inventory.
3. 3 H atoms (from HCl) and 3 H atoms (from $Al(OH)_3$) would give 6 H atoms needed on both sides of the equation—so a coefficient of 3 for water is needed to provide that balance.

Products: aluminum hydroxide and hydrochloric acid.

{Recall that *aqueous* HCl is a "hydro"-chlor-"ic" acid. Can you think of at least one other example ? Hint: Examine Group 7A !}

c. $2\ NaNO_3\ (s) + H_2SO_4\ (\ell) \rightarrow$ $\mathbf{Na_2SO_4}$ $(s) + 2\ HNO_3\ (g)$

1. Note the need for at least 2 Na and 1 S and 4 O atoms.
2. $2\ NaNO_3$ will provide the proper Na atom inventory.
3. The coefficient of 2 in front of $NaNO_3$ requires a coefficient of 2 for HNO_3 —providing a balance for N atoms.
4. The implied coeffienct of 1 for Na_2SO_4 suggests a similar coefficient for H_2SO_4 —to balance the S atom inventory.
5. O atom inventory is done "automatically" when we balanced N and S inventories.

Products: sodium sulfate and nitric acid [....although nitric acid typically exists as an aqueous solution.]

d. $\mathbf{NiCO_3}$ $(s) + 2\ HNO_3\ (aq) \rightarrow Ni(NO_3)_2\ (aq) + CO_2\ (g) + H_2O\ (\ell)$

1. Note the need for at least 1 Ni atom on both sides. This inventory will mandate 2 NO_3 groups on the right—and also on the left. Since these come from HNO_3 molecules, we'll need 2 HNO_3 on the left.
2. The 2 H from the acid and the CO_3 from nickel carbonate, provide 2H, 1 C and 3 O atoms. 1 H_2O takes care of the 2H, and **one** of the O atoms, 1 CO_2 consumes the 1 C and the remaining 2 O atoms.

Products: nickel(II) nitrate, carbon dioxide, and water

10. Balance:

a. the synthesis of urea:

$CO_2(g) + 2\ NH_3(g) \rightarrow$ $\mathbf{CO(NH_2)_2}$$(s) + H_2O(\ell)$

1. Note the need for two NH_3 in each molecule of urea, so multiply NH_3 by 2.
2. 2 NH_3 provides the two H atoms for a molecule of H_2O.
3. Each CO_2 provides the O atom for a molecule of H_2O.

b. the synthesis of uranium(VI) fluoride

$UO_2(s) + 4\ HF(aq) \rightarrow$ $\mathbf{UF_4}$$(s) + 2\ H_2O(aq)$

$UF_4(s) + F_2(g) \rightarrow UF_6(s)$

1. The 4 F atoms in UF_4 requires 4 F atoms from HF. (equation 1)
2. The H atoms in HF produce 2 molecules of H_2O. (equation 1)
3. The 1:1 stoichiometry of $UF_6 : UF_4$ provides a simple balance. (equation 2)

c. synthesis of titanium metal from TiO_2:

$TiO_2(s) + 2\,Cl_2(g) + 2\,C(s) \rightarrow TiCl_4(\ell) + 2\,CO(g)$

$TiCl_4(\ell) + 2\,Mg(s) \rightarrow Ti(s) + 2\,MgCl_2(s)$

1. The O balance mandates 2 CO for each TiO_2. (equation 1)
2. A coefficient of 2 for C provides C balance. (equation 1)
3. The Ti balance ($TiO_2 : TiCl_4$) requires 4 Cl atoms, hence 2 Cl_2 (equation 1)
4. The Cl balance requires 2 $MgCl_2$, hence 2 Mg. (equation 2)

General Stoichiometry

12. Moles of oxygen needed to reach with 6.0 mol of Al:

$$4\,Al\,(s) + 3\,O_2\,(g) \rightarrow 2\,Al_2O_3\,(s)$$

$$6.0 \text{ mol Al} \bullet \frac{3 \text{ mol } O_2}{4 \text{ mol Al}} = 4.5 \text{ mol } O_2$$

What mass of Al_2O_3 should be produced ?

$$6.0 \text{ mol Al} \bullet \frac{2 \text{ mol } Al_2O_3}{4 \text{ mol Al}} \bullet \frac{102 \text{ g } Al_2O_3}{1 \text{ mol } Al_2O_3} = 310 \text{ g } Al_2O_3 \text{ (to 2 sf)}$$

14. For the reaction of methane with oxygen:

a. the products of the reaction:

$$CH_4\,(g) + 2\,O_2(g) \rightarrow CO_2\,(g) + 2\,H_2O\,(g)$$

b. the balanced equation for the reaction is given above

c. Mass of oxygen, in grams, needed to completely consume the 16.04 g CH_4.

$$16.04 \text{ g } CH_4 \bullet \frac{1 \text{ mol } CH_4}{16.04 \text{ g } CH_4} \bullet \frac{2 \text{ mol } O_2}{1 \text{ mol } CH_4} \bullet \frac{31.999 \text{ g } O_2}{1 \text{ mol } O_2} = 63.99 \text{ g } O_2 .$$

d. total mass of products expected:

This could be solved in several ways. The simplest is to recognize that—according to the Law of Conservation of Matter, mass is conserved in a reaction. If 63.99 g of oxygen react with exactly 16.04 g of methane, the total products would also have a mass of (63.99 + 16.04)g or 80.03 g.

16. Quantity of Br_2 to react with 2.56 g of Al:

According to the balanced equation 2 mol of Al react with 3 mol of Br_2 .

Calculate the # of moles of Al, then multiply by 3/2 to obtain # mol of Br_2 required.

$$2.56 \text{ g Al} \cdot \frac{1 \text{ mol Al}}{26.98 \text{ g Al}} \cdot \frac{3 \text{ mol Br}_2}{2 \text{ mol Al}} \cdot \frac{159.8 \text{ g Br}_2}{1 \text{ mol Br}_2} = 22.7 \text{ g Br}_2$$

Mass of Al_2Br_6 expected:

This could be solved in several ways. The simplest is to recognize that—according to the Law of Conservation of Matter, mass is conserved in a reaction. If 22.7 g of bromine react with exactly 2.56 g of aluminum, the total products would also have a mass of (22.7 g+ 2.56 g) or 25.3 g Al_2Br_6.

18. The reaction of iron with oxygen to given iron(III) oxide:

a. The balanced equation for the reaction:

$$4 \text{ Fe (s)} + 3 \text{ O}_2 \text{ (g)} \rightarrow 2 \text{ Fe}_2\text{O}_3\text{(s)}$$

b. Mass of Fe_2O_3 produced when 2.68 g Fe react:

$$2.68 \text{ g Fe} \cdot \frac{1 \text{ mol Fe}}{55.85 \text{ g Fe}} \cdot \frac{2 \text{ mol Fe}_2\text{O}_3}{4 \text{ mol Fe}} \cdot \frac{159.7 \text{ g Fe}_2\text{O}_3}{1 \text{ mol Fe}_2\text{O}_3} = 3.83 \text{ g Fe}_2\text{O}_3$$

c. Mass of oxygen required:

This could be solved in several ways. The simplest is to recognize that—according to the Law of Conservation of Matter, mass is conserved in a reaction. If 3.83 g Fe_2O_3 are produced when 2.68 g of iron react, the total oxygen required would have a mass of (3.83 g - 2.68 g) or 1.15 g O_2.

20. Mass of acetone available from decomposition of 125 mg of acetoacetic acid (AA)

	$CH_3COCH_2CO_2H$	$\rightarrow$	$CH_3COCH_3 + CO_2$
molar mass:	102.09 g/mol		58.08 g/mol

$$125 \text{ mg AA} \cdot \frac{1 \text{ mol AA}}{102.09 \text{ g AA}} \cdot \frac{1 \text{ mol acetone}}{1 \text{ mol AA}} \cdot \frac{58.08 \text{ g acetone}}{1 \text{ mol acetone}} = 71.2 \text{ mg acetone}$$

Note that it **is not necessary** to convert milligrams of acetoacetic acid into grams of acetoacetic acid, since the mass of product (acetone) is also expressed in milligrams.

Limiting Reactants

22. For the reaction of aluminum with chlorine:

$$2\ Al(s) + 3\ Cl_2(g) \rightarrow 2\ AlCl_3(s)$$

a. Limiting reagent:

$$2.70\ g\ Al \cdot \frac{1\ mol\ Al}{26.98\ g\ Al} = 0.100\ mol\ Al$$

$$4.05\ g\ Cl_2 \cdot \frac{1\ mol\ Cl_2}{70.91\ g\ Cl_2} = 0.0571\ mol\ Cl_2$$

Since each mol of Al requires 1.5 mol Cl_2, we clearly have a deficiency of Cl_2, so **Cl_2 is the Limiting Reagent.**

b. Mass of $AlCl_3$ possible:

$$0.0571\ mol\ Cl_2 \cdot \frac{2\ mol\ AlCl_3}{3\ mol\ Cl_2} \cdot \frac{133.3\ g\ AlCl_3}{1\ mol\ AlCl_3} = 5.08\ g\ AlCl_3$$

c. Mass of Al remaining:

$$0.0571\ mol\ Cl_2 \cdot \frac{2\ mol\ Al}{3\ mol\ Cl_2} \cdot \frac{26.98\ g\ Al}{1\ mol\ Al} = 1.03\ g\ Al\ consumed$$

Remaining = (2.70 - 1.03) = 1.67 g Al

24. The limiting reagent can be determined by calculating the moles-available and moles-required ratios for the equation:

$$CO\ (g) + 2\ H_2\ (g) \rightarrow CH_3OH\ (\ell)$$

Moles of each reactant present:

$$12.0\ g\ H_2 \cdot \frac{1\ mol\ H_2}{2.016\ g\ H_2} = 5.95\ mol\ H_2$$

$$74.5\ g\ CO \cdot \frac{1\ mol\ CO}{28.01\ g\ CO} = 2.66\ mol\ CO$$

moles-required ratio: $\frac{2\ mol\ H_2}{1\ mol\ CO}$

moles-available ratio: $\frac{5.95\ mol\ H_2}{2.66\ mol\ CO} = \frac{2.24\ mol\ H_2}{1\ mol\ CO}$

Since we require 2 moles of hydrogen per mole of CO, and we have available 2.24 moles of hydrogen per mole of CO, **CO is the limiting reagent** and **H_2 is present in excess.**

To determine the mass of excess reagent remaining after the reaction is complete, calculate the amount needed:

$$2.66 \text{ mol CO} \cdot \frac{2 \text{ mol } H_2}{1 \text{ mol CO}} \cdot \frac{2.016 \text{ g } H_2}{1 \text{ mol } H_2} = 10.7 \text{ g } H_2 \text{ needed}$$

Excess H_2 = 12.0 - 10.7 = 1.3 g

We calculate the theoretical yield of CH_3OH from the amount of limiting reagent:

$$2.66 \text{ mol CO} \cdot \frac{1 \text{ mol } CH_3OH}{1 \text{ mol CO}} \cdot \frac{32.04 \text{ g } CH_3OH}{1 \text{ mol } CH_3OH} = 85.2 \text{ g } CH_3OH$$

26. The limiting reagent can be determined by calculating the mole-available and mole-needed ratios for the equation:

$$CaO\ (s) + 2\ NH_4Cl\ (s) \rightarrow 2\ NH_3\ (g) + H_2O\ (g) + CaCl_2\ (s)$$

Calculate the moles of CaO and of NH_4Cl :

$$112 \text{ g CaO} \cdot \frac{1 \text{ mol CaO}}{56.08 \text{ g CaO}} = 2.00 \text{ mol CaO}$$

$$224 \text{ g } NH_4Cl \cdot \frac{1 \text{ mol } NH_4Cl}{53.49 \text{ g } NH_4Cl} = 4.19 \text{ mol } NH_4Cl$$

moles-required ratio: $\frac{2 \text{ mol } NH_4Cl}{1 \text{ mol CaO}}$

moles-available ratio: $\frac{4.19 \text{ mol } NH_4Cl}{2.00 \text{ mol CaO}} = \frac{2.10 \text{ mol } NH_4Cl}{1.00 \text{ mol CaO}}$

CaO is the limiting reagent, and will determine the maximum amount of products obtainable:

$$112 \text{ g CaO} \cdot \frac{1 \text{ mol CaO}}{56.08 \text{ g CaO}} \cdot \frac{2 \text{ mol } NH_3}{1 \text{ mol CaO}} \cdot \frac{17.03\text{g } NH_3}{1 \text{ mol } NH_3} = 68.0 \text{ g } NH_3$$

The balanced equation shows that for each mole of CaO, 2 moles of NH_4Cl are required. So 2.00 mol of CaO would require 4.00 mol of NH_4Cl, leaving (4.19 - 4.00) 0.19 mol of NH_4Cl in excess.

This number of moles would have a mass of:

$$0.19 \text{ mol } NH_4Cl \cdot \frac{53.49 \text{ g } NH_4Cl}{1 \text{ mol } NH_4Cl} = 10. \text{ g } NH_4Cl$$

Percent Yield

28. Percent yield of NH_3:

$$\frac{\text{actual}}{\text{theoretical}} = \frac{103 \text{ g } NH_3}{136 \text{ g } NH_3} \cdot 100 = 75.7 \text{ \% yield}$$

30. Theoretical yield of $ZnCl_2$:

Given the fact that the only $ZnCl_2$ is formed in this reaction, we can calculate the % of Zn in $ZnCl_2$: $\frac{65.39 \text{ g Zn}}{136.30 \text{ g } ZnCl_2}$

This means that we can calculate the **theoretical amount of $ZnCl_2$** that can be formed from 35.5 g Zn:

$$35.5 \text{ g Zn} \cdot \frac{136.30 \text{ g } ZnCl_2}{65.39 \text{ g Zn}} = 74.0 \text{ g } ZnCl_2 \text{ (theoretical yield)}$$

% yield of $ZnCl_2$:

$$\frac{\text{actual}}{\text{theoretical}} = \frac{65.2 \text{ g } ZnCl_2}{74.0 \text{ g } ZnCl_2} \cdot 100 = 88.1 \text{ \% yield}$$

Chemical Analysis

32. Weight percent of $CuSO_4 \cdot 5\,H_2O$ in the mixture:

Mass of H_2O = 1.245 g - 0.832 g = 0.413 g H_2O

Since this water was a part of the hydrated salt, let's calculate the mass of that salt present: In 1 mol of $CuSO_4 \cdot 5\,H_2O$ there are 90.08 g H_2O and 159.61 g $CuSO_4$ or 249.69 g $CuSO_4 \cdot 5\,H_2O$. So:

$$0.413 \text{ g } H_2O \cdot \frac{249.69 \text{ g } CuSO_4 \cdot 5\,H_2O}{90.08 \text{ g } H_2O} = 1.14 \text{ g hydrated salt}$$

$$\text{\% hydrated salt} = \frac{1.14 \text{ g hydrated salt}}{1.245 \text{ g mixture}} \cdot 100 = 91.9\%$$

34. The moles of benzene present are:

$$0.951 \text{ g } C_6H_6 \bullet \frac{1 \text{ mol } C_6H_6}{78.11 \text{ g } C_6H_6} = 1.22 \times 10^{-2} \text{ mol } C_6H_6$$

The amount of $Al(C_6H_5)_3$ present is:

$$1.22 \times 10^{-2} \text{ mol } C_6H_6 \bullet \frac{1 \text{ mol } Al(C_6H_5)_3}{3 \text{ mol } C_6H_6} \bullet \frac{258.30 \text{ g } Al(C_6H_5)_3}{1 \text{ mol} Al(C_6H_5)_3} = 1.05 \text{ g } Al(C_6H_5)_3$$

The percent of $Al(C_6H_5)_3$ in the sample is: $\frac{1.05 \text{ g } Al(C_6H_5)_3}{1.25 \text{ g sample}} \bullet 100 = 83.9\%$

Determination of Empirical Formulas

36. The basic equation is:

$$C_xH_y + O_2 \rightarrow x\,CO_2 + \frac{y}{2} H_2O$$

Without balancing the equation, one can see that all the C in CO_2 comes from the styrene as does all the H in H_2O. Let's use the percentage of C in CO_2 to determine the mass of C in styrene, and the percentage of H in H_2O to provide the mass of H in styrene.

$$1.481 \text{ g } CO_2 \bullet \frac{12.01 \text{ g C}}{44.01 \text{ g } CO_2} = 0.404 \text{ g C}$$

Similarly :

$$0.303 \text{ g } H_2O \bullet \frac{2.02 \text{ g H}}{18.02 \text{ g } H_2O} = 0.0340 \text{ g H}$$

Alternatively, the mass of H could be determined by subtracting the mass of C from the 0.438 g styrene

$$\text{Mass H} = 0.438 \text{ g styrene} - 0.404 \text{ g C}$$

Establish the ratio of C atoms to H atoms

$$0.0340 \text{ g H} \bullet \frac{1 \text{ mol H}}{1.008 \text{ g H}} = 0.0337 \text{ mol H}$$

$$0.404 \text{ g C} \bullet \frac{1 \text{ mol C}}{12.011 \text{ g C}} = 0.0336 \text{ mol C}$$

This number of H and C atoms indicates an empirical formula for styrene of 1:1 or C_1H_1.

38. From combustion data:

$$\underset{95.6 \text{ mg}}{C_xH_yO_z} + O_2 \rightarrow \underset{269 \text{ mg}}{CO_2} + \underset{110 \text{ mg}}{H_2O}$$

To determine the empirical formula, we must determine the amount of C and O in the combustion products:

$$\% \text{ C in } CO_2 = \frac{12.01 \text{ g C}}{44.01 \text{ g } CO_2} \qquad \% \text{ H in } H_2O = \frac{2.02 \text{ g H}}{18.02 \text{ g } H_2O}$$

Establishing the ratio of C atoms to H atoms and the amount present

$$269 \text{ mg } CO_2 \bullet \frac{12.01 \text{ g C}}{44.01 \text{ g } CO_2} = 73.41 \text{ mg C}$$

$$110 \text{ mg } H_2O \bullet \frac{2.02 \text{ g H}}{18.02 \text{ g } H_2O} = 12.33 \text{ mg H}$$

Now we can determine the amount of O present in the sample:

$$95.6 \text{ mg} = 73.41 \text{ mg C} + 12.33 \text{ mg H} + x \text{ mg O}$$

$$\text{mg O} = 9.86 \text{ mg}$$

Finding the number of moles of each element:

$$73.41 \text{ mg} \bullet \frac{1 \text{ mol C}}{12.01 \text{ g C}} = 6.11 \text{ mol C}$$

$$12.33 \text{ mg H} \bullet \frac{1 \text{ mol H}}{1.008 \text{ g H}} = 12.23 \text{ mol H}$$

$$9.68 \text{ mg O} \bullet \frac{1 \text{ mol O}}{16.00 \text{ g O}} = 0.616 \text{ mol O}$$

This gives ratios of $C_{9.9}H_{19.85}O_1$ for the smallest whole numbers: $C_{10}H_{20}O$

40. Calculate the mass of Si in the compounds; using the fraction of Si in SiO_2:

$$11.64 \text{ g } SiO_2 \bullet \frac{28.09 \text{ g Si}}{60.084 \text{ g } SiO_2} = 5.442 \text{ g Si or } 0.1937 \text{ mol Si}$$

and the mass of H using the fraction of water that is hydrogen :

$$6.980 \text{ g } H_2O \bullet \frac{2.016 \text{ g H}}{18.02 \text{ g } H_2O} = 0.7809 \text{ g H or } 0.7747 \text{ mol H}$$

The ratio of Si:H is $\frac{0.7747 \text{ mol H}}{0.1937 \text{ mol Si}} = 4$

The empirical formula is then SiH_4.

General Questions

42. Mass of oxygen formed when 125 mg of glucose burn:

$$C_6H_{12}O_6 + 6\,O_2 \rightarrow 6\,CO_2 + 6\,H_2O$$

molar mass: 180.16 g/mol 32.00 g/mol 44.01 g/mol 18.02 g/mol

Quantity of O_2 required:

$$125 \text{ mg glucose} \cdot \frac{1 \text{ mol glucose}}{180.16 \text{ g glucose}} \cdot \frac{6 \text{ mol } O_2}{1 \text{ mol glucose}} \cdot \frac{32.00 \text{ g } O_2}{1 \text{ mol } O_2} = 133 \text{ mg } O_2$$

Quantity of CO_2 formed:

$$125 \text{ mg glucose} \cdot \frac{1 \text{ mol glucose}}{180.16 \text{ g glucose}} \cdot \frac{6 \text{ mol } CO_2}{1 \text{ mol glucose}} \cdot \frac{44.01 \text{ g } CO_2}{1 \text{ mol } CO_2} = 183 \text{ mg } CO_2$$

Quantity of H_2O formed:

$$125 \text{ mg glucose} \cdot \frac{1 \text{ mol glucose}}{180.16 \text{ g glucose}} \cdot \frac{6 \text{ mol } H_2O}{1 \text{ mol glucose}} \cdot \frac{18.02 \text{ g } H_2O}{1 \text{ mol } H_2O} = 75.0 \text{ mg } H_2O$$

Note that it **is not necessary** to convert milligrams of a substance into grams of that substance, since the mass of products are also expressed in milligrams.

44. For the reaction of aluminum hydroxide with hydrochloric acid:

$$Al(OH)_3 + 3\,HCl \rightarrow AlCl_3 + 3\,H_2O$$

molar mass: 78.00 g/mol 36.46 g/mol 133.3 g/mol 18.02 g/mol

What reactant is the limiting reactant if 15.5 g of $Al(OH)_3$ is combined with 20.5 g of HCl?

Calculate the # of moles available and the # of moles required:

a. Moles available:

$$15.5 \text{ g } Al(OH)_3 \cdot \frac{1 \text{ mol } Al(OH)_3}{78.00 \text{ g } Al(OH)_3} = 0.199 \text{ mol } Al(OH)_3$$

$$20.5 \text{ g HCl} \cdot \frac{1 \text{ mol HCl}}{36.46 \text{ g HCl}} = 0.562 \text{ mol HCl}$$

moles-required ratio: $\frac{3 \text{ mol HCl}}{1 \text{ mol } Al(OH)_3}$

moles-available ratio: $\dfrac{0.562 \text{ mol HCl}}{0.199 \text{ mol Al(OH)}_3} = \dfrac{2.83 \text{ mol HCl}}{1.00 \text{ mol Al(OH)}_3}$

Since the moles-available ratio for $HCl/Al(OH)_3$ is less than the moles-required ratio, **HCl is the limiting reactant**, and **$Al(OH)_3$ is the reactant in excess.**

b. Mass of water that can be formed:

Since HCl is the limiting reactant, use the # of mole of HCl to determine the mass of water.

$$0.562 \text{ mol HCl} \cdot \frac{3 \text{ mol } H_2O}{3 \text{ mol HCl}} \cdot \frac{18.02 \text{ g } H_2O}{1 \text{ mol } H_2O} = 10.1 \text{ g } H_2O$$

46. a. Balanced equation: $2\,Fe(s) + 3\,Cl_2(g) \rightarrow 2\,FeCl_3(s)$

b. 1. Mass of Cl_2 to react with 10.0 g iron:

$$10.0 \text{ g Fe} \cdot \frac{1 \text{ mol Fe}}{55.85 \text{ g Fe}} \cdot \frac{3 \text{ mol } Cl_2}{2 \text{ mol Fe}} \cdot \frac{70.91 \text{ g } Cl_2}{1 \text{ mol } Cl_2} = 19.0 \text{ g } Cl_2$$

2. Amount of $FeCl_3$ produced :

$$10.0 \text{ g Fe} \cdot \frac{1 \text{ mol Fe}}{55.85 \text{ g Fe}} \cdot \frac{2 \text{ mol } FeCl_3}{2 \text{ mol Fe}} = 0.179 \text{ mol } FeCl_3$$

$$0.179 \text{ mol } FeCl_3 \cdot \frac{162.2 \text{ g } FeCl_3}{1 \text{ mol } FeCl_3} = 29.0 \text{ g } FeCl_3$$

c. Percent yield of $FeCl_3$:

$$\frac{\text{Actual}}{\text{Theoretical}} \cdot 100 \text{ or } \frac{18.5 \text{ g } FeCl_3}{29.0 \text{ g } FeCl_3} \cdot 100 = 63.8\ \%$$

48. The reaction of titanium(IV) chloride with water:

$$TiCl_4\,(\ell) + 2\,H_2O\,(g) \rightarrow TiO_2(s) + 4\,HCl\,(g)$$

a. names: titanium(IV) chloride; water; titanium(IV) oxide; hydrogen chloride

b. Mass of water obtainable from 14.0 mL of $TiCl_4$:

Begin by asking the question :"How many moles of $TiCl_4$ are in 14.0 mL of $TiCl_4$?

$$14.0 \text{ mL } TiCl_4 \cdot \frac{1.73 \text{ g } TiCl_4}{1 \text{ mL}} \cdot \frac{1 \text{ mol } TiCl_4}{189.7 \text{ g } TiCl_4} = 0.128 \text{ mol } TiCl_4$$

The relationship between moles of $TiCl_4$ and moles of water is:

$$0.128 \text{ mol } TiCl_4 \cdot \frac{18.02 \text{ g } H_2O}{1 \text{ mol } H_2O} = 4.60 \text{ g } H_2O$$

c. What mass of TiO_2 and HCl will result ?

$$0.128 \text{ mol } TiCl_4 \cdot \frac{1 \text{ mol } TiO_2}{1 \text{ mol } TiCl_4} \cdot \frac{79.90 \text{ g } TiO_2}{1 \text{ mol } TiO_2} = 10.2 \text{ g } TiO_2$$

$$0.128 \text{ mol } TiCl_4 \cdot \frac{4 \text{ mol HCl}}{1 \text{ mol } TiCl_4} \cdot \frac{36.46 \text{ g HCl}}{1 \text{ mol HCl}} = 18.6 \text{ g HCl}$$

As an aside, note that the **total** mass of the reactants was (24.2 g $TiCl_4$ + 4.60 g H_2O) = 28.8 grams. Summing the masses of products we also obtain:

(10.2 g TiO_2 + 18.6 g H_2O) 28.8 grams.

Conceptual Questions

50. The equation for the reaction shows that molecular iodine reacts with molecular chlorine in a (1 mol I_2 : 3 mol Cl_2) ratio. The illustration shows 4 molecules of EACH gas. Since 4 molecules of I_2 would require 12 molecules of Cl_2, chlorine is the LIMITING REACTANT. So the question is transformed into asking "How many molecules of ICl_3 can one make with 4 molecules of Cl_2? Since each ICl_3 molecule requires THREE Cl atoms, one can quickly see that the MAXIMUM NUMBER OF ICl_3 molecules is TWO, leaving two chlorine atoms (or 1 chlorine molecule) unreacted. Having used ONE molecule of I_2 to make TWO ICl_3 molecules, three molecules I_2 remain. Choice (b) has this composition.

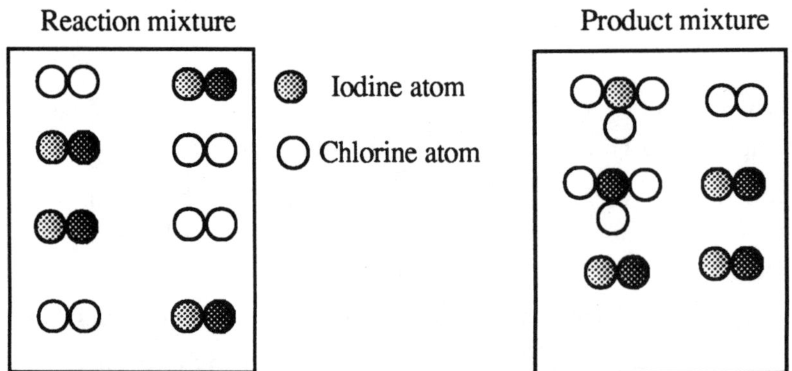

52. The evolution of CO_2 ceases when sufficient acetic acid is no longer available. Earlier in the chapter, we see the equation for the reaction:

$$CH_3CO_2H + NaHCO_3 \rightarrow NaCH_3CO_2 + CO_2 + H_2O$$

Calculate the moles of $NaHCO_3$ in 5.0 g of $NaHCO_3$:

$$5.0 \text{ g } NaHCO_3 \cdot \frac{1 \text{ mol } NaHCO_3}{84.0 \text{ g } NaHCO_3} = 0.060 \text{ mol } NaHCO_3$$

The balanced equation indicates a 1:1 mol ratio between the acetic acid and the $NaHCO_3$:

$$0.060 \text{ mol } NaHCO_3 \cdot \frac{1 \text{ mol } CH_3CO_2H}{1 \text{ mol } NaHCO_3} \cdot \frac{60.1 \text{ g } CH_3CO_2H}{1 \text{ mol } CH_3CO_2H} = 3.57 \text{ g } CH_3CO_2H$$

(3.6 g $NaHCO_3$—2 sf)

Challenging Questions

54. 1.056 g MCO_3 produced MO + 0.376 g CO_2

The MO had a mass of (1.056 - 0.376) 0.680 g

According to the equation given, CO_2 and MO are produced in equimolar amounts.

$$0.376 \text{ g } CO_2 \cdot \frac{1 \text{ mol } CO_2}{44.0 \text{ g } CO_2} = 8.54 \times 10^{-3} \text{ mol } CO_2$$

So the metal oxide (0.680 g) must correspond to 8.54×10^{-3} mol of metal oxide.

The molar mass of metal oxide is then:

$$\frac{0.680 \text{ g}}{8.54 \times 10^{-3} \text{ mol}} = 79.6 \text{ g/ mol}$$

If the oxide contains one mol of O atoms per mol of M atoms, we can deduce the molar mass of M

MO = 79.6 g/mol

79.6 = M g/mol + 16.0 g O/mol

63.6 g/mol = M

The metal with the atomic weight close to 63.6 is (b) Cu.

56. $TiO_2 + H_2 \rightarrow H_2O + Ti_xO_y$

1.598 g 1.438 g

The moles of TiO_2 initially present:

$$1.598 \text{ g } TiO_2 \cdot \frac{1 \text{ mol } TiO_2}{79.879 \text{ g } TiO_2} = 0.02000 \text{ mol } TiO_2 \text{ (and Ti)}$$

The mass of Ti present in the TiO_2 :

$$1.598 \text{ g } TiO_2 \cdot \frac{47.88 \text{ g Ti}}{79.879 \text{ g } TiO_2} = 0.9579 \text{ g Ti}$$

Since all the Ti in the unknown compound, Ti_xO_y, originates in the TiO_2, 0.9579 g of the 1.438 g of the new oxide is Ti, leaving (1.438 - 0.9579) g of O. The number of moles of O is:

$$0.480 \text{ g O} \cdot \frac{1 \text{ mol O}}{16.00 \text{ g O}} = 0.03000 \text{ mol O}$$

The new oxide is then $Ti_{0.02000}O_{0.03000}$ or Ti_2O_3

58. Balancing the following equations:

a. Ag (s) + H_2SO_4 (aq) $\rightarrow$ **Ag_2SO_4** (s)+ SO_2 (g)+ H_2O (ℓ)

[The emboldened substance represents a good starting place for balancing.]

1. The subscript 2 on Ag indicates that 2 Ag (s) will be necessary for the Ag inventory
2. Note that the "SO_4 + SO_2" will require AT LEAST 2 S atoms (and 6 O atoms) hence a coefficient of 2 for H_2SO_4
3. The 4 H atoms (2 H_2SO_4) will mandate $2H_2O$. So the balanced equation is:

2 Ag (s) + 2 H_2SO_4 (aq) $\rightarrow$ **Ag_2SO_4** (s) + SO_2 (g) + 2 H_2O (ℓ)

b. Au (s) + HNO_3 (aq) + HCl (aq) $\rightarrow$ **$HAuCl_4$** (aq) + NO (g) + H_2O (ℓ)

1. 4 Cl atoms in $HAuCl_4$ will mandate 4 HCl (for the Cl inventory).
2. 4 HCl suggests a coefficient of 2 for water: 2 H_2O - giving 5 H atoms on both sides (Did you remember the H in HNO_3?)
3. The 2 H_2O and the 1 NO gives 3 O atoms on the right so 1 HNO_3 completes the O inventory. The balanced equation is then:

Au (s) + HNO_3 (aq) + 4 HCl (aq) $\rightarrow$ **$HAuCl_4$** (aq) + NO (g) + 2 H_2O (ℓ)

c. Zn (s) + NaOH (aq) + H_2O (ℓ) $\rightarrow$ **$Na_2Zn(OH)_4$** + H_2 (g)

1. $Na_2Zn(OH)_4$ mandates a coefficient of 2 for NaOH.
2. The 4 O contained in $Na_2Zn(OH)_4$ requires 4 O on the left. 2 O atoms in NaOH will require 2 H_2O.
3. The 4 H atoms in 2 H_2O will require 2 H_2 (on the right). The balanced equation is:

Zn (s) + 2 NaOH (aq) + 2 H_2O (ℓ) $\rightarrow$ **$Na_2Zn(OH)_4$** + 2 H_2 (g)

Summary Questions

60. The graph representing the mass of iron consumed with a **fixed** mass of bromine shows that for the mass of bromine used, the amount of iron consumed maximizes at 2.0 g of Fe (with the production of 10.6 grams of compound Fe_xBr_y.

   a. According to the graph, the mass of bromine used when 2.0 g of Fe is consumed is maximized at 8.6 grams of bromine. Any excess iron that is added does not result in additional product.

   $$2.0 \text{ g Fe} \cdot \frac{1 \text{ mol Fe}}{55.85 \text{ g Fe}} = 0.036 \text{ mol Fe}$$

   The amount of bromine contained would then be: (10.6 g - 2.0 g) or 8.6 g Br

   $$8.6 \text{ g Br} \cdot \frac{1 \text{ mol Br}}{79.9 \text{ g Br}} = 0.108 \text{ mol Br}$$

   b. The ratio of $\frac{Br}{Fe}$ is $\frac{0.108}{0.036}$ or 3. (to 1 sf)

   c. The empirical formula is $FeBr_3$.

   d. The balanced equation: $2 \text{ Fe (s)} + 3 \text{ Br}_2 (\ell) \rightarrow 2 \text{ FeBr}_3 \text{ (s)}$

   e. The product is $FeBr_3$: iron(III) bromide

   f. Statement (i) is correct. *Note that the addition of iron in excess of 1 gram, results in the formation of more product*. This product formation maximizes at about 2 grams of iron.

Chapter 5
Reactions in Aqueous Solution

Review Questions

6. Water soluble: $Ni(NO_3)_2$, $NiCl_2$ — nitrates and chlorides are soluble
   Water insoluble: $NiCO_3$, $Ni_3(PO_4)_2$ — carbonates and phosphates are insoluble

8. a. CuS is copper(II) sulfide
   b. $CaCO_3$ is calcium carbonate
   c. AgI is silver iodide

10. One example of a **gas-forming reaction** is the chemical puzzler at the introduction to this chapter.

$NaHCO_3$ (s) +	H^+ (aq)	→	Na^+ (aq) +	CO_2 (g) +	H_2O (ℓ)
sodium hydrogen carbonate	hydrogen ion (from citric acid)		sodium ion	carbon dioxide	water

Another example is the reaction occurring when bread rises. That reaction can be written:

$C_4H_6O_6$ (aq) +	HCO_3^- (aq)	→	$C_4H_5O_6^-$ (aq) +	CO_2 (g) +	H_2O (ℓ)
tartaric acid	hydrogen carbonate ion		tartrate ion	carbon dioxide	water

12.

Cl_2 (g)	+	2 NaBr (aq)	→	2 NaCl (aq)	+	Br_2 (ℓ)
reduced		oxidized				
<u>oxidizing agent</u>		<u>reducing agent</u>				

oxidation numbers: Cl in Cl_2: 0 Cl in NaCl: -1 chlorine **gains** electrons
Br in NaBr: -1 Br in Br_2 : 0 bromine **loses** electrons

14. A 0.500 M solution of KCl contains 0.500 mol of KCl per LITER of solution. One would place 1/4 of the KCl (0.125 mol of KCl) into the volumetric flask, add water until the KCl had dissolved. The distilled water should then be added to the volumetric flask until the level of solution in the flask reaches the calibration mark on the flask—indicating that the flask contains 250 mL. Upon proper swirling to achieve mixing, the flask would then contain 0.125mol/0.250 L or a 0.500 M solution of KCl.

16. Larger concentration of ions: 0.20 M $BaCl_2$ or 0.25 M NaCl.

Examine the concentrations of the constitutent ions in each of the solutions.
When 0.20 M $BaCl_2$ dissolves, it produces 0.20 M Ba^{2+} ions and (2 x 0.20) 0.40 M Cl^- .
When 0.25 M NaCl dissolves, it produces 0.25 M Na^+ ions and 0.25 M Cl^- The **total** concentration of ions in the $BaCl_2$ solution is greater.

PROPERTIES OF AQUEOUS SOLUTIONS

Solubility of Compounds

18. a. $FeCl_2$ is expected to be soluble.
    b. $AgNO_3$ is soluble. AgI and Ag_3PO_4 are not soluble.
    c. Li_2CO_3 , NaCl and $KMnO_4$ are soluble. In general, salts of the alkali metals are soluble.

20. a. NaCl is a soluble chloride.
    b. $Cu(OH)_2$ is an insoluble hydroxide.
    c. $BaCO_3$ is an insoluble carbonate.
    d. $NaNO_3$ is a soluble nitrate.

22.

Compound	Cation	Anion
a. KOH	K^+	OH^-
b. K_2SO_4	K^+	SO_4^{2-}
c. $NaNO_3$	Na^+	NO_3^-
d. $(NH_4)_2SO_4$	NH_4^+	SO_4^{2-}

24.

Compound	Water Soluble	Cation	Anion
a. Na_2CO_3	yes	Na^+	CO_3^{2-}
b. $CuSO_4$	yes	Cu^{2+}	SO_4^{2-}
c. NiS	no		
d. $CaBr_2$	yes	Ca^{2+}	Br^-

26. Anions to produce a soluble Cu^{2+} salt: SO_4^{2-} , NO_3^-, $C_2H_3O_2^-$,ClO_3^-, ClO_4^-, halides (F^-, Cl^-, Br^-, I^-),

Anions to produce an insoluble Cu^{2+} salt: S^{2-}, O^{2-}, $C_2O_4^{2-}$, SO_3^{2-}, OH^-, CO_3^{2-}

TYPES OF CHEMICAL REACTIONS

Precipitation Reactions

28. $CdCl_2(aq) + 2\ NaOH(aq) \rightarrow Cd(OH)_2\ (s) + 2\ NaCl\ (aq)$
Net ionic equation: $Cd^{2+}(aq) + 2\ OH^-(aq) \rightarrow Cd(OH)_2\ (s)$

30. Balanced equations for precipitation reactions:

a. $NiCl_2(aq) + (NH_4)_2S(aq) \rightarrow NiS(s) + 2\ NH_4Cl(aq)$

b. $3\ Mn(NO_3)_2(aq) + 2\ Na_3PO_4(aq) \rightarrow Mn_3(PO_4)_2(s) + 6\ NaNO_3(aq)$

32. $Pb(NO_3)_2(aq) + 2\ KOH(aq) \rightarrow Pb(OH)_2(s) + 2\ KNO_3(aq)$

$Pb(NO_3)_2(aq)$	$2\ KOH(aq)$	$Pb(OH)_2(s)$	$2\ KNO_3(aq)$
lead(II) nitrate	potassium hydroxide	lead (II) hydroxide	potassium nitrate

Writing Net Ionic Equations

34. a. $(NH_4)_2CO_3\ (aq) + Cu(NO_3)_2\ (aq) \rightarrow CuCO_3\ (s) + 2\ NH_4NO_3\ (aq)$
(net) $CO_3^{2-}\ (aq) + Cu^{2+}\ (aq) \rightarrow CuCO_3\ (s)$
b. $Pb(OH)_2\ (s) + 2\ HCl\ (aq) \rightarrow PbCl_2(s) + 2\ H_2O\ (\ell)$
(net) $Pb(OH)_2\ (s) + 2\ H^+(aq) + 2\ Cl^-\ (aq) \rightarrow PbCl_2(s) + 2\ H_2O\ (\ell)$
c. $BaCO_3\ (s) + 2\ HCl\ (aq) \rightarrow BaCl_2\ (aq) + H_2O\ (\ell) + CO_2\ (g)$
(net) $BaCO_3\ (s) + 2\ H^+\ (aq) \rightarrow Ba^{2+}(aq) + H_2O\ (\ell) + CO_2\ (g)$

36. a. $ZnCl_2\ (aq) + 2\ KOH\ (aq) \rightarrow 2\ KCl\ (aq) + Zn(OH)_2\ (s)$
(net) $Zn^{2+}(aq) + 2\ OH^-\ (aq) \rightarrow Zn(OH)_2\ (s)$
b. $AgNO_3\ (aq) + KI\ (aq) \rightarrow AgI(s) + KNO_3\ (aq)$
(net) $Ag^+\ (aq) + I^-\ (aq) \rightarrow AgI(s)$
c. $2\ NaOH\ (aq) + FeCl_2(aq) \rightarrow Fe(OH)_2\ (s) + 2\ NaCl\ (aq)$
(net) $2\ OH^-\ (aq) + Fe^{2+}(aq) \rightarrow Fe(OH)_2\ (s)$

Acid-Base Reactions

38. HNO_3 (aq) + H_2O (ℓ) → H_3O^+ (aq) + NO_3^- (aq)
alternatively HNO_3 (aq) → H^+ (aq) + NO_3^- (aq)

40. $H_2C_2O_4$(aq) → H^+(aq) + $HC_2O_4^-$(aq)
$HC_2O_4^-$(aq) → H^+(aq) + $C_2O_4^{2-}$(aq)

42. MgO(s) + H_2O(ℓ) → $Mg(OH)_2$(aq)

44. Complete and Balance

a. 2 CH_3CO_2H(aq) + $Mg(OH)_2$(s) → $Mg(CH_3CO_2)_2$(aq) + 2 H_2O(ℓ)

acetic acid	magnesium hydroxide	magnesium acetate	water

b. $HClO_4$(aq) + NH_3(aq) → NH_4ClO_4(aq)

perchloric acid	ammonia	ammonium perchlorate

46. Write and balance the equation:

$Ba(OH)_2$ (s) + 2 HNO_3 (aq) → $Ba(NO_3)_2$ (aq) + 2 H_2O (ℓ)

barium hydroxide	nitric acid	barium nitrate

Gas-Forming Reactions

48. $FeCO_3$ (s) + 2 HNO_3 (aq) → $Fe(NO_3)_2$ (aq) + H_2O (ℓ) + CO_2(g)

iron(II) carbonate	nitric acid	iron(II) nitrate	water	carbon dioxide

Types of Reactions in Aqueous Solution

50. Acid-Base (AB), Precipitation (PR), or Gas-Forming (GF)

a. $Ba(OH)_2$ (s) + 2 HCl (aq) → $BaCl_2$(aq) + 2 H_2O (ℓ) AB

b. 2 HNO_3 (aq) + $CoCO_3$ (s) → $Co(NO_3)_2$ (aq) + H_2O (ℓ) + CO_2 (g) GF

c. 2 Na_3PO_4 (aq) + 3 $Cu(NO_3)_2$ (aq) → $Cu_3(PO_4)_2$ (s) + 6 $NaNO_3$ (aq) PR

52. Acid-Base (AB) , Precipitation (PR), or Gas-Forming (GF)

a. $MnCl_2(aq) + Na_2S(aq) \rightarrow MnS(s) + 2\,NaCl(aq)$ PR
b. $K_2CO_3(aq) + ZnCl_2(aq) \rightarrow ZnCO_3(s) + 2\,KCl(aq)$ PR
c. $K_2CO_3(aq) + 2\,HClO_4(aq) \rightarrow 2\,KClO_4(aq) + CO_2(g) + H_2O(\ell)$ GF

Net ionic equations:
a. $Mn^{2+}(aq) + S^{2-}(aq) \rightarrow MnS(s)$
b. $CO_3^{2-}(aq) + Zn^{2+}(aq) \rightarrow ZnCO_3(s)$
c. $CO_3^{2-}(aq) + 2\,H^+(aq) \rightarrow CO_2(g) + H_2O(\ell)$

Oxidation-Reduction Reactions

54. For questions on oxidation number, read the symbol (x) as "the oxidation number of x."

a. BrO_3^- (Br) + 3(O) = -1

Since oxygen almost always has an oxidation number of -2, we can substitute this value and solve for the oxidation number of Br.

(Br) + 3(-2) = -1
(Br) = +5

b. $C_2O_4^{2-}$
2 (C) + 4 (O) = -2
2 (C) + 4 (-2) = -2
2 (C) + -8 = -2
2 (C) = +6
(C) = +3

c. F_2 The oxidation number for any free element is zero.

d. CaH_2
(Ca) + 2 (H) = 0
(Ca) + 2 (-1) = 0
(Ca) = +2

e. H_4SiO_4
4(H) + (Si) + 4(O) = 0
4(+1) + (Si) + 4(-2) = 0
(Si) = +4

f. SO_4^{2-}
(S) + 4(O) = -2
(S) + 4(-2) = -2
(S) = +6

56. a. Oxidation-Reduction: $Zn(s)$ has an oxidation number of 0, while $Zn^{2+}(aq)$ has an oxidation number of +2—hence Zn is being oxidized. N in NO_3^- has an oxidation number of +5, while N in NO_2 has an oxidation number of +4—hence N is being reduced.

b. Acid-Base reaction: There is no change in oxidation number for any of the elements in this reaction—hence it is NOT an oxidation-reduction reaction. H_2SO_4 is an acid, and $Zn(OH)_2$ acts as a base.

c. Oxidation-Reduction: $Ca(s)$ has an oxidation number of 0, while $Ca^{2+}(aq)$ has an oxidation number of +2—hence Ca is being oxidized. H in H_2O has an oxidation number of +1, while H in H_2 has an oxidation number of 0—hence H is being reduced.

58. Determine which reactant is oxidized and which is reduced:

a. $2\ Mg\ (s) + O_2\ (g) \rightarrow 2\ MgO\ (s)$

	ox. number			
specie	before	after	has experienced	functions as the
Mg	0	+2	oxidation	(Mg) reducing agent
O	0	-2	reduction	(O_2) oxidizing agent

b. $C_2H_4\ (g) + 3\ O_2\ (g) \rightarrow 2\ CO_2\ (g) + 2\ H_2O\ (g)$

	ox. number			
specie	before	after	has experienced	functions as the
C	-2	+4	oxidation	(C_2H_4) reducing agent
H	+1	+1	no change	
O	0	-2	reduction	(O_2) oxidizing agent

c. $Si\ (s) + 2\ Cl_2\ (g) \rightarrow SiCl_4\ (\ell)$

	ox. number			
specie	before	after	has experienced	functions as the
Si	0	+4	oxidation	(Si) reducing agent
Cl	0	-1	reduction	(Cl_2) oxidizing agent

Solution Concentration

60. Molarity of Na_2CO_3 solution:

$$6.73 \text{ g } Na_2CO_3 \cdot \frac{1 \text{ mol } Na_2CO_3}{106.0 \text{ g } Na_2CO_3} = 0.0635 \text{ mol } Na_2CO_3$$

$$\text{Molarity} \equiv \frac{\# \text{ mol}}{\text{L}} = \frac{0.0635 \text{ mol } Na_2CO_3}{0.250 \text{ L}} = 0.254 \text{ M } Na_2CO_3$$

Concentration of Na^+ and CO_3^{2-} ions:

$$\frac{0.254 \text{ mol } Na_2CO_3}{\text{L}} \cdot \frac{2 \text{ mol } Na^+}{1 \text{ mol } Na_2CO_3} = 0.508 \text{ M } Na^+$$

$$\frac{0.254 \text{ mol } Na_2CO_3}{\text{L}} \cdot \frac{1 \text{ mol } CO_3^{2-}}{1 \text{ mol } Na_2CO_3} = 0.254 \text{ M } CO_3^{2-}$$

62. Mass of $KMnO_4$:

$$\frac{0.0125 \text{ mol } KMnO_4}{\text{L}} \cdot \frac{0.250 \text{ L}}{1} \cdot \frac{158.0 \text{ g } KMnO_4}{1 \text{ mol } KMnO_4} = 0.494 \text{ g } KMnO_4$$

64. Volume of 0.123 M NaOH to contain 25.0 g NaOH:

Calculate moles of NaOH in 25.0 g:

$$\frac{25.0 \text{ g NaOH}}{1} \cdot \frac{1 \text{ mol NaOH}}{40.00 \text{ g NaOH}} = 0.625 \text{ mol NaOH}$$

The volume of 0.123 M NaOH that contains 0.625 mol NaOH:

$$0.625 \text{ mol NaOH} \cdot \frac{1 \text{ L}}{0.123 \text{ mol NaOH}} \cdot \frac{1 \times 10^3 \text{ mL}}{1\text{L}} = 5.08 \times 10^3 \text{ mL}$$

66. Molarity of Cu(II) sulfate in the diluted solution:

We can calculate the molarity if we know the number of moles of $CuSO_4$ in the 10.0 mL solution

1. Moles of $CuSO_4$ in 4.00 mL of 0.0250 M $CuSO_4$:

$$\text{M x V} = \frac{0.0250 \text{ mol } CuSO_4}{\text{L}} \cdot \frac{4.00 \times 10^{-3} \text{ L}}{1} = 1.00 \times 10^{-4} \text{ mol } CuSO_4$$

2. When that number of moles is distributed in 10.0 mL:

$$\frac{1.00 \times 10^{-4} \text{ mol } CuSO_4}{10.0 \times 10^{-3} \text{ L}} = 0.0100 \text{ M } CuSO_4$$

Perhaps a shorter way to solve this problem is to note the number of moles (found by multiplying the original molarity times the volume) is distributed in a given volume, resulting in the diluted molarity. Mathematically: $M_1 \times V_1 = M_2 \times V_2$

68. Using the definition of molarity we can calculate the concentration of H_2SO_4 in the following:

b. $\frac{6.00 \text{ mol } H_2SO_4}{L} \cdot 20.8 \times 10^{-3} \text{ L} = 0.125 \text{ moles of } H_2SO_4$

When amount of acid is diluted to 1.00 L, a solution of 0.125 M H_2SO_4 results.

70. Identity and concentration of ions in each of the following solutions:
   a. 0.25 M $(NH_4)_2SO_4$ gives rise to (2 x 0.25) 0.50 M NH_4^+ ions and 0.25 M SO_4^{2-} ions
   b. 0.056 M HNO_3 gives rise to 0.056 M H^+ ions and 0.056 M NO_3^- ions.
   c. 0.123 M Na_2CO_3 gives rise to (2 x 0.123) 0.246 M Na^+ ions and 0.123 M CO_3^{2-}.
   d. 0.00124 M $KClO_4$ gives rise to 0.00124 M K^+ ions and 0.00124 M ClO_4^- ions

Stoichiometry of Reactions in Solution

72. Volume of 0.125 M HNO_3 to react with 1.30 g of $Ba(OH)_2$:

Need several steps:

1. calculate mol of barium hydroxide in 1.30 g
2. calculate mole of HNO_3 needed to react with that # of mol of barium hydroxide
3. calculate volume of 0.125 M HNO_3 that contains that # of mol of nitric acid.

$$1.30 \text{ g } Ba(OH)_2 \cdot \frac{1 \text{ mol } Ba(OH)_2}{171.3 \text{ g } Ba(OH)_2} \cdot \frac{2 \text{ mol } HNO_3}{1 \text{ mol } Ba(OH)_2} \cdot \frac{1 \text{ L}}{0.125 \text{ mol } HNO_3}$$

and multiplying by the fraction $\frac{1000 \text{ mL}}{1 \text{ L}}$ gives 121 mL.

74. Mass of NaOH formed from 10.0 L of 0.15 M NaCl:

$$\frac{0.15 \text{ mol NaCl}}{1 \text{ L}} \cdot \frac{10.0 \text{ L}}{1} \cdot \frac{2 \text{ mol NaOH}}{2 \text{ mol NaCl}} \cdot \frac{40.0 \text{ g NaOH}}{1 \text{ mol NaOH}} = 60. \text{ g NaOH}$$

Mass of Cl_2 obtainable:

$$\frac{0.15 \text{ mol NaCl}}{1 \text{ L}} \cdot \frac{10.0 \text{ L}}{1} \cdot \frac{1 \text{ mol } Cl_2}{2 \text{ mol NaCl}} \cdot \frac{70.9 \text{ g } Cl_2}{1 \text{ mol } Cl_2} = 53 \text{ g } Cl_2$$

76. Calculate

1. mol of AgBr
2. mol of $Na_2S_2O_3$ needed to react (balanced equation)
3. volume of 0.0138 M $Na_2S_2O_3$ containing that number of moles.

$$0.250 \text{ g AgBr} \cdot \frac{1 \text{ mol AgBr}}{187.8 \text{ g AgBr}} \cdot \frac{2 \text{ mol } Na_2S_2O_3}{1 \text{ mol AgBr}} \cdot \frac{1 \text{ L}}{0.0138 \text{ mol } Na_2S_2O_3}$$

$$\cdot \frac{1000 \text{ mL}}{1 \text{ L}} = 193 \text{ mL } Na_2S_2O_3$$

78. The balanced equation:

$$Pb(NO_3)_2 + 2\,NaCl \rightarrow PbCl_2 + 2\,NaNO_3$$

Volume of 0.750 M $Pb(NO_3)_2$ needed:

$$\frac{2.25 \text{ mol NaCl}}{1 \text{ L}} \cdot \frac{1.00 \text{ L}}{1} \cdot \frac{1 \text{ mol } Pb(NO_3)_2}{2 \text{ mol NaCl}} \cdot \frac{1 \text{ L}}{0.750 \text{ mol } Pb(NO_3)_2} \cdot \frac{1000 \text{ mL}}{1 \text{ L}}$$

$$= 1500 \text{ mL or } 1.50 \times 10^3 \text{ mL}$$

80. To determine the possible amount of excess calcium carbonate, calculate # of moles of each reactant:

$$2.56 \text{ g } CaCO_3 \cdot \frac{1 \text{ mol } CaCO_3}{100.1 \text{ g } CaCO_3} = 0.0256 \text{ mol } CaCO_3 \text{ and}$$

$$0.250 \text{ L} \cdot \frac{0.125 \text{ mol HCl}}{1 \text{ L}} = 0.0313 \text{ mol HCl}$$

The balanced equation indicates that 2 mol HCl react PER MOLE of calcium carbonate, so the # of mol of $CaCO_3$ that can react is:

$$0.0313 \text{ mol HCl} \cdot \frac{1 \text{ mol } CaCO_3}{2 \text{ mol HCl}} = 0.0156 \text{ mol } CaCO_3.$$

So (0.0256 mol - 0.0156 mol) $CaCO_3$ remain. This amount would have a mass of:

$$0.010 \text{ mol} \cdot \frac{100.1 \text{ g } CaCO_3}{1 \text{ mol } CaCO_3} = 1.00 \text{ g of } CaCO_3$$

Mass of $CaCl_2$ produced:

The balanced equation indicates that we get 1 mol $CaCl_2$ for 1 mol of $CaCO_3$

$$0.0156 \text{ mol } CaCO_3 \cdot \frac{1 \text{ mol} CaCl_2}{1 \text{ mol } CaCO_3} \cdot \frac{111.0 \text{ g } CaCl_2}{1 \text{ mol } CaCl_2} = 1.73 \text{ g } CaCl_2$$

Titrations

82. To calculate the volume of HCl needed, we calculate the moles of NaOH in 1.33 g, then use the stoichiometry of the balanced equation:

$$HCl\ (aq) + NaOH\ (aq) \rightarrow NaCl\ (aq) + H_2O\ (\ell)$$

$$1.33\ g\ NaOH \cdot \frac{1\ mol\ NaOH}{40.00\ g\ NaOH} \cdot \frac{1\ mol\ HCl}{1\ mol\ NaOH} \cdot \frac{1\ L}{0.812\ mol\ HCl} \cdot \frac{1000\ mL}{1\ L} = 40.9\ mL\ HCl$$

84. Calculate:
   1. moles of Na_2CO_3 corresponding to 2.152 g Na_2CO_3
   2. moles of HCl that react with that number of moles (using the balanced equation)
   3. the volume of HCl containing that number of moles of HCl

   The balanced equation is:

$$Na_2CO_3\ (aq) + 2\ HCl\ (aq) \rightarrow 2\ NaCl\ (aq) + H_2O\ (\ell) + CO_2\ (g)$$

$$2.152\ g\ Na_2CO_3 \cdot \frac{1\ mol\ Na_2CO_3}{106.0\ g\ Na_2CO_3} \cdot \frac{2\ mol\ HCl}{1\ mol\ Na_2CO_3} \cdot \frac{1\ L}{0.955\ mol\ HCl} \cdot \frac{1000\ mL}{1\ L} = 42.5\ mL\ HCl$$

86. Mass of citric acid per 100. mL of soft drink:

   Calculate:
   1. mol NaOH used in the neutralization
   2. mol citric acid that react with that amount of NaOH (balanced equation)
   3. mass of citric acid corresponding to that number of moles.

$$\frac{0.0102\ mol\ NaOH}{1\ L} \cdot \frac{0.03351\ L}{1} \cdot \frac{1\ mol\ citric\ acid}{3\ mol\ NaOH} \cdot \frac{192.1\ g\ citric\ acid}{1\ mol\ citric\ acid} = 0.0219\ g\ citric\ acid$$

88. To determine the acid's identity, calculate the moles of each acid corresponding to 0.956 g.

$$0.956 \text{ g} \cdot \frac{1 \text{ mol citric acid}}{192.1 \text{ g citric acid}} = 4.98 \times 10^{-3} \text{ mol citric acid}$$

$$0.956 \text{ g} \cdot \frac{1 \text{ mol tartaric acid}}{150.1 \text{ g tartaric acid}} = 6.37 \times 10^{-3} \text{ mol tartaric acid}$$

Calculate the number of moles of NaOH present:

$$\frac{0.513 \text{ mol NaOH}}{1 \text{ L}} \cdot \frac{0.0291 \text{ L}}{1} = 1.49 \times 10^{-2} \text{ mol NaOH}$$

From the balanced equations note that each mole of citric acid requires 3 moles NaOH while each mole of tartaric acid requires 2 moles NaOH. Using the moles of citric acid and tartaric acid calculated above, calculate the amount of NaOH needed for each of the acids.

$$4.98 \times 10^{-3} \text{ mol citric acid} \cdot \frac{3 \text{ mol NaOH}}{1 \text{ mol citric acid}} = 1.49 \times 10^{-2} \text{ mol NaOH}$$

$$6.37 \times 10^{-3} \text{ mol tartaric acid} \cdot \frac{2 \text{ mol NaOH}}{1 \text{ mol tartaric acid}} = 1.27 \times 10^{-2} \text{ mol NaOH}$$

From this latter calculation we see that the solid acid is **citric acid.**

90. Mass of vitamin C in a 1.00 g tablet :

$$\frac{0.102 \text{ mol } Br_2}{1 \text{ L}} \cdot \frac{0.02785 \text{ L}}{1} \cdot \frac{1 \text{ mol } C_6H_8O_6}{1 \text{ mol } Br_2} \cdot \frac{176.1 \text{ g vitamin C}}{1 \text{ mol } C_6H_8O_6}$$

$$= 0.500 \text{ g vitamin C}$$

General Questions

92. $MgCO_3 (s) + 2 H^+(aq) + 2 Cl^-(aq) \rightarrow CO_2(g) + Mg^{2+}(aq) + 2 Cl^-(aq) + H_2O (\ell)$

1. Write species as they exist in aqueous solution, and name the spectator ions.

$MgCO_3 (s) + 2 H^+(aq) + 2 Cl^-(aq) \rightarrow CO_2(g) + Mg^{2+}(aq) + 2 Cl^-(aq) + H_2O (\ell)$

spectator ion (chloride ion)

2. Remove any species which appear in exactly the same form on both sides of the equation. For this reaction: 2 Cl^- (aq).

$MgCO_3 (s) + 2 H^+(aq) \rightarrow CO_2(g) + Mg^{2+}(aq) + H_2O (\ell)$

3. This is a gas-forming reaction.

94. a. A balanced equation:

$$(NH_4)_2S(aq) + Hg(NO_3)_2(aq) \rightarrow HgS(s) + 2\ NH_4NO_3(aq)$$

b. compounds named:

ammonium	mercury(II)	mercury(II)	ammonium
sulfide	nitrate	sulfide	nitrate

c. This is a **precipitation** reaction.

96. a. The reaction is a **gas-forming** reaction.

b. Calculating the # of moles of citric acid:

$$0.100 \text{ g of citric acid} \cdot \frac{1 \text{ mol citric acid}}{192.1 \text{ g citric acid}} = 5.21 \times 10^{-4} \text{ mol citric acid}$$

Note that citric acid can have THREE H^+ removed per molecule, so for 1 mol of citric acid there will need to be THREE moles of $NaHCO_3$. The mass of $NaHCO_3$ needed is:

$$5.21 \times 10^{-4} \text{ mol citric acid} \cdot \frac{3 \text{ mol } NaHCO_3}{1 \text{ mol citric acid}} \cdot \frac{84.01 \text{ g } NaHCO_3}{1 \text{ mol } NaHCO_3}$$

$$\cdot \frac{1000 \text{ mg } NaHCO_3}{1 \text{ g } NaHCO_3} = 131 \text{ mg } NaHCO_3$$

98. Diluting 10.0 mL of 2.56 M HCl to 250. mL results in a concentration of HCl of:

$$\text{moles HCl}: \frac{2.56 \text{ mol HCl}}{1 \text{ L}} \cdot 0.0100 \text{ L} = 2.56 \times 10^{-2} \text{ mol HCl}$$

$$\text{Molarity HCl} = \frac{2.56 \times 10^{-2} \text{ mol HCl}}{0.250 \text{ L}} \quad \text{or} \quad 0.102 \text{ M HCl (to 3 sf)}$$

Since HCl is a strong acid, we anticipate total dissociation into H^+ and Cl^- ions.

So 0.102 M HCl $\Rightarrow$ 0.102 M $H^+(aq)$ and 0.102 M Cl^- (aq).

100. The weight percent of $Na_2S_2O_3$ in the material:

1. Determine the moles of iodine in the solution (V x M)
2. Use the balanced equation to determine the # of moles of thiosulfate present
   The equation shows 1 mol I_2 reacts with 2 mol $Na_2S_2O_3$
3. Determine the mass of thiosulfate present, and from the total mass of sample—the percent of $Na_2S_2O_3$ in the sample.

$$\frac{0.246 \text{ mol } I_2}{1 \text{ L}} \cdot \frac{0.04021 \text{ L}}{1} \cdot \frac{2 \text{ mol } Na_2S_2O_3}{1 \text{ mol } I_2} \cdot \frac{158.1 \text{ g } Na_2S_2O_3}{1 \text{ mol } Na_2S_2O_3} = 3.13 \text{ g } Na_2S_2O_3$$

$$\% \ Na_2S_2O_3 = \frac{3.13 \text{ g } Na_2S_2O_3}{3.232 \text{ g sample}} \cdot 100 = 96.8 \ \% \ Na_2S_2O_3$$

Conceptual Questions

102. Prepare zinc chloride by:

   a. an acid-base reaction:

      $Zn(OH)_2 \text{ (s)} + 2 \text{ HCl (aq)} \rightarrow ZnCl_2 \text{ (aq)} + 2 \text{ } H_2O \text{ } (\ell)$

      To obtain the salt, you would need to heat the solution to drive off the water, leaving zinc chloride.

   b. a gas-forming reaction

      $ZnCO_3 \text{ (s)} + 2 \text{ HCl (aq)} \rightarrow ZnCl_2 \text{ (aq)} + H_2O \text{ } (\ell) + CO_2 \text{ (g)}$

   c. an oxidation-reduction reaction

      $Zn \text{ (s)} + 2 \text{ HCl (aq)} \rightarrow ZnCl_2 \text{ (aq)} + H_2 \text{ (g)}$

104. To decide the relative concentrations, calculate the dilutions. Let's assume that the HCl has a concentration of 0.100 M. Then calculate the diluted concentrations in each case.

Student 1 :

20.0 mL of 0.100 M HCl is diluted to 40.0 mL total

The diluted molarity is then :

$$\frac{0.100 \text{ mol HCl}}{1 \text{ L}} \cdot 0.0200 \text{ L} = 0.400 \text{ L} \cdot \text{M}$$

$$0.050 \frac{\text{mol HCl}}{\text{L}} = \text{M}$$

Student 2 :

20.0 mL of 0.100 M HCl is diluted to 80.0 mL total

The diluted molarity is :

$$\frac{0.100 \text{ mol HCl}}{1\text{L}} \cdot 0.0200 \text{ L} = 0.0800 \text{ L} \cdot \text{M}$$

$$0.025 \frac{\text{mol HCl}}{\text{L}} = \text{M}$$

So the second student's molarity is half the concentration of the first student's. Noting that the volumes of the two solutions differ by a factor of 2, and their concentrations differ by a factor of 2, when the two students calculate **the concentration of the original**

solution, they will find identical concentrations (e)—since they both used equal volumes (and equal number of moles) of the original HCl solution.

Challenging Questions

106. Since the entire sample of dye (1.0 g) was eventually dissolved in the swimming pool, the ratio of **moles of dye/volume of pool** is 4.1×10^{-8} mol/L. Recalling the relationship: # moles = M • V we can calculate the volume of solution that would produce this concentration with the # of moles of dye in 1.0 g.

$$1.0 \text{ g dye} \cdot \frac{1 \text{ mol dye}}{319.85 \text{ g dye}} = 3.1 \times 10^{-4} \text{ mol dye.}$$

Substituting this value into the equation gives:

$$3.1 \times 10^{-4} \text{ mol dye} = 4.1 \times 10^{-8} \text{ mol dye/L} \cdot V$$, and rearranging gives

$\frac{3.1 \times 10^{-4} \text{ mol dye}}{4.1 \times 10^{-8} \text{ mol dye/L}} = V = 76{,}000$ L. Converting this to the units of gallons we get:

$$76{,}000 \text{ L} \cdot \frac{1.056 \text{ quarts}}{1 \text{ gallon}} \cdot \frac{1 \text{ gallon}}{4.00 \text{ quarts}} = 20{,}000 \text{ gallons}$$

108. The precipitation reaction may be written:

$$FeCl_3 \text{ (aq)} + 3\ NaOH \rightarrow Fe(OH)_3 \text{ (s)} + 3\ NaCl \text{ (aq)}$$

Calculate the # of moles of each reactant present:

$$\frac{0.234 \text{ mol } FeCl_3}{1 \text{ L}} \cdot \frac{0.0250 \text{ L}}{1} = 5.85 \times 10^{-3} \text{ mol } FeCl_3$$

$$\frac{0.453 \text{ mol NaOH}}{1 \text{ L}} \cdot \frac{0.0425 \text{ L}}{1} = 1.93 \times 10^{-2} \text{ mol NaOH}$$

If we take the moles of iron(III) chloride and multiply by 3 we get: 1.76×10^{-2} mol. Since we have **more moles of NaOH** than this amount, the $FeCl_3$ is the **limiting reactant—and NaOH is present in excess.**

The maximum mass of $Fe(OH)_3$ obtainable is:

$$5.85 \times 10^{-3} \text{ mol } FeCl_3 \cdot \frac{1 \text{ mol } Fe(OH)_3}{1 \text{ mol } FeCl_3} \cdot \frac{106.9 \text{ g } Fe(OH)_3}{1 \text{ mol } Fe(OH)_3} = 0.625 \text{ g } Fe(OH)_3$$

The concentration of NaOH remaining is found by subtracting the NaOH used **from** the NaOH available, and dividing by the **total volume** (of 25.0 + 42.5 mL):

$$\frac{(0.0193 - 0.0176)\text{ mol NaOH}}{0.0675\text{ L}} = 0.025\text{ M}$$

110. Calculate the # of moles of HCl used (and # moles of NH_3)

$$\frac{1.500\text{ mol HCl}}{1\text{L}} \bullet \frac{0.02363\text{ L}}{1} \bullet \frac{1\text{ mol } NH_3}{1\text{ mol HCl}} = 0.03545\text{ mol } NH_3$$

To determine the formula of the compound, we need to determine the number of moles of ammonia per mole of compound. We can do that if we know the # of moles of compound present. We can calculate that by dividing the mass of compound by its molar mass.

mol compound = $\frac{1.580\text{ g compound}}{\text{MM}}$. We know the MM of the compound is equal to the sum of all the moles of Co(III) chloride and the moles of ammonia. We can represent that as:

$$\frac{1.580\text{ g compound}}{\frac{1\text{ mol } CoCl_3}{1} \bullet \frac{165.29\text{ g } CoCl_3}{1\text{ mol } CoCl_3} + \frac{x\text{ mol } NH_3}{1} \bullet \frac{17.03\text{ g } NH_3}{1\text{ mol } NH_3}}$$

simplified that can be written as: $\frac{1.580\text{ g compound}}{165.29 + x \bullet 17.03\text{ g}}$

What is also true is that we can express the # of moles of compound as:

mol compound $= \frac{1\text{ mol compound}}{x\text{ mol } NH_3} \bullet 0.03545\text{ mol } NH_3$

Since both these fraction are equal to the # of moles of compound, we can set them equal to one another:

$$\frac{1.580\text{ g compound}}{165.29 + x \bullet 17.03\text{ g}} = \frac{1\text{ mol compound}}{x\text{ mol } NH_3} \bullet 0.03545\text{ mol } NH_3\ .$$

Solving for x we get:

$$0.03545 \bullet (165.29 + x \bullet 17.03\text{ g}) = 1.580x$$

$$5.8596 + 0.6037x = 1.580x$$

$$5.8596 = 0.976x \text{ and } x = 6.$$

So the formula for the compound is: $Co(NH_3)_6Cl_3$

Summary Question

112. a. The balanced equation for the formation of cisplatin:

$(NH_4)_2PtCl_4 + 2\ NH_3 \rightarrow Pt(NH_3)_2Cl_2 + 2\ NH_4Cl$

b. Volume of 0.125 M NH_3 required to make 12.50 g of cisplatin:

$$12.5 \text{ g } Pt(NH_3)_2Cl_2 \cdot \frac{1 \text{ mol } Pt(NH_3)_2Cl_2}{300.0 \text{ g } Pt(NH_3)_2Cl_2} \cdot \frac{2 \text{ mol } NH_3}{1 \text{ mol } Pt(NH_3)_2Cl_2}$$

or 0.0833 mol NH_3

Calculating the volume that contains that number of moles of NH_3

$$0.0833 \text{ mol } NH_3 \cdot \frac{1 \text{ L}}{0.125 \text{ mol } NH_3} = 0.667 \text{ L or } 667 \text{ mL}$$

c. Determine the value of x in the compound $Pt(NH_3)_2Cl_2(C_5H_5N)_x$

Determine the # of moles of cisplatin in 1.50 g:

$$1.50 \text{ g cisplatin} \cdot \frac{1 \text{ mol } Pt(NH_3)_2Cl_2}{300.0 \text{ g } Pt(NH_3)_2Cl_2} = 0.00500 \text{ mol cisplatin}$$

Determine the **total** number of moles of pyridine:

$$1.50 \text{ mL pyridine} \cdot \frac{0.979 \text{ g pyridine}}{1 \text{ mL pyridine}} \cdot \frac{1 \text{ mol pyridine}}{79.10 \text{ g pyridine}} = 0.0186 \text{ mol pyridine}$$

Once some of the pyridine has reacted, the amount of pyridine **remaining** can be calculated by titrating with HCl. The balanced equation shows that 1 mol of HCl reacts with 1 mol pyridine. The # moles HCl (and therefore the # moles pyridine) is:

$$\frac{0.475 \text{ mol HCl}}{1\text{L}} \cdot 0.0370 \text{ L} = 0.0176 \text{ mol HCl} = 0.0176 \text{ mol pyridine remaining}$$

The pyridine that reacted is then (0.0186 mol - 0.0176 mol) = 0.0010 mol pyridine.

The # of moles of pyridine PER mol of cisplatin is then : $\frac{0.0010 \text{ mol pyridine}}{0.00500 \text{ mol cisplatin}} = 2$

The formula is then: $Pt(NH_3)_2Cl_2(C_5H_5N)_2$.

Chapter 6
Principles of Reactivity: Energy and Chemical Reactions

NUMERICAL QUESTIONS

Energy Units

10. Express 1200 Calories/day in Joules:

$$\frac{1200 \text{ Cal}}{\text{day}} \bullet \frac{1000 \text{ calorie}}{1 \text{ Cal}} \bullet \frac{4.184 \text{ J}}{1 \text{ cal}} = 5.0 \times 10^6 \text{ Joules/day}$$

12. Energy lost per day through the glass door:

$$\frac{1.0 \times 10^6 \text{ J}}{1 \text{ hr}} \bullet \frac{1 \text{ kJ}}{1000 \text{ J}} \bullet \frac{24 \text{ hr}}{1 \text{ day}} = \frac{2.4 \times 10^4 \text{ kJ}}{\text{day}}$$

$$\frac{2.4 \times 10^4 \text{ kJ}}{\text{day}} \bullet \frac{1000 \text{ J}}{1 \text{ kJ}} \bullet \frac{1 \text{ kwh}}{3.61 \times 10^6 \text{ J}} = \frac{6.6 \text{ kwh}}{\text{day}}$$

Specific Heat

14. Heat energy to warm 168g nickel from -15.2 °C to 23.6 °C:

Heat $=$ mass x heat capacity x ΔT

Heat $= (168 \text{ g})(\frac{0.445 \text{ J}}{\text{g} \bullet \text{K}})(296.8 \text{ K} - 258.0 \text{ K}) = 2.90 \times 10^3 \text{ J}$

Note that we have converted 23.6 °C and -15.2 °C to their Kelvin equivalents, by adding 273.2.

16. Heat energy to warm nickel from -15.2 °C to 23.6 °C:

Heat $=$ mass x heat capacity x ΔT

For water $= (50.0 \text{ g})(\frac{4.184 \text{ J}}{\text{g} \bullet \text{K}})(63 \text{ K})$

$= 1.32 \times 10^4$ J or 13 kJ (2 sf)

For aluminum $= (200. \text{ g})(\frac{0.902 \text{ J}}{\text{g} \bullet \text{K}})(63 \text{ K})$

$= 1.14 \times 10^4$ J or 11 kJ (2 sf)

Note that the temperature difference is 63 K whether you calculate the temperature difference in Celsius (85 °C - 22 °C) or Kelvin (358 K - 295 K).

18. Using the specific heat for aluminum (0.902 $\frac{J}{g \cdot K}$) and the change in T (255 °C - 25 °C), calculate the energy added:

$$q_{Al} = \text{(mass)(heat capacity)}(\Delta T)$$
$$= (485\ g)(0.902\ \frac{J}{g \cdot K})(230.\ K)$$
$$= 100{,}600\ J \text{ or } 101\ kJ$$

20. While you could calculate the heat energy needed using the equation, $q = \text{(mass)(heat capacity)}(\Delta T)$, you can easily answer this question by noting the relative magnitudes of the heat capacity for ethylene glycol (2.42 $\frac{J}{g \cdot K}$) and water (4.184 $\frac{J}{g \cdot K}$). **Water would require more heat energy**. See question 16 for another example of this problem.

22. Final T of copper-water mixture:

We must **assume** that **no energy** will be transferred to or from the beaker containing the water. Then the **magnitude** of energy lost by the hot copper and the energy gained by the cold water will be equal (but opposite in sign).

$$q_{copper} = -q_{water}$$

Using the heat capacities of H_2O and copper, and expressing the temperatures in Kelvin we can write:

$$(192\ g)(0.385\ \frac{J}{g \cdot K})(T_{final} - 373.2\ K) = -\overbrace{(750.\ mL)(1.00\frac{g}{mL})}^{\text{mass of water}}(4.184\ \frac{J}{g \cdot K})(T_{final} - 277.2\ K)$$

Simplifying each side gives:

$$73.92\frac{J}{K} \cdot T_{final} - 27{,}600\ J = -3138\frac{J}{K} \cdot T_{final} + 870{,}000\ J$$
$$3212\frac{J}{K} \cdot T_{final} = 897600\ J$$
$$T_{final} = 279\ K \text{ or } 6\ °C$$

Don't forget: **Round numbers only at the end**. To show the steps, I have rounded some of these intermediate numbers.

24. Initial temperature of gold sample:

This problem is solved almost exactly like question 22. The difference is that we know T_{final} for the gold.

$$q_{gold} = -q_{water}$$

$$(182 \text{ g})(0.128 \frac{\text{J}}{\text{g} \cdot \text{K}})(300.7 \text{ K} - T_{initial}) = -(22.1 \text{ g})(4.184 \frac{\text{J}}{\text{g} \cdot \text{K}})(2.5 \text{ K})$$

$$7005. \text{ J} - 23.3 \frac{\text{J}}{\text{K}} \cdot T_{initial} = -231 \text{ J}$$

rearranging:

$$7236 \text{ J} = 23.5 \text{ J/K} \cdot T_{initial}$$

$$311 \text{ K} = T_{initial}$$

or 37 °C

26. Example 6.3 in your text is a good template for this problem.

$$q_{metal} = -q_{water}$$

$$(150.0 \text{ g})(C_{metal})(296.5 \text{ K} - 353.2 \text{ K}) = -(150.0 \text{ g})(4.184 \frac{\text{J}}{\text{g} \cdot \text{K}})(296.5 \text{ K} - 293.2 \text{ K})$$

$$-8505 \text{ g} \cdot \text{K}(C_{metal}) = -2071.1 \text{ J}$$

$$C_{metal} = 0.24 \frac{\text{J}}{\text{g} \cdot \text{K}}$$

Changes of State

28. Quantity of energy to melt 16 ice cubes at 0 °C.

The mass of ice involved:

$$16 \text{ ice cubes} \cdot \frac{62.0 \text{ g ice}}{1 \text{ ice cube}} = 992 \text{ g ice}$$

To melt 992 g ice:

$$992 \text{ g ice} \cdot \frac{333 \text{ J}}{1.000 \text{ g ice}} = 330. \times 10^3 \text{ J or } 330. \text{ kJ}$$

30. To calculate the quantity of heat for the process described, think of the problem in three steps:

1. melt ice at 0°C to liquid water at 0°C
2. warm liquid water from 0°C to 100°C
3. convert liquid water at 100 °C to gaseous water at 100 °C

1. The energy to melt 60.1 g of ice at 0 °C is:

$$60.1 \text{ g ice} \cdot \frac{333 \text{ J}}{\text{g ice}} = 2.00 \times 10^4 \text{ J}$$

2. The energy required to warm the liquid water from 0°C to 100 °C (ΔT = 100 K) is:

$$(4.18 \text{ J/g} \cdot \text{K}) \cdot 60.1 \text{ g} \cdot 100 \text{ K} = 2.51 \times 10^4 \text{ J}$$

3. To convert liquid water at 100 °C to gaseous water at 100 °C:

$$2260 \text{ J/g} \cdot 60.1 \text{ g} = 13.6 \times 10^4 \text{ J}$$

The total energy required is: $[2.00 \times 10^4 + 2.51 \times 10^4 \text{ J} + 13.6 \times 10^4 \text{J}] = 1.81 \times 10^5 \text{ J}$

or 181 kJ

32. To accomplish the process, one must:

1. heat the tin from 25.0 °C to 231.9 °C ($\Delta T = 206.9$ K)
2. melt the tin at 231.9 °C

Using the specific heat for tin, the energy for the first step is:

$$(0.227 \frac{\text{J}}{\text{g} \cdot \text{K}})(454 \text{ g})(206.9 \text{ K}) = 21{,}300 \text{ J}$$

To melt the tin at 231.9 °C, we need:

$$59.2 \frac{\text{J}}{\text{g}} \cdot 454 \text{ g} = 26{,}900 \text{ J}$$

The total heat energy needed (in J) is (21,300 + 26,900) = 48,200 J or 48.2 kJ

Enthalpy

Note that in this chapter, I have left negative signs with the value for heat released

(heat released = - ; heat absorbed = +)

34. For a process in which the $\Delta H°$ is negative, that process is **exothermic**.

To calculate heat released when 1.25 g NO react, note that the energy shown (-114.1 kJ) is released when **2** moles of NO react, so we'll need to account for that:

$$1.25 \text{ g NO} \cdot \frac{1 \text{ mol NO}}{30.01 \text{ g NO}} \cdot \frac{-114.1 \text{ kJ}}{2 \text{ mol NO}} = -2.38 \text{ kJ}$$

36. The combustion of isooctane (IO) is **exothermic**. The molar mass of IO is: 114.2 g/mol.

The heat evolved is:

$$1.00 \text{ L of IO} \cdot \frac{0.6878 \text{ g IO}}{1\text{mL}} \cdot \frac{1 \times 10^3 \text{ mL}}{1 \text{ L}} \cdot \frac{1 \text{ mol IO}}{114.2 \text{ g IO}} \cdot \frac{10922 \text{ kJ}}{2 \text{ mol IO}} = 32{,}900 \text{ kJ}$$

38. The molar mass of CH_3OH is 32.04 g/mL

$$0.0466 \text{ g } CH_3OH \cdot \frac{1 \text{ mol } CH_3OH}{32.04 \text{ g } CH_3OH} = 1.45 \times 10^{-3} \text{ mol } CH_3OH$$

The enthalpy change will be:

$$\frac{-1110 \text{ J}}{1.45 \times 10^{-3} \text{ mol } CH_3OH} = -763{,}000 \frac{\text{J}}{\text{mol}} \text{ or } -763 \frac{\text{kJ}}{\text{mol}} \text{ (molar heat of combustion)}$$

The ΔH°rxn for the equation shown would be -1526 kJ (or -1530 to 3 significant figures).

Hess's Law

40. The molar enthalpy of formation for benzene may be calculated by describing **several processes** that **when added** are:

$$6\ C(s) + 3\ H_2(g) \rightarrow C_6H_6\ (\ell)$$

Begin with the equation given for the combustion of benzene.

Reversing it gives:

$12\ CO_2(g) + 6\ H_2O(\ell) \rightarrow 2\ C_6H_6(\ell) + 15\ O_2(g)$ $\Delta H° = +6534.8$ kJ

Use the equation for the formation of CO_2, noting that we need 12 $CO_2(g)$

$12\ C(s) + 12\ O_2(g) \rightarrow 12\ CO_2(g)$ $\Delta H° = (-393.509)(12)$ kJ

and the equation for the formation of H_2O, noting that we need 6 $H_2O(\ell)$

$6\ H_2(g) + 3\ O_2(g) \rightarrow 6\ H_2O(\ell)$ $\Delta H° = (-285.830)(6)$ kJ

Adding these 3 equations gives

$$12C(s) + 6\ H_2(g) \rightarrow 2\ C_6H_6\ (\ell) \qquad \Delta H° = +97.9\ \text{kJ}$$

Halving all the coefficients provides the desired equation with a $\Delta H = +49.0$ kJ

42. The desired equation is: $Pb(s) + 1/2\ O_2(g) \rightarrow PbO(s)$

1.	$Pb(s) + CO(g) \rightarrow PbO(s) + C(s)$	$\Delta H° = -106.8$ kJ
1/2 x 2.	$C(s) + 1/2\ O_2(g) \rightarrow CO(g)$	$\Delta H° = (-221.0\ \text{kJ})(0.5)$
	$Pb(s) + 1/2\ O2(g) \rightarrow PbO(s)$	$\Delta H° = -217.3$ kJ

The process is **exothermic**. The heat evolved when 250 g Pb react:

$$250\ \text{g Pb} \cdot \frac{1\ \text{mol Pb}}{207.2\ \text{g Pb}} \cdot \frac{-217.3\text{kJ}}{1\ \text{mol Pb}} = -260\ \text{kJ (2 sf)}$$

Standard Enthalpies of Formation

44. The equation requested requires that we form **one** mol of product (chromium(III) oxide) from its elements—each in their standard state.

Begin by writing a balanced equation:

$$4\ Cr(s) + 3\ O_2(g) \rightarrow 2\ Cr_2O_3(s)$$

Now express the reaction so that you form one mole of Cr_2O_3—divide all coefficients by 2

$$2\ Cr(s) + 3/2 O_2(g) \rightarrow Cr_2O_3(s)$$

46. a. The ΔH° is **negative**; the formation of glucose from its elements is **exothermic**.

b. The balanced equation:

$$6\ C(graphite) + 3\ O_2(g) + 6\ H_2(g) \rightarrow C_6H_{12}O_6(s)$$

48. Use Hess's Law to obtain the overall $\Delta H°_{rxn}$: Note that we'll need to **reverse the second equation** shown. The result of reversing the second equation is to change the sign of ΔH° for that equation.

$Pb(s) + 2\ Cl_2(g) \rightarrow PbCl_4(\ell)$	$\Delta H°_f$ = -329.3 kJ
$PbCl_4(\ell) \rightarrow PbCl_2(s) + Cl_2(g)$	ΔH° = - 30.1 kJ
$Pb(s) + Cl_2(g) \rightarrow PbCl_2(s)$	ΔH° = -359.4 kJ

50. The enthalpic change for SO_3 formation:

$$SO_2(g) + 1/2\ O_2(g) \rightarrow SO_3(g)$$

$$\begin{aligned}\Delta H°_{rxn} &= \Delta H°_f\ SO_3 - [\Delta H°_f\ SO_2 + 1/2\ \Delta H°_f\ O_2] \\ &= (-395.7\ kJ/mol)(1\ mol) - [(-296.8\ kJ/mol)(1\ mol) + (0\ kJ/mol)(1/2\ mol)] \\ &= -395.7\ kJ + 296.8\ kJ \\ &= -98.9\ kJ\end{aligned}$$

Since the enthalpic change is negative, the reaction is **exothermic**.

52. a. The enthalpy change for the reaction:

	$4\ NH_3(g)$	$+\ 5\ O_2(g)$	$\rightarrow 4\ NO(g)$	$+\ 6\ H_2O(g)$
$\Delta H°_f$(kJ/mol)	-46.1	0	+90.3	-241.8

$$\begin{aligned}\Delta H°_{rxn} &= [\ (4\ mol)(+90.3\ \frac{kJ}{mol}) + (6\ mol)(-241.8\ \frac{kJ}{mol})\ ] - [\ (4\ mol)(-46.1\ \frac{kJ}{mol}) + (5\ mol)(0)] \\ &= (-1089.6\ kJ) - (-184.4\ kJ) \\ &= -905.2\ kJ\end{aligned}$$

The reaction is **exothermic**.

b. Heat evolved when 10.0 g NH_3 react:

The balanced equation shows that 4 mol NH_3 result in the release of 905.2 kJ.

$$10.0 \text{ g } NH_3 \cdot \frac{1 \text{ mol } NH_3}{17.03 \text{ g } NH_3} \cdot \frac{-905.2 \text{ kJ}}{4 \text{ mol } NH_3} = -133 \text{ kJ}$$

54. a. The enthalpy change for the reaction:

$$WO_3(s) + 3\, H_2(g) \rightarrow W(s) + 3\, H_2O(\ell)$$

$\Delta H°_f$(kJ/mol)	-842.9	0	0	-285.8

$$\Delta H°_{rxn} = [\,(3 \text{ mol})(-285.8 \tfrac{\text{kJ}}{\text{mol}})] - [(1 \text{ mol})(-842.9 \tfrac{\text{kJ}}{\text{mol}})]$$

$$= (-857.4 \text{ kJ}) - (-842.9 \text{ kJ})$$

$$= -14.5 \text{ kJ}$$

b. The heat **evolved** when 1.00 g WO_3 reacts:

$$1.00 \text{ g } WO_3 \cdot \frac{1 \text{ mol } WO_3}{231.8 \text{ g } WO_3} \cdot \frac{-14.5 \text{ kJ}}{1 \text{ mol } WO_3} = -0.0626 \text{ kJ}$$

or - 62.6 J

56. The molar enthalpy of formation of naphthalene can be calculated since we're given the enthalpic change for the reaction:

$$C_{10}H_8\,(s) + 12\, O_2(g) \rightarrow 10\, CO_2(g) + 4\, H_2O(\ell)$$

$\Delta H°_f$(kJ/mol)	?	0	-393.5	-285.8

$$\Delta H°_{rxn} = \Sigma \Delta H°_f \text{ products} - \Sigma \Delta H°_f \text{ reactants}$$

$$-5156.1 \text{ kJ} = [(10 \text{ mol})(-393.5 \tfrac{\text{kJ}}{\text{mol}}) + (4 \text{ mol})(-285.8 \tfrac{\text{kJ}}{\text{mol}})] - [\Delta H°_f\, C_{10}H_8]$$

$$-5156.1 \text{ kJ} = (-5078.2 \text{ kJ}) - \Delta H°_f\, C_{10}H_8$$

$$-77.9 \text{ kJ} = -\Delta H°_f\, C_{10}H_8$$

$$77.9 \text{ kJ} = \Delta H°_f\, C_{10}H_8$$

Calorimetry

58. Heat evolved = Heat absorbed by bomb and water

$$-q_{rxn} = q_{bomb} + q_{water}$$

$$q_{bomb} = C \cdot \Delta t = (650 \text{ J/ K})(3.33 \text{ K}) = 2.16 \times 10^3 \text{ J}$$

(where C = heat capacity of bomb)

$$q_{water} = S \bullet m \bullet \Delta t = (4.184\ J/g \bullet K)(320.\ g)(3.33\ K) = 4.46 \times 10^3\ J$$

(where S = specific heat of water)

$$-q_{rxn} = 2.16 \times 10^3\ J + 4.46 \times 10^3\ J = 6.62 \times 10^3\ J$$

$$q_{rxn} = -6.62 \times 10^3\ J \text{ or } -6.62\ kJ$$

60. Heat evolved by reaction = - Heat absorbed by surroundings

= - (Heat absorbed by bomb and water)

= - (q_{bomb} + q_{water})

Δt = (27.38°C - 25.00°C) = 2.38 °C or 2.38 K

q_{bomb} = (893 J/K) • 2.38 K = 2130 J

q_{water} = (775 g)(4.18 J/g •K) • 2.38 K = 7710 J

Heat absorbed by bomb + water = 9840 J or 9.84 kJ

The heat evolved by reaction of 0.300 g C is then: -9.84 kJ

Expressing this on a molar basis:

$$\frac{-9.84\ kJ}{0.300\ g\ C} \bullet \frac{12.01\ g\ C}{1\ mol\ C} = \frac{-394\ kJ}{1\ mol\ C}$$

62. 100.0 mL of 0.200 M CsOH and 50.0 mL of 0.400 M HCl each supply 0.0200 moles of base and acid respectively. If we assume the specific heat capacities of the solutions are 4.2 J/g • K, the **heat evolved** for 0.0200 moles of CsOH is:

q = (4.2 J/g • K)(150. g)(24.28 °C - 22.50°C) [and since 1.78°C = 1.78 K]

q = (4.2 J/g • K)(150. g)(1.78 K)

q = 1120 J

The molar enthalpy of neutralization is : $\frac{-1120\ J}{0.0200\ mol\ CsOH}$ = - 56000 J/mol (2 sf)

or -56 kJ/mol

64. Heat absorbed by the ice : $\frac{333\ J}{1.00\ g\ ice}$ • 7.33 g ice = 2440 J (to 3 sf)

[NOTE the data 333 J/g ice is provided in problem 28 as well as elsewhere.]

Since this energy (2440 J) is released by the metal, we can calculate the heat capacity of the metal:

heat = heat capacity • mass • ΔT

-2440 J = C • 50.0 g • (273.2 K - 373 K) [Note that ΔT is negative!]

$$0.489 \frac{J}{g \cdot K} = C$$

Note that the heat released (left side of equation) has a negative sign to indicate the **directional flow** of the energy.

GENERAL QUESTIONS

66. Greater quantity of heat released :

To calculate heat transferred:

$$q = \text{(mass)(heat capacity)}(\Delta T)$$

$$q_{H_2O} = (50.0 \text{ g})(\frac{4.184 \text{ J}}{\text{g} \cdot \text{K}})(283 \text{ K} - 323 \text{ K})$$

$$= -8400 \text{ J} \quad \text{(to 2 sf)}$$

$$q_{C_2H_5OH} = (100.0 \text{ g})(\frac{2.46 \text{ J}}{\text{g} \cdot \text{K}})(283 \text{ K} - 323 \text{ K})$$

$$= -9800 \text{ J} \quad \text{(to 2 sf)}$$

Ethanol releases (q is negative) more heat.

68. The enthalpy change for the reaction:

	$Mg(s)$ +	$2 H_2O(\ell)$ →	$Mg(OH)_2(s)$ +	$H_2(g)$
$\Delta H°_f$(kJ/mol)	0	-285.8	-924.5	0

$$\Delta H°_{rxn} = (1 \text{ mol})(-924.5 \frac{\text{kJ}}{\text{mol}}) - (2 \text{ mol})(-285.8 \frac{\text{kJ}}{\text{mol}})$$

$$= -352.9 \text{ kJ} \quad \text{or } -3.529 \times 10^5 \text{ J}$$

Each mole of magnesium releases 352.9 kJ of heat energy.

Calculate the heat required to warm 25 mL of water from 25 to 85 °C.

$$\text{heat} = \text{heat capacity} \times \text{mass} \times \Delta T$$

$$= (4.184 \frac{\text{kJ}}{\text{mol}})(25 \text{ mL})(\frac{1.00 \text{ g}}{1 \text{ mL}})(60 \text{ K})$$

$$= 6276 \text{ or } 6300 \text{ J} \quad \text{or } 6.3 \text{ kJ} \quad \text{(to 2 sf)}$$

Magnesium required:

$$6.3 \text{ kJ} \cdot \frac{1 \text{ mol Mg}}{352.9 \text{ kJ}} \cdot \frac{24.3 \text{ g Mg}}{1 \text{ mol Mg}} = 0.43 \text{ g Mg}$$

70. Calculate the $\Delta H°_f$ for $N_2H_4(\ell)$

	$N_2H_4(\ell)$ +	$O_2(g)$ →	$N_2(g)$ +	$2 H_2O(g)$
$\Delta H°_f$(kJ/mol)	?	0	0	-241.8

Given that the ΔH°_{rxn} is -534 kJ.

$$\Delta H^\circ_{rxn} = [(2\text{ mol})(-241.8\ \frac{kJ}{mol})] - \Delta H^\circ_f\ N_2H_4(\ell)$$

$$-534\text{ kJ} = -483.6\text{ kJ} - \Delta H^\circ_f\ N_2H_4(\ell)$$

$$-50.\text{ kJ} = -\Delta H^\circ_f\ N_2H_4(\ell)$$

$$50.\text{ kJ} = \Delta H^\circ_f\ N_2H_4(\ell)$$

72. Enthalpy change for:

	$C(s)$ +	$H_2O(g)$ →	$CO(g)$ +	$H_2(g)$
ΔH°_f(kJ/mol)	0	-241.818	-110.525	0

$$\Delta H^\circ_{rxn} = [(1\text{ mol})(-110.525\ \frac{kJ}{mol}) + 0] - [0 + (1\text{ mol})(-241.818\ \frac{kJ}{mol})]$$

$= +131.293$ kJ (The process is **endothermic**.)

Heat involved when 1.0 metric ton (1000.0 kg) of C is converted to coal gas:

$$1000.0\text{ kg C} \bullet \frac{1000\text{ g C}}{1\text{ kg C}} \bullet \frac{1\text{ mol C}}{12.0\text{ g C}} \bullet \frac{+131.293\text{ kJ}}{1\text{ mol C}} = 1.0931 \times 10^7\text{ kJ}$$

74. a. Enthalpy change for the decomposition of ammonium nitrate: 6.78 IM

	$NH_4NO_3(s)$ →	$2\ H_2O(g)$ +	$N_2O(g)$
ΔH°_f(kJ/mol)	-365.6	-241.8	82.1

$$\Delta H^\circ_{rxn} = [(2\text{ mol})(-241.8\ \frac{kJ}{mol}) + (1\text{ mol})(82.1\ \frac{kJ}{mol})] - [(1\text{ mol})(-365.6\ \frac{kJ}{mol})]$$

$= -35.9$ kJ

b. Quantity of heat evolved when 8.00 kg of ammonium nitrate decomposes:

$$8.00 \times 10^3\text{ g} \bullet \frac{1\text{ mol } NH_4NO_3}{80.04\text{ g } NH_4NO_3} \bullet \frac{-35.9\text{ kJ}}{1\text{ mol } NH_4NO_3} = -3590\text{ kJ}$$

76.

	$C_6H_{12}O_6(s)$ +	$6\ O_2(g)$ →	$6\ CO_2(g)$ +	$6\ H_2O(\ell)$
ΔH°_f(kJ/mol)	-1273.3	0	-393.5	-187.8

$$\Delta H^\circ_{rxn} = [(6\text{ mol})(-393.5\ \frac{kJ}{mol}) + (6\text{ mol})(-187.8\ \frac{kJ}{mol})] - [(1\text{ mol})(-1273.3\ \frac{kJ}{mol})$$

$= (-3487.8\text{ kJ}) - (-1273.3\text{ kJ})$ for glucose

$= -2214.5$ kJ for glucose. Substituting (-1265.6 kJ) for fructose gives - 2222.2 kJ for fructose.

Since the ΔH°_{rxn} for fructose is more negative **and** since the two sugars have identical molar masses, **fructose** produces **more energy/gram.**

78. q_{metal} = heat capacity x mass x ΔT

$q_{metal} = C_{metal} \times 27.3\text{ g} \times (-69.03\text{ K})$

Note that ΔT is negative, since T_{final} of the metal is LESS THAN $T_{initial}$]

and $q_{water} = 15.0\text{ g} \cdot \frac{4.184\text{ J}}{\text{g} \cdot \text{K}} \cdot (303.02\text{ K} - 298.15\text{ K}) = 305.6\text{ J}$

Setting $q_{metal} = -q_{water}$

$C_{metal} \times 27.3\text{ g} \times (-69.03\text{ K}) = -305.6\text{ J}$ and solving for C gives:

$C_{metal} = 0.162 \frac{\text{J}}{\text{g} \cdot \text{K}}$

Conceptual Questions

80. The molar heat capacities for Al, Fe, Cu, and Au are:

$$0.902 \frac{\text{J}}{\text{g} \cdot \text{K}} \cdot \frac{26.98\text{ g Al}}{1\text{ mol Al}} = 24.3 \frac{\text{J}}{\text{mol} \cdot \text{K}}$$

$$0.451 \frac{\text{J}}{\text{g} \cdot \text{K}} \cdot \frac{55.85\text{ g Fe}}{1\text{ mol Fe}} = 25.2 \frac{\text{J}}{\text{mol} \cdot \text{K}}$$

$$0.385 \frac{\text{J}}{\text{g} \cdot \text{K}} \cdot \frac{63.55\text{ g Cu}}{1\text{ mol Cu}} = 24.5 \frac{\text{J}}{\text{mol} \cdot \text{K}}$$

$$0.128 \frac{\text{J}}{\text{g} \cdot \text{K}} \cdot \frac{197.0\text{ g Au}}{1\text{ mol Au}} = 25.2 \frac{\text{J}}{\text{mol} \cdot \text{K}}$$

Since all molar heat capacities are approximately $25 \frac{\text{J}}{\text{mol} \cdot \text{K}}$ we can calculate for Ag:

$$25 \frac{\text{J}}{\text{mol} \cdot \text{K}} \cdot \frac{1\text{mol Ag}}{107.9\text{ g Ag}} = 0.23 \frac{\text{J}}{\text{g} \cdot \text{K}}$$

82. One can determine (using standard tables or via an experiment) the value for the ΔH of the reaction in which C(graphite) reacts with oxygen to form CO_2 (g). Combining this with the ΔH of combustion of one mole of diamond. Reversing the second equation and adding the two equations (Hess' Law) we get:

1. C (graphite) + O_2 (g) $\rightarrow$ CO_2 (g) $\Delta H^\circ = -393.5$ kJ (appendix)
2. CO_2 (g) $\rightarrow$ C (diamond) + O_2 (g) $\Delta H^\circ = 395.4$ kJ (appendix)

---

giving : C (graphite) $\rightarrow$ C (diamond) $\Delta H^\circ = 1.9$ kJ

84. This is a losing battle. To extract heat from the inside of the refrigerator, work has to be done. That work (by the condenser and motor) releases heat to the environment (your room). So while the temporary relief of cool air from the inside of the refrigerator is pleasant, the motor has to do work—and heats your room.

Challenging Questions

86. Let's represent the total mass of the ethanol-water mixture be **e** grams.

What is true is that the energy absorbed by the ethanol PLUS the energy absorbed by the water must be 3.6 kJ (or 3.6×10^3 J).

The enthalpy of vaporizations of ethanol and water indicate how much energy is required to vaporize ONE GRAM of the respective substance. Multiplying the mass of ethanol x it's enthalpy of vaporization would give an indication of the energy absorbed as the ethanol vaporized. We could write that as 850 J/g • Me (where Me is the mass of ethanol), and the similar term for water would be 2360 J/g • Mw (where Mw is the mass of water). We know the **relative** masses of each substance— ethanol is 10 % of the mixture and water 90%. Using **e** grams as the mass of the mixture we can write:

Me = 0.10 **e** and Mw = 0.90 **e**

Collecting we have: energy absorbed by the ethanol = 850 J/g • Me = 850 J/g • 0.10 **e**

and the energy absorbed by the water = 2360 J/g • Mw = 2360 J/g • 0.90 **e**

3.6×10^3 J = 850 J/g • 0.10 **e** + 2360 J/g • 0.90 **e**

3.6×10^3 J = (85 J/g • **e** + 2124 J/g • **e**) = 2209J/g **e** and

e = 3.6×10^3 J/2209J/g = 1.63 grams of mixture (about 2 grams)

88. For the combustion of C_8H_{18}:

$$C_8H_{18}(\ell) + 25/2\ O_2(g) \rightarrow 8\ CO_2(g) + 9\ H_2O(\ell)$$

$$\Delta H^\circ_{rxn} = [(8\text{ mol})(-393.5\frac{\text{kJ}}{\text{mol}}) + (9\text{ mol})(-285.8\frac{\text{kJ}}{\text{mol}})] - [(1\text{ mol})(-259.2\frac{\text{kJ}}{\text{mol}}) + 0]$$

$$\Delta H^\circ_{rxn} = -5461.0\text{ kJ}$$

Expressed on a gram basis

$$-5461.0\frac{\text{kJ}}{\text{mol}} \bullet \frac{1\text{ mol }C_8H_{18}}{114.2\text{ g }C_8H_{18}} = -47.81\text{ kJ/g}$$

For the combustion of CH_3OH:

$$2\ CH_3OH(\ell) + 3\ O_2\ (g) \rightarrow 4\ H_2O(\ell) + 2\ CO_2\ (g)$$

ΔH°_{rxn} = [(- 393.5 kJ/mol)(2 mol) + (- 285.8 kJ/mol)(4 mol)] - [(- 238.66 kJ/mol)(2 mol) + 0]

= [(- 787.0) + (- 967.2)] + 477.4 kJ

= - 1452.9 kJ

Express this on a per mol and per gram basis:

$$\frac{-\ 1452.9\ \text{kJ}}{2\ \text{mol}\ CH_3OH} \bullet \frac{1\ \text{mol}\ CH_3OH}{32.04\ \text{g}\ CH_3OH} = -\ 22.67\ \text{kJ/g}$$

On a per gram basis, **isooctane liberates the greater amount** of heat energy.

90. a. The Enthalpy diagram for the isomers of butene:

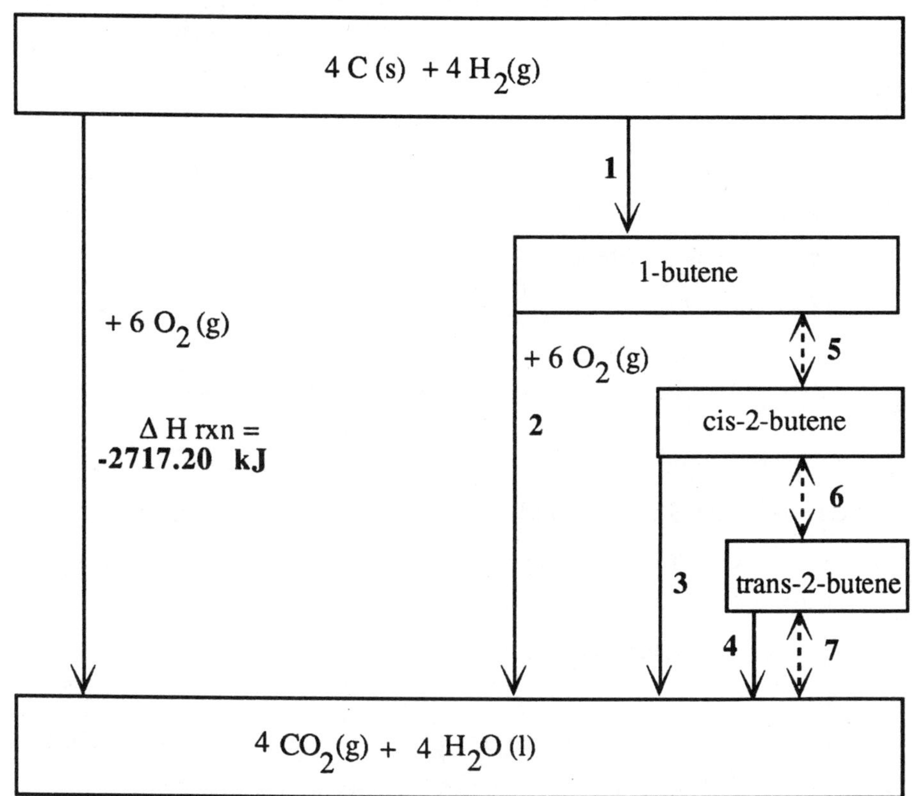

Step	ΔH (in kJ/mol)
1	- 20.5
2	-2696.7
3	-2687.5
4	-2684.2

b. The enthalpy change from *cis*-2-butene to *trans*-2-butene:

This energy difference corresponds to step 6 in the diagram. Note that this difference corresponds to the difference in the Enthalpies of Combustion of steps 3-4 or: $\Delta H = (-2684.2) - (-2687.5) = -3.3$ kJ

c. The enthalpies of formation for both *cis*-2-butene and *trans*-2-butene:

Step 1 corresponds to the $\Delta H°_f$ for 1-butene. Step 1 + step 5 corresponds to the $\Delta H°_f$ for *cis*-2-butene. Step 5 corresponds to the difference in the Enthalpies of Combustion of steps 2-3 or:

$\Delta H = (-2687.5) - (-2696.7) = -9.2$ kJ

Step 1 + Step 5 = $(-20.5) + (-9.2) = -29.7$ kJ ($\Delta H°_f$ for *cis*-2-butene)

Using the same logic as above, step 1 + step 5 + step 6 corresponds to the $\Delta H°_f$ for *trans*-2-butene. The magnitude for step 6 corresponds to the difference in the Enthalpies of Combustion of steps 3-4, as calculated in part b above, or:

$\Delta H = (-2684.2) - (-2687.5) = -3.3$ kJ

Step 1 + Step 5 + Step 6 = $(-20.5) + (-9.2) + (-3.3) = -33.0$ kJ

($\Delta H°_f$ for *trans*-2-butene)

Note: There may have been an error in the first printing of the book in which the data and pictures were swapped. The data in the diagram for steps 2,3, and 4 are correct for the $\Delta H_{combustion}$ for 1-butene, *cis*-2-butene, and *trans*-2-butene respectively.

92. One can arrive at the desired answer if you recall the **definition** of $\Delta H°_f$. The definition is the enthalpy change associated with the formation of **one mole** of the substance (in this case B_2H_6) from its elements—each in their standard state (s for boron and g for hydrogen). Note that the 1st equation given uses **four** moles of B as a reactant — and we'll need only 2, so divide the first equation by 2 to give:

$2\,B\,(s) + 3/2\,O_2\,(g) \rightarrow B_2O_3\,(s)$ $\Delta H = 1/2(-2543.8 \text{ kJ}) = -1271.9$ kJ

The formation of 1 mole of B_2H_6 will require the use of 6 moles of H (or 3 moles of H_2), so multiply the second equation by 3/2 to give:

$3\,H_2\,(g) + 3/2\,O_2\,(g) \rightarrow 3\,H_2O\,(g)$ $\Delta H = 3/2(-484 \text{ kJ}) = -726$ kJ

Finally the third equation given has B_2H_6 as a **reactant and not a product**. So let's reverse the third equation to give:

$B_2O_3\,(s) + 3\,H_2O\,(g) \rightarrow B_2H_6\,(g) + 3\,O_2\,(g)$ $\Delta H = -(-2032.9 \text{ kJ}) = +2032.9$ kJ

Adding the three equations gives the equation:

$$2\ B\ (s) + 3\ H_2(g) \rightarrow B_2H_6\ (g) \text{ with a } \Delta H = (+2032.9 - 1271.9 - 726) \text{ or } + 36\ kJ$$

The literature value is 35.6 kJ/mol.

94. The desired equation is: $CH_4\ (g) + 3\ Cl_2\ (g) \rightarrow 3\ HCl\ (g) + CHCl_3\ (\ell)$

Begin with equation 1 (the combustion of methane)

$$CH_4\ (g) + 2\ O_2\ (g) \rightarrow 2\ H_2O\ (\ell) + CO_2\ (g) \qquad \Delta H = -890.4\ kJ = -890.4\ kJ$$

The second equation has HCl as one of the reactants. We need to **reverse** it and (to adjust the coefficient of HCl to 3), multiply by 3/2 to give:

$$3/2\ H_2\ (g) + 3/2\ Cl_2\ (g) \rightarrow 3\ HCl \qquad \Delta H = -3/2(+184.6)\ kJ = -\ 276.9\ kJ$$

Note that CO_2 (formed in equation 1) doesn't appear in the overall equation, so let's use equation 4 (reversed) to "consume" the CO_2:

$$CO_2\ (g) \rightarrow C\ (graphite) + O_2\ (g) \qquad \Delta H = -1(-393.5)\ kJ = +\ 393.5\ kJ$$

Noting also that equation 1 produces 2 water molecules, let's "consume" them by using equation 5 (reversed)—multiplied by 2:

$$2\ H_2O\ (\ell) \rightarrow 2\ H_2\ (g) + O_2\ (g) \qquad \Delta H = -2(-285.8)\ kJ = +\ 571.6\ kJ$$

and finally we need to produce $CHCl_3$, which we can do with the equation that represents the $\Delta H°_f$ for $CHCl_3$:

$$C(graphite) + 1/2\ H_2\ (g) + 3/2\ Cl_2\ (g) \rightarrow CHCl_3\ (\ell) \qquad \Delta H = -134.47\ kJ$$

The overall enthalpy change would then be:

$$\Delta H = -890.4\ kJ\ - 276.9\ kJ + 393.5\ kJ + 571.6\ kJ\ -134.47\ kJ = -336.7\ kJ$$

Summary Question

95. a. Mass of SO_2 contained in 440 million gallons of wine that is 100. ppm in SO_2:

$$440 \times 10^6\ gal \bullet \frac{4\ qt}{1\ gal} \bullet \frac{0.9463\ L}{1\ qt} \bullet \frac{1 \times 10^3\ cm^3}{1\ L} \bullet \frac{1.00\ g\ wine}{1\ cm^3} \bullet \frac{100.\ g\ SO_2}{1 \times 10^6\ g\ wine} = 1.7 \times 10^8\ g\ SO_2$$

The number of moles to which this corresponds is:

$$1.7 \times 10^8\ g\ SO_2 \bullet \frac{1\ mol\ SO_2}{64.06\ g\ SO_2} = 2.6 \times 10^{10}\ mol\ SO_2$$

b. For the consumption of 21 million tons of SO_2 by MgO:

$$21. \times 10^6 \text{ T } SO_2 \bullet \frac{1 \text{ mol } SO_2}{64.06 \text{ g } SO_2} \bullet \frac{1 \text{ mol MgO}}{1 \text{ mol } SO_2} \bullet \frac{40.30 \text{ g MgO}}{1 \text{ mol MgO}}$$

$$= 1.3 \times 10^7 \text{ T MgO } (1.2 \times 10^{13} \text{ g MgO})$$

The quantity of $MgSO_4$ produced:

$$1.3 \times 10^7 \text{ T MgO} \bullet \frac{120.4 \text{ g } MgSO_4}{40.30 \text{ g MgO}} = 3.9 \times 10^7 \text{ T } MgSO_4 \ (3.6 \times 10^{13} \text{ g } MgSO_4)$$

Note : For the two previous calculations in (b), the answers in parentheses are arrived at by multiplying the # of T of the substance by 2000 (converts T to pounds) and then by 454 (converts pounds to grams). Alternately one could **begin** the calculation by converting T of SO_2 or MgO into grams (using the two stoichiometric factors).

c. The heat energy change per mole of $MgSO_4$ by the reaction:

$$MgO\text{ (s)} + SO_2\text{ (g)} + \frac{1}{2} O_2\text{ (g)} \rightarrow MgSO_4\text{ (s)}$$

$\Delta H_f°$ - 601.70 - 296.8 0 - 2817.5

$$\begin{aligned} \Delta H_{rxn}° &= \Sigma \Delta H_f° \text{ products} - \Sigma \Delta H_f° \text{ reactants} \\ &= (-2817.5 \text{ kJ/mol})(1 \text{ mol}) \\ &\quad - [(-601.70 \text{ kJ/mol})(1 \text{ mol}) + (-296.8 \text{ kJ/mol})(1 \text{ mol}) + 0] \\ &= -1919.0 \text{ kJ} \quad \textbf{Heat evolved} \end{aligned}$$

d. Heat evolved by production of 750. T H_2SO_4 per day.

The overall ΔH_{rxn} is the sum of the three reactions given:

$$\Delta H_{overall} = (-296.8 \text{ kJ}) + (-98.9 \text{ kJ}) + (-130.0 \text{ kJ}) = -525.7 \text{ kJ/mol } H_2SO_4$$

The number of moles of H_2SO_4 corresponding to 750. T is:

$$750 \text{ T } H_2SO_4 \bullet \frac{9.08 \times 10^5 \text{ g } H_2SO_4}{1 \text{ T } H_2SO_4} \bullet \frac{1 \text{ mol } H_2SO_4}{98.08 \text{ g } H_2SO_4} = 6.94 \times 10^6 \text{ mol}$$

The overall energy produced per day is then:

$$6.94 \times 10^6 \text{ mol } H_2SO_4 \bullet \frac{-525.7 \text{ kJ}}{\text{mol } H_2SO_4} = 3.65 \times 10^9 \text{ kJ}$$

Chapter 7
Atomic Structure

NUMERICAL AND OTHER QUESTIONS

Electromagnetic Radiation

20. a **Red, orange, and yellow** light have less energy than green light.

    b. **Blue** light photons are of greater energy than yellow light.

    c. **Blue** light has greater frequency than green light.

22. Since frequency • wavelength = speed of light (c)

$$88.9 \times 10^{6}\ s^{-1} \cdot l = 3.0 \times 10^{8}\ m/s$$

$$l = \frac{\text{speed of light}}{\text{frequency}} = \frac{3.0 \times 10^{8}\ m/s}{88.9 \times 10^{6}\ s^{-1}} = 3.37\ m$$

24. $$\text{frequency} = \frac{\text{speed of light}}{\text{wavelength}} = \frac{2.9979 \times 10^{8}\ m/s}{4.10 \times 10^{2}\ nm} \cdot \frac{1.00 \times 10^{9}\ nm}{1.00\ m}$$

$$= 7.3 \times 10^{14}\ s^{-1} \quad \text{(to 2 sf)}$$

The energy of a photon of this light may be determined :

$$E = h\upsilon$$

Planck's constant, h, has a value of $6.626 \times 10^{-34}\ J \cdot s \cdot photons^{-1}$

$$E = (6.626 \times 10^{-34}\ J \cdot s \cdot photons^{-1})(7.3 \times 10^{14}\ s^{-1})$$
$$= 4.8 \times 10^{-19}\ J \cdot photon^{-1} \quad \text{(to 2 sf)}$$

Energy of 1.00 mol of photons=

$$4.8 \times 10^{-19}\ J \cdot photon^{-1} \cdot \frac{6.0221 \times 10^{23}\ photons}{1.00\ mol\ photons} \cdot \frac{1\ kJ}{1000\ J} = 290\ kJ/mol\ photons$$

Compare energies of photons of violet light with those of red light::

From Example 7.2 we find that the energy of a mol of photons of red light = 175 kJ/mol so

$$\frac{E_{violet}}{E_{red}} = \frac{290\ kJ/mol}{175\ kJ/mol} = 1.7$$ Violet light is more energetic by a factor of 1.8!

26. The frequency of the line at 396.15 nm:

$$\text{frequency} = \frac{\text{speed of light}}{\text{wavelength}} = \frac{2.9979 \times 10^{8}\ \text{m/s}}{3.9615 \times 10^{2}\ \text{nm}} \bullet \frac{1.00 \times 10^{9}\ \text{nm}}{1.00\ \text{m}}$$

$$= 7.5676 \times 10^{14}\ \text{s}^{-1}$$

The energy of a photon of this light may be determined : $E = h\upsilon$

Planck's constant, h, has a value of 6.626×10^{-34} J • s • photons^{-1}

$$E = (6.626 \times 10^{-34}\ \text{J} \bullet \text{s} \bullet \text{photons}^{-1})(7.5676 \times 10^{14}\ \text{s}^{-1})$$
$$= 5.0144 \times 10^{-19}\ \text{J} \bullet \text{photon}^{-1}$$

$$\text{Energy of 1.00 mol of photons} = 5.0144 \times 10^{-19}\ \text{J} \bullet \text{photon}^{-1} \bullet \frac{6.0221 \times 10^{23}\ \text{photons}}{1.00\ \text{mol photons}}$$

$$= 3.0197 \times 10^{5}\ \text{J/mol photon or 302 kJ/mol photons}$$

28. a. The **most energetic light** would be represented by the light of **shortest wavelength** (253.652 nm).

b. The frequency of this light is :

$$\frac{2.9979 \times 10^{8}\ \text{m/s}}{253.652\ \text{nm}} \bullet \frac{1.00 \times 10^{9}\ \text{nm}}{1.00\ \text{m}} = 1.18190 \times 10^{15}\ \text{s}^{-1}$$

The energy of 1 photon with this wavelength is:

$$E = h\upsilon = (6.62608 \times 10^{-34}\ \frac{\text{J} \bullet \text{s}}{\text{photon}})(1.18190 \times 10^{15}\ \text{s}^{-1})$$
$$= 7.83139 \times 10^{-19}\ \frac{\text{J}}{\text{photon}}$$

c. The line emission spectrum of mercury shows the visible region between ≈ 400 and 750 nm. The lines at 404 and 436 nm are present while the lines at 253 nm, 365 nm and 1013 nm lie outside the visible region. The 404 nm line and the 436 nm line are in the violet end of the spectrum.

30. Since energy is proportional to frequency ($E = h\nu$), we can arrange the radiation in order of increasing energy per photon by listing the types of radiation in increasing frequency (or decreasing wavelength).

→ Energy increasing →			
FM music	microwave	yellow light	x-rays

→ Frequency (ν) increasing →

← Wavelength (λ) increasing ←

Photoelectric Effect

32. $$\text{Energy} = 2.0 \times 10^{2}\ \text{kJ/mol} \cdot \frac{1\ \text{mol}}{6.0221 \times 10^{23}\ \text{photons}} \cdot \frac{1.00 \times 10^{3}\ \text{J}}{1.00\ \text{kJ}}$$

$$= 3.3 \times 10^{-19}\ \text{J} \cdot \text{photons}^{-1}$$

What wavelength of light would provide this energy ?

$$E = h\nu = \frac{hc}{\lambda} \text{ or } \lambda = \frac{hc}{E} = \frac{(6.626\text{x}10^{-34}\ \text{J} \cdot \text{s} \cdot \text{photons}^{-1})(2.9979 \times 10^{8}\ \text{m} \cdot \text{s}^{-1})}{3.3 \times 10^{-19}\ \text{J} \cdot \text{photons}^{-1}}$$

$$= 6.0 \times 10^{-7}\ \text{m or } 6.0 \times 10^{2}\ \text{nm}$$

Radiation of this wavelength--in the **visible** region of the electromagnetic spectrum--would appear **orange**.

Atomic Spectra and The Bohr Atom

34. a.

Transitions from	to
n = 5	n = 4,3,2, or 1
n = 4	n = 3, 2, or 1
n = 3	n = 2 or 1
n = 2	n = 1

Ten transitions are possible from these five quantum levels, providing 10 emission lines.

b. Photons of the lowest frequency (lowest energy) will be emitted in a transition from the level with **n = 5** to the level **n = 4**. This is easily seen with the aid of the equation:

$$\Delta E = Rhc(\frac{1}{n^2_f} - \frac{1}{n^2_i}).$$

Since R, h, and c are constant for any transition, inspection shows that the largest change in energy results if $n_f = 4$ and $n_i = 5$.

c. The emission line having the **shortest wavelength** also has the **highest frequency**. A transition from $n_i = 5$ to $n_f = 1$ would provide the shortest wavelength line.

d. The emission line having the **highest energy** corresponds to a transition from the level with **n = 5** to the level with **n = 1** .

36. One can calculate the wavelength of any transition by applying the relationship:

$$\frac{1}{\lambda} = R(\frac{1}{n^2_f} - \frac{1}{n^2_i}).$$

For any line with a wavelength greater than 102.6 nm, the left side of the equation $(\frac{1}{\lambda})$ would be smaller than that for the 102.6 nm line. This will be true if the term $(\frac{1}{n^2_f} - \frac{1}{n^2_i})$ is smaller than 8/9.

For $n_f = 1$ and $n_i = 3$: $\frac{1}{1^2} - \frac{1}{3^2} = 8/9$ or 0.89

Substituting the suggested transitions into the term $(\frac{1}{n^2_f} - \frac{1}{n^2_i})$ yields:

	transition between	$(\frac{1}{n^2_f} - \frac{1}{n^2_i})$
a.	2 and 4	0.19
b.	1 and 4	0.94
c.	1 and 5	0.96
d.	3 and 5	0.07

The transition between **2 and 4** and between **3 and 5** would produce a line of longer wavelength than 102.6 nm (lower energy). The remaining transitions would produce lines of shorter wavelength (and higher energy) .

38. The wavelength of emitted light for the transition n = 3 to n = 1.

$$\frac{1}{\lambda} = R\left(\frac{1}{1^2} - \frac{1}{3^2}\right) = (1.0974 \times 10^7\ m^{-1})(8/9) = 9.7547 \times 10^6\ m^{-1}$$

$\lambda = 1.026 \times 10^{-7}$ m or approximately 103 nm (far ultraviolet)

40. The energy needed to move an electron from n = 1 to n = 5 will be the same as the amount emitted as the electron relaxed from n = 5 to n = 1, i.e. 2.093×10^{-18} J.

De Broglie and Matter Waves

42. Mass of an electron: 9.11×10^{-31} kg

Planck's constant: 6.626×10^{-34} J • s • photon^{-1}

Velocity of the electron: 2.5×10^8 cm • s^{-1} or 2.5×10^6 m • s^{-1}

$$\lambda = \frac{h}{m \cdot v} = \frac{6.626 \times 10^{-34}\ J \cdot s}{(9.11 \times 10^{-31}\ kg \cdot 2.5 \times 10^6\ m \cdot s^{-1})}$$

$= 2.9 \times 10^{-10}$ m = 2.9 Angstroms = 0.29 nm

44. The wavelength can be determined exactly as in Question 42:

$$\lambda = \frac{h}{m \cdot v} = \frac{6.626 \times 10^{-34}\ J \cdot s}{(1.0 \times 10^{-1}\ kg \cdot 30.\ m \cdot s^{-1})}$$

$= 2.2 \times 10^{-34}$ m or 2.2×10^{-25} nm

What velocity is needed to have a λ of 5.6×10^{-3} nm ?

First express the wavelength in meters:

$$5.6 \times 10^{-3}\ nm \cdot \frac{1\ m}{1 \times 10^9\ nm} = 5.6 \times 10^{-12}\ m$$

Now calculate the velocity needed to have a wavelength of 5.6×10^{-12} m.

Rearranging the equation above to solve for velocity:

$$v = \frac{h}{m \cdot \lambda} = \frac{6.626 \times 10^{-34}\ J \cdot s}{(1.0 \times 10^{-1}\ kg \cdot 5.6 \times 10^{-12}\ m)} = 1.2 \times 10^{-21}\ \frac{m}{s}$$

Quantum Mechanics

46. Complete the following table: (Answers are emboldened.)

ATOMIC PROPERTY	QUANTUM NUMBER
orbital size	**n**
relative orbital orientation	**m_ℓ**
orbital shape	**ℓ**

48. a. n = 4 possible ℓ values = 0,1,2,3 (ℓ = 0,1,... (n - 1))
b. ℓ = 2 possible m_ℓ values = -2,-1,0,+1,+2 (-ℓ ..., 0,....+ℓ)
c. orbital = 4s n = 4; ℓ = 0; m_ℓ = 0
d. orbital = 4f n = 4; ℓ = 3; m_ℓ = -3,-2,-1,0,+1,+2,+3

50. An electron in a 4p orbital must have n = 4 and ℓ = 1. The possible m_ℓ values give rise to the following sets of n, ℓ, and m_ℓ

n	ℓ	m_ℓ
4	1	-1
4	1	0
4	1	+1

Note that the **three values** of m describe **three orbital orientations.**

52. Subshells in the electron shell with n = 4 :
There are 4 : s, p, d, and f sublevels corresponding to ℓ = 0, 1, 2, and 3 respectively.

54. Explain why each of the following is not a possible set of quantum numbers for an electron in an atom.
a. n = 2, ℓ = 2, m_ℓ = 0 For n = 2, maximum value of ℓ is one (1).
b. n = 3, ℓ = 0, m_ℓ = -2 For ℓ = 0, possible values of m_ℓ are ± ℓ and 0.
c. n = 6, ℓ = 0, m_ℓ = 1 For ℓ = 0, possible values of m_ℓ are ± ℓ and 0.

56.

	quantum number designation	maximum number of orbitals
a.	n = 3; ℓ = 0; m_ℓ = +1	none; for ℓ = 0, the only possible value of m_ℓ = 0
b.	n = 5; ℓ = 1	3 ("**p**" orbitals)

	quantum number designation	maximum number of orbitals
c.	$n = 7;\ \ell = 5$	eleven; the # of orbitals is "$2\ell + 1$"
d.	$n = 4;\ \ell = 2;\ m_\ell = -2$	1 (one of the three 4 "**p**" orbitals)

58. The number of planar nodes possessed by each of the following:

	orbital	number of planar nodes
a.	2s	0 (because $\ell = 0$)
b.	5d	2 (because $\ell = 2$)
c.	5f	3 (because $\ell = 3$)

60. Which of the following orbitals cannot exist and why:

2s exists	$n = 2$	permits ℓ values as large as 1 ($\ell = 0$ is an s sublevel)
2d cannot exist	$\ell = 2$	is not permitted for $n < 3$ ($\ell = 2$ is a d sublevel)
3p exists	$n = 3$	permits ℓ values as large as 2 ($\ell = 1$ is a p sublevel)
3f cannot exist	$\ell = 3$	is not permitted for $n < 4$ ($\ell = 3$ is an f sublevel)
4f exists	$\ell = 4$	permits ℓ values as large as 3 ($\ell = 3$ is an f sublevel)
5s exists	$n = 5$	permits ℓ values as large as 4 ($\ell = 0$ is an s sublevel)

62. The complete set of quantum numbers for :

		n	ℓ	m_ℓ	
a.	2p	2	1	-1, 0, +1	(3 orbitals)
b.	3d	3	2	-2, -1, 0, +1, +2	(5 orbitals)
c.	4f	4	3	-3, -2, -1, 0, +1, +2, +3	(7 orbitals)

64. All orbitals except **the s orbital** may have magnetic quantum numbers, $m_\ell = -1$.
The s orbital has as its only value for the magnetic quantum number, $m_\ell = 0$.

General Questions

66. The energy (in kJ/mol photons) of light with a wavelength of 305 nm (or 305×10^{-9} m)

$$\text{Energy} = \frac{hc}{\lambda} = \frac{6.626 \times 10^{-34}\ \text{J} \cdot \text{s}}{\text{photon}} \cdot \frac{2.9979 \times 10^{8}\ \text{m} \cdot \text{s}^{-1}}{305 \times 10^{-9}\ \text{m}} \cdot \frac{6.022 \times 10^{23}\ \text{photons}}{1\ \text{mol}} \cdot \frac{1\ \text{kJ}}{1000\ \text{J}} = 392\ \text{kJ/mol}$$

68. Number of photons (λ = 12 cm) to raise the temperature of the eye by 3.0 °C:

1. Determine the energy needed to change the temperature :

$$\text{Energy} = \text{mass} \cdot \text{heat capacity} \cdot \Delta t$$
$$= (11\text{ g})(4.0\frac{\text{J}}{\text{g} \cdot \text{K}})(3.0\text{ K})$$
$$= 132\text{ J (or 130 to 2 sf)}$$

2. Determine the energy of a photon of light with λ = 12 cm: (or 12 x 10^{-2} m)

$$\text{Energy} = \frac{hc}{\lambda} = \frac{(6.626 \times 10^{-34}\text{ J} \cdot \text{s} \cdot \text{photon}^{-1})(2.9979 \times 10^{8}\text{ m} \cdot \text{s}^{-1})}{12 \times 10^{-2}\text{ m}}$$
$$= \frac{1.66 \times 10^{-24}\text{ J}}{\text{photon}}$$

3. The number of photons needed to furnish 130 J:

$$1.3 \times 10^{2}\text{ J} \cdot \frac{1\text{ photon}}{1.66 \times 10^{-24}\text{ J}} = 7.8 \times 10^{25}\text{ photons (2 sf)}$$

and dividing by Avogadro's number, this equals 130 mol photons!

70. Time for Sojourner's signal to travel 7.8 x 10^7 km:

If light travels at 2.9979 x 10^8 m • s^{-1}, we can calculate the time:

$$\frac{7.8 \times 10^{7}\text{ km}}{1} \cdot \frac{1 \times 10^{3}\text{ m}}{1\text{ km}} \cdot \frac{1\text{ s}}{2.9979 \times 10^{8}\text{ m}} = 260\text{ s or 4.3 minutes}$$

72. Example 7.3 in your text illustrates the calculation of the ionization energy for H's electron

$$E = \frac{-Z^2Rhc}{n^2} = -2.179 \times 10^{-18}\text{ J/atom} \Rightarrow -1312\text{ kJ/mol}$$

For He^+ the calculation yields:

$$E = \frac{-(2)^2(1.097 \times 10^{7}\text{ m}^{-1})(6.626 \times 10^{-34}\text{ J} \cdot \text{s})(2.998 \times 10^{8}\text{ m} \cdot \text{s}^{-1})}{(1)^2}$$

$$= -8.717 \times 10^{-18}\text{ J/ion and expressing this energy for a mol of ions}$$

$$= \frac{-8.717 \times 10^{-18}\text{ J}}{\text{ion}} \cdot \frac{6.0221 \times 10^{23}\text{ atoms}}{\text{mol}} \cdot \frac{1\text{ kJ}}{1000\text{ J}} = -5248\frac{\text{kJ}}{\text{mol}}$$

The energy to remove the electron is 5248 kJ/mol of ions. Note that this energy is four times that for H.

74. (i) The photon with the smallest energy would be produced by a transition between the **closest levels**, or (b) n = 7 to n=6.

(ii) The photon with the highest frequency (and with the highest energy) would be produced by a transition between the **two most distant levels**, (a) n = 7 to n= 1.

(iii) The photon with the shortest wavelength (and therefore with the highest energy) would be produced by a transition between the **two most distant levels**, (a) n = 7 to n= 1.

76. Orbitals ranked in order of **increasing energy** :

1s < 2s, 2p< 3s , 3p , 3d < 4s

Note that for the Hydrogen atom, the Bohr equation deals with the **principal quantum number**, n-- so the 2s and 2p have the same energy, and the 3s, 3p and 3d all have the same energy.

78. a. The quantum number n describes the **size (and energy)** of an atomic orbital .

b. The shape of an atomic orbitals is given by the quantum number ℓ.

c. A photon of orange light has **less energy** than a photon of yellow light.

d. The maximum number of orbitals that may be associated with the quantum numbers n= **4**, ℓ = **3** is **seven.** (corresponding to m_ℓ values of ± 3, ± 2, ± 1 and 0)

e. The maximum number of orbitals that may be associated with the quantum numbers n= **3**, ℓ = **2**, and m_ℓ = -2 is **one**.

f. The orbital on the left is a **p orbital** and the one on the right is a **d orbital**.

g. When n = 5, the possible values of ℓ are 0, 1, 2, 3 and 4. (Range is 0 ..... (n-1))

h. The maximum number of orbitals that can be assigned to the n = 4 shell is **16**.

n = 4	ℓ = 0	m_ℓ = 0	1 orbital
n = 4	ℓ = 1	m_ℓ = -1,0,+1	3 orbitals
n = 4	ℓ = 2	m_ℓ = -2,-1,0,+1,+2	5 orbitals
n = 4	ℓ = 3	m_ℓ = -3,-2,-1,0,+1,+2,+3	7 orbitals
			16 orbitals

Conceptual Questions

80. For the first three electron shells:

N = 1	N = 2	N = 3
L = 1	L = 2	L = 3
M = ±1,0	M = ±1,0	M = ±1,0

There would be 3 orbitals in each of the 3 shells for a **total of 9 orbitals.**

82. The wave-particle duality describes the behavior of subatomic particles which sometimes behave like particles (and have particulate properties like **mass**) and sometimes behave as waves (and exhibit wave-like properties like **diffraction**). Our modern view of atomic structure pictures the electrons behaving as waves in the atom.

84. Observables (measurable):

(b)—can observe the light emitted

(e),(f)—diffraction patterns produced by electrons and light

(g)—can measure the energy absorbed

(h),(i)—atoms and molecules can now be seen with scanning tunneling microscopes.

(j)—water waves are visible

86. The sources of colored light from certain salts in burning organic liquids and neon signs are the **energy transitions** as excited electrons return to the ground state, releasing the energy corresponding to the **difference** in the energy levels of the ground and excited states.

Challenging Questions

88. The $2p_z$ orbital is oriented along the z axis, with minimal density along the x (and y) axis. The density 0.05 nm from the nucleus on the x axis would be **lower than** the density at the same distance from the nucleus along the z axis.

90. a. The spherical symmetry of the 1s orbital means that the probability of finding the electron at y = d is **equal** to the probability of finding the electron at x = d(0.01).

b. The $2p_x$ orbital is oriented along the x axis. Since the maximum density of the p_x orbital is oriented along that axis, the probability of finding the electron at a distance on the y axis (y = d) should be less than 0.001(zero).

CD-ROM Questions

Screen 7.4 The Electromagnetic Spectrum

a1. As one "moves" from the blue region to the red region, **wavelength increases**, and since wavelength and frequency are inversely related, as the wavelength increases **the frequency decreases.**

a2. The blue region of light is around the 450 nm region. This would correspond to 450 x 10^{-9} m (or 4.50 x 10^{-7} m)

a3. The orange region of light is around 650 nm. This would correspond to 650 x 10^{-9} m (or 6.50 x 10^{-7} m)

a4. **Red light** has the **longest wavelength** , while **blue (or violet)** has the **shortest**.

b. Since **blue** (or violet) has the shortest wavelength, it has the **highest frequency.**

c. The frequency of radiation in a microwave over is **higher** than that of FM radio stations.

d. The wavelength of x-rays is **shorter** than that of ultraviolet light.

Screen 7.6 Atomic Line Spectra

a. Line spectra are so-called owing to the existence of "lines" of light that arise during the electronic transitions from "excited" or higher-energy states to lower energy states. This release of energy is manifest in the production of light.

b. As n increases to values of 3 and higher, the **wavelength** of the light emitted **decreases.** The electronic transitions (with certain amounts of energy) result in the emission of light in a part of the electromagnetic spectrum that produces lines of varying colors.

Screen 7.7 Bohr's Model of the Hydrogen Atom

a. Movements of electrons from higher energy levels to lower levels is accompanied by the emission of electromagnetic radiation (sometimes in the visible region).

b. As an electron is moved to an energy level with n = infinity, the electron is "removed" for the atom. We sometimes refer to this as "ionization", since the loss of an electron from a neutral atom creates a charged specie—an ion.

c. The movement of electrons from lower energy levels to higher energy levels is accompanied by the **absorption of energy**. Many dual-beam spectrometers use this principle to detect the absorption of energy. When a beam of light is split, if part of the beam is passed through the sample (our "atoms of interest") while a part of the beam is not passed through the sample, if we compare the energy of the two beams at some part, the energy of the "sample beam" will be reduced—when compared to our "reference beam" owing to the absorption of energy by the "atoms of interest".

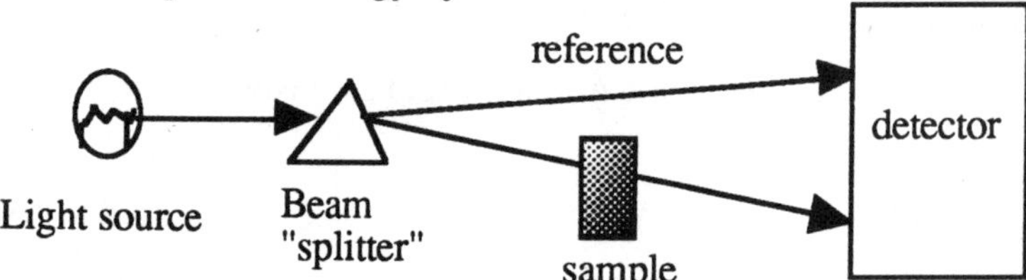

The diagram above provides a simple schematic for such a spectrometer.

d. The pickle glows since the materials in the pickle are being "excited" by the addition of the energy (electric current). Since the pickle has been soaked in brine (NaCl), the electrons in the sodium atom are excited and release energy as they "return" to lower energy states, providing "yellow" light. The same kind of light is visible in many street lamps.

Screen 7.11 Shells, Subshells, and Orbitals

a. A $4p_z$ orbital is located in the **fourth shell**, the **"p" subshell** ($\ell = 1$), and the **z** orbital.

b. The **number of subshells in a given level** is equal to the quantum number n for that level (or shell).

n = 1 has 1 subshell

n = 2 has 2 subshells.......etc

c. There is **no subshell with four orbitals**. Since the possible # of orbitals in a given subshell is $2\ell+1$ (where ℓ is the angular momentum quantum number), the number of orbitals is always an **odd** number.

Screen 7.13 Shapes of Atomic Orbitals

a. The general shape of s orbitals is **spherical** . S orbitals differ from each other in their "radius" , as n becomes larger, the sphere becomes larger. They share the common shape (boundary surfaces), and they also share the common "center" for the sphere—namely the nucleus of the atom.

b. The **p** orbitals are thought to be "peanut" or "dumbbell" shaped. Whle they share a **common nodal location**—the nucleus of the atom, they **differ** one from the other in the relative location of their greatest electron densities. The p_x orbital is oriented along the "x-axis" while the p_y is oriented 90 ° with respect to it (along the "y-axis"), and the p_z orbital along the "z-axis" mutually perpendicular to both the x and y orbitals.

c. The 3d orbitals are frequently described as "p orbitals that are crossed"—almost like a 3-dimensional clover-leaf.

d. The **number of nodal planes** for an orbital is **equal to the ℓ quantum number.** Hence if $\ell = 0$, there are 0 nodal planes, if $\ell = 1$, there is 1 nodal plane, etc.

Chapter 8
Atomic Electron Configurations and Chemical Periodicity

Writing Electron Configurations

10. Electron configurations for Al and P using both the orbital box and spectroscopic notations:

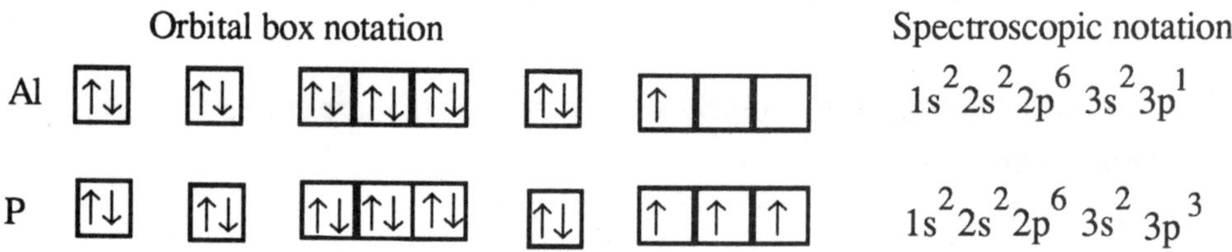

Note that Al is in group 3A (13) indicating that there are THREE electrons in the outer shell, while P is in group 5A (15) indicating that there are FIVE electrons in the outer shell. Both Al and P are on the "right side" of the periodic table—where elements have their "outermost" electrons in p subshells.

12. Vanadium's electron configuration: $1s^2\ 2s^2\ 2p^6\ 3s^2\ 3p^6\ 3d^3\ 4s^2$. As listed in Table 8.2 the configuration is [Ar] $3d^3\ 4s^2$, in which the argon "core" represents the "first" 18 electrons.

14. Arsenic's electron configuration: (33 electrons)
    Spectroscopic notation: $1s^2\ 2s^2\ 2p^6\ 3s^2\ 3p^6\ 3d^{10}\ 4s^2\ 4p^3$
    Noble gas notation: [Ar] $3d^{10}\ 4s^2\ 4p^3$

16. Electron configurations, using the Noble gas notation:

    a. Sr: [Kr] $5s^2$
    b. Zr: [Kr] $4d^2\ 5s^2$
    c. Rh: [Kr] $4d^8\ 5s^1$
    d. Sn: [Kr] $4d^{10}\ 5s^2\ 5p^2$

    The spectroscopic notation for each of these is obtained by replacing [Kr] with $1s^2\ 2s^2\ 2p^6\ 3s^2\ 3p^6\ 3d^{10}\ 4s^2\ 4p^6$

18. Electron configuration for:
    a. Sm: $1s^2\ 2s^2\ 2p^6\ 3s^2\ 3p^6\ 3d^{10}\ 4s^2\ 4p^6\ 4d^{10}\ 4f^6\ 5s^2\ 5p^6\ 6s^2$ or [Xe] $4f^6\ 6s^2$
    Predicted: [Xe] $4f^5\ 5d^1 6s^2$

b. Yb: $1s^2\ 2s^2\ 2p^6\ 3s^2\ 3p^6\ 3d^{10}\ 4s^2\ 4p^6\ 4d^{10}\ 4f^{14}\ 5s^2\ 5p^6\ 6s^2$ or [Xe] $4f^{14}\ 6s^2$
Predicted: [Xe] $4f^{13}\ 5d^1 6s^2$
Note two things: 1) The f subshell is filling as we progress from samarium to ytterbium.
2) The outermost shell is the "6th"—corresponding to the "row" in the periodic table in which these elements are located.

20. Electron configuration for:

a. Pu: $1s^2\ 2s^2\ 2p^6\ 3s^2\ 3p^6\ 3d^{10}\ 4s^2\ 4p^6\ 4d^{10}\ 4f^{14}\ 5s^2\ 5p^6\ 5d^{10}\ 5f^6\ 6s^2\ 6p^6\ 7s^2$
or [Rn] $5f^6\ 7s^2$
Predicted: [Rn] $5f^5 6d^1 7s^2$

b. Cm: $1s^2\ 2s^2\ 2p^6\ 3s^2\ 3p^6\ 3d^{10}\ 4s^2\ 4p^6\ 4d^{10}\ 4f^{14}\ 5s^2\ 5p^6\ 5d^{10}\ 5f^7\ 6s^2\ 6p^6 6d^1 7s^2$
[Rn] $5f^7 6d^1 7s^2$
Note that the predicted and actual are **both** :[Rn] $5f^7 6d^1 7s^2$

22. The orbital box representations for the following ions:

Orbital box notation

Ion	1s	2s	2p	3s	3p
a. Mg^{2+}	[↑↓]	[↑↓]	[↑↓][↑↓][↑↓]		
b. K^+	[↑↓]	[↑↓]	[↑↓][↑↓][↑↓]	[↑↓]	[↑↓][↑↓][↑↓]
c. Cl^-	[↑↓]	[↑↓]	[↑↓][↑↓][↑↓]	[↑↓]	[↑↓][↑↓][↑↓]
d. O^{2-}	[↑↓]	[↑↓]	[↑↓][↑↓][↑↓]		

24. Electron configurations of:

		3d	4s	
a. V	[Ar]	[↑][↑][↑][ ][ ]	[↑↓]	[Ar] $3d^3\ 4s^2$
b. V^{2+}	[Ar]	[↑][↑][↑][ ][ ]	[ ]	[Ar] $3d^3$
c. V^{5+}	[Ar]	[ ][ ][ ][ ][ ]	[ ]	[Ar]

Note that the V^{2+} ion contains unpaired electrons, and is therefore paramagnetic.

26. Element 25 is manganese.

			3d	4s	
a.	Mn	[Ar]	↑ ↑ ↑ ↑ ↑	↑↓	[Ar] $3d^5 4s^2$
b.	Mn^{2+}	[Ar]	↑ ↑ ↑ ↑ ↑	☐	[Ar] $3d^5$

c. Having unpaired electrons, Mn^{2+} is **paramagnetic.**

d. Mn^{2+} has five (5) unpaired electrons.

28. For ruthenium :

a. The predicted and actual electron configuration are:

Predicted : Ru [Kr] $4d^6 5s^2$ or $1s^2 2s^2 2p^6 3s^2 3p^6 3d^{10} 4s^2 4p^6 4d^6 5s^2$

Actual : Ru [Kr] $4d^7 5s^1$ or $1s^2 2s^2 2p^6 3s^2 3p^6 3d^{10} 4s^2 4p^6 4d^7 5s^1$

b. The orbital box and spectroscopic notation for the Ru^{3+} ion:

		4d	5s	
Ru^{3+}	[Kr]	↑ ↑ ↑ ↑ ↑	☐	[Kr] $4d^5$

The formation of the Ru^{3+} ion can be arrived at (from the predicted configuration) by loss of the two 5s electrons and one 4d electron. From the actual configuration, we see the loss of two 4d electrons and the 5s electron.

30. The electron configuration for U and U^{4+} :

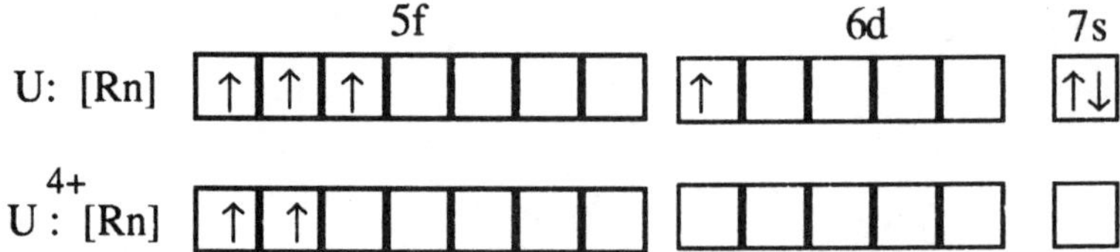

Both species have unpaired electrons and are therefore **paramagnetic**.

32. The electron configuration for Ti^{2+} is : [Ar] $3d^2$ hence this ion has 2 unpaired electrons. The electron configuration for Mn^{3+} is : [Ar] $3d^4$ giving this ion 4 unpaired electrons. Both ions therefore are paramagnetic (contain unpaired electrons).

34. The 2+ ions for Ti through Zn are:

Ion	Unpaired electrons	Ion	Unpaired electrons
Ti^{2+}	2	Co^{2+}	3
V^{2+}	3	Ni^{2+}	2
Cr^{2+}	4	Cu^{2+}	1
Mn^{2+}	5 (greatest number)	Zn^{2+}	0 **diamagnetic**
Fe^{2+}	4		

Electron Configurations and Quantum Numbers

36. The electron configuration for Mg using the orbital box method:

	1s	2s	2p	3s	
Mg:	[↑↓]	[↑↓]	[↑↓][↑↓][↑↓]	[↑↓]	[Ne] $3s^2$
Electron number				11 12	

Electron number:	n	ℓ	m_ℓ	m_s
11	3	0	0	+ 1/2
12	3	0	0	- 1/2

38. The electron configuration for Ga using the orbital box method:

	3d	4s	4p	
Ga [Ar]	[↑↓][↑↓][↑↓][↑↓][↑↓]	[↑↓]	[↑][ ][ ]	[Ar] $3d^{10} 4s^2 4p^1$

A possible set of quantum numbers for the highest energy electron:

n	ℓ	m_ℓ	m_s
4	1	-1	+ 1/2

40. Maximum number of electrons associated with the following sets of quantum numbers:

	Characterized as	Maximum number
a. n = 4 and ℓ = 3	4f electrons	14
b. n = 6, ℓ = 1, m_ℓ = -1	6p electrons	2
c. n = 3, ℓ = 3, m_ℓ = -3	[Maximum ℓ = n-1]	none
d. n = 2, ℓ = 1, m_ℓ = 1, and m_s = +1/2	2p orbital	1 (4 q. n. completely describe 1 electron)

42. Explain why the following sets of quantum numbers are not valid:

a. $n = 4$, $\ell = 2$, $m_\ell = 0$, $m_s = 0$:

The possible values of m_s can only be +1/2 or -1/2

b. $n = 3$, $\ell = 1$, $m_\ell = -3$, $m_s = -1/2$:

The possible values for m_ℓ is $-\ell$......0.....$+\ell$. Changing m_ℓ to -1, 0, +1 would give a valid set of quantum numbers.

c. $n = 3$, $\ell = 3$, $m_\ell = -1$, $m_s = +1/2$

The maximum value of ℓ is (n-1). Changing ℓ to 2 would provide a valid set of quantum numbers.

Periodic Properties

44. The Xe-F bond length can be estimated by adding the two radii:

Xe (131 pm) + F (72 pm) = 203 pm

46. Elements arranged in order of increasing size: C < B < Al < Na < K

Radii from Figure 8.10 (in pm): 77 < 85 < 143 < 186 < 227

48. The specie in each pair with the larger radius:

a. Cl^- is larger than Cl—The ion has more electrons/proton than the atom.

b. Al is larger than O—Al is in period 3, while O is in period 2.

c. In is larger than I—Atomic radii decrease, in general, across a period.

50. The group of elements with correctly ordered increasing ionization energy (IE):

c. Li < Si < C < Ne.

Neon would have the greatest IE. Silicon, being slightly larger in atomic radius than carbon, has a lesser IE. Lithium, the largest atom of this group, would have the smallest IE.

52. For the elements : Li, K, C, and N

Increasing ionization energy: K < Li < C < N

Ionization energy varies **inversely** with atomic radius—the smaller the atom, the greater the ionization energy.

54. For the elements Mg, Na, O, and P:

a. The largest atomic radius: Na

The greater the period number, the larger the atom. So Mg, Na, and P are larger than O. Atomic radii also decrease across a period.

b. The most positive electron affinity: O

Nonmetals have a more positive EA than metals. In general, EA increases across a period and up a group, so oxygen—being the element (of this group) closer to the "upper right" corner of the periodic table—is expected to have the most positive EA of these four elements.

c. Increasing ionization energy: Na < Mg < P < O

The ionization energy varies inversely with the atomic radius. The smaller the atom, the greater the IE.

56. a. Increasing ionization energy: S < O < F

Ionization energy is inversely proportional to atomic size.

b. Largest ionization energy of O, S, or Se: O

Oxygen is the smallest of these Group 6A elements, and hence has the largest IE.

c. Most positive electron affinity of Se, Cl, or Br: Cl

Chlorine is the smallest of these three elements. EA tends to increase on a diagonal from the lower left of the periodic table to the upper right.(See Figure 8.13).

d. Largest radius of O^{2-}, F^-, F: O^{2-}

The oxide ion has the largest electron : proton ratio. If one considers the attraction of the nuclear species (protons) for the extranuclear species (electrons), the greater the number of protons/electron the smaller the specie—owing to an increased electron-proton attraction. So the oxide ion (with 10 electrons and 8 protons) has the fewest electrons (of these three species) per proton.

General Questions

58. Using the spectroscopic notation the atom described would have an electron configuration:

$1s^2\ 2s^2\ 2p^6\ 3s^2\ 3p^6\ 4s^2$

a. Adding the electrons gives a sum of 20. Since this is a **neutral** atom, the # of protons and electrons would be equal. Twenty protons gives this element an atomic number = 20.

b. There are 2 s electrons in each of 4 shells : 8 s electrons total

c. There are 6 p electrons in each of 2 shells : 12 p electrons total

d. There are 0 d electrons

e. With its outer electrons in an "s" sublevel--specifically the 4s sublevel, the element is a **metal**-- specifically elemental **Calcium**.

60. The set of quantum numbers which is incorrect is (b). The maximum value of ℓ is (n- 1). With an ℓ =1, this value could not exceed 0.

62. The number of complete electron shells in element 71 is four (4). Element 71 is Lu, which has all subshells of the fourth shell filled.

64. a. The element with electron configuration $1s^2\ 2s^2\ 2p^6\ 3s^2\ 3p^3$ is: P (atomic # = 15)

b. The element with smallest atomic radius of the alkaline earth metals is Be—atomic radius increases as one goes down the group.

c. The element with the largest IE in Group 5A would also be the element with the smallest atomic radius—N

d. The element whose 2+ ion would have the configuration: [Kr] $4d^5$: Tc

e. The element with the most positive EA in Group 7A : Cl

f. The element with electron configuration: [Ar] $3d^{10}\ 4s^2$ is: Zn (atomic # = 30)

66. For element A = [Kr] $5s^1$ and B = [Ar] $3d^{10}4s^24p^4$

a. Element A is a metal (one electron in the s sublevel).

b. Element B has the greater IE. B's atomic radius would be smaller than that of A.

c. Element B has the greater EA. In general EA for nonmetals is greater than that of the metals in the same period

d. A would have the larger atomic radius—with its outermost electrons in the 5th shell.

68. Ions not likely to be found include: In^{4+}, Fe^{6+}, and Sn^{5+}

- Indium would form a cation by losing **three** electrons, forming the In^{3+} cation.
- Iron can lose either 2 or 3 electrons, forming Fe^{2+} and Fe^{3+} ions. Loss of more electrons would necessitate loss of d electrons from a half-filled d subshell.
- Tin can lose up to 4 electrons (2 s and 2 p) to form the 4+ cation. Loss of more electrons would require removal of electrons from a filled d subshell.

70. a. Element with the largest atomic radius: Se has the largest radius—being in period 4, group 6A. S would be smaller. Cl would be smaller than S. Recall the fact that atomic radius decreases across a period.

b. Cl^- is larger than Cl—having more electrons per proton.

c. Na would have the largest **difference** between the 1st and 2nd IE. Removing one electron would provide the stable 1+ cation and provide a specie with a filled shell.

d. Element with the largest ionization energy: N (IE is inversely proportional to atomic radius).

e. Largest radius: N^{3-} (radius 171 pm) is largest, owing to the repulsion of the 3 "added" electrons. While Xe is in period 5, its radius is cited as only 131 pm.

72. For the elements Na, B, Al, and C:

a. The largest atomic radius: Na (186 pm)

b. The largest electron affinity: C

c. Increasing ionization energy: Na < Al < B < C

74. a. The element (containing 27 electrons) is Cobalt

b. The sample contains unpaired electrons—so it is paramagnetic

c. The 3+ ion of cobalt would be formed by the loss of the two 4s electrons and 1 of the d electrons— leaving four unpaired electrons.

Conceptual Questions

76. The decrease in atomic size across a period is attributable to the increasing nuclear charge with an increasing number of protons. Given that the electrons are in the same outer energy level, the nuclear attraction for the electrons results in a diminishing atomic radius.

78. The slight decrease in atomic radius of the transition metals is a result of increased repulsions of (n-1)d electrons for (n)s electrons. Thus repulsion reduces the effects of the increasing nuclear charge across a period.

80. The element with the greatest difference between the first and second IE's is lithium. Of the four elements shown, Li is in the first group. The loss of the first electron results in an ion with a filled outer shell—a very stable configuration. Removal of a second electron would require a much larger amount of energy.

82. Ions likely to be formed:

Cs^+ and Se^{2-}; Both these ions have the noble gas configuration.

84. Since Ca is smaller than K, we would expect the first IE of Ca to be greater than that of K. Once K has lost its "first" electron, it possess an [Ar] core. Removal of an additional electron (the second) requires much energy. Ca, on the other hand, can lose a "second" electron to obtain the stable noble gas configuration with a much smaller amount of energy (smaller IE).

86. For the reaction to be: $2\ Ca\ (s) + 3\ F_2\ (g) \rightarrow 2\ CaF_3\ (s)$, calcium would have to form a 3+ cation. Since Calcium (in Group 2A) forms 2+ cations (and produces an ion that has the noble gas configuration), the formation of the 3+ cation is highly unlikely.

Challenging Questions

88. A plot of the atomic radii of the elements K—V shows a decrease. With the mass of these elements increasing from K through V (as more protons, neutrons, and electrons are added), the density is expected to increase.

90. The ionizations energies for the four elements are quite easily understood when two things are taken into account:

a. As one "crosses" a period (Si,P,S,Cl), the atoms decrease in radius. This decrease results in an increased nuclear-electron attraction, making removal of an electron successively more difficult (greater energy required).

b. The seeming anomaly (with P) is accounted for by remembering that with P the element has achieved a "half-filled" p subshell. Elemental S would then have **four** electrons in the p subshell—and removing one would create that "half-filled" subshell—and so less energy is required to remove the "fourth" electron—since removing that electron would reduce electron-electron repulsion in the p sublevel.

Summary Question

92. a. Orbital box notation for sulfur:

1s	2s	2p	3s	3p
[↑↓]	[↑↓]	[↑↓][↑↓][↑↓]	[↑↓]	[↑↓][↑][↑]

b. Quantum numbers for the "last electron": $n = 3$, $\ell = 1$, $m_\ell = +1$ (or 0 or -1), $m_s = -1/2$

c. Element with the smallest ionization energy : S

Element with the smallest radius: O

d. S: Negative ions are always larger than the element from which they are derived.

e. Grams of Cl_2 to make 675 g of $OSCl_2$:

$$675 \text{ g } OSCl_2 \cdot \frac{1 \text{ mol } OSCl_2}{119.0 \text{ g } OSCl_2} \cdot \frac{2 \text{ mol } Cl_2}{1 \text{ mol } OSCl_2} \cdot \frac{70.91 \text{ g } Cl_2}{1 \text{ mol } Cl_2} = 804 \text{ g } Cl_2$$

f. The theoretical yield of $OSCl_2$ if 10.0 g of SO_2 and 20.0 g of Cl_2 are used:

$$\text{Moles of } SO_2 = 10.0 \text{ g } SO_2 \cdot \frac{1 \text{ mol } SO_2}{64.06 \text{ g } SO_2} = 0.156 \text{ moles } SO_2$$

$$\text{Moles of } Cl_2 = 20.0 \text{ g } Cl_2 \cdot \frac{1 \text{ mol } Cl_2}{70.91 \text{ g } Cl_2} = 0.282 \text{ moles } Cl_2$$

$$\text{Moles-available ratio: } \frac{0.156 \text{ moles } SO_2}{0.282 \text{ moles } Cl_2} = \frac{0.553 \text{ moles } SO_2}{1 \text{ mol } Cl_2}$$

$$\text{Moles-required ratio: } \frac{1 \text{ mol } SO_2}{2 \text{ mol } Cl_2} = \frac{0.5 \text{ mol } SO_2}{1 \text{ mol } Cl_2}$$

Cl_2 is the limiting reagent.

$$0.282 \text{ moles } Cl_2 \cdot \frac{1 \text{ mol } OSCl_2}{2 \text{ moles } Cl_2} \cdot \frac{119.0 \text{ g } OSCl_2}{1 \text{ mol } OSCl_2} = 16.8 \text{ g } OSCl_2$$

g. For the reaction:

$$SO_2 (g) + 2 Cl_2 (g) \rightarrow OSCl_2 (g) + Cl_2O (g) \qquad \Delta H^\circ_{rxn} = +164.6 \text{ kJ}$$

$$\Delta H^\circ_{rxn} = [1 \cdot \Delta H^\circ_f OSCl_2 (g) + 1 \cdot \Delta H^\circ_f Cl_2O (g)] - [1 \cdot \Delta H^\circ_f SO_2(g) + 2 \cdot \Delta H^\circ_f Cl_2 (g)]$$

$$+164.6 \text{ kJ} = [\Delta H^\circ_f OSCl_2 (g) + 80.3 \text{ kJ}] - [-296.8 \text{ kJ} + 2 \cdot (0)]$$

$$+164.6 \text{ kJ} = \Delta H^\circ_f OSCl_2 (g) + 377.1 \text{ kJ}$$

$$-212.5 \text{ kJ} = \Delta H^\circ_f OSCl_2 (g)$$

Chapter 9
Bonding and Molecular Structure: Fundamental Concepts

Review Questions

4. Prediction on the bonding in:

KI	ionic	bonding between metal and nonmetal
MgS	ionic	bonding between metal and nonmetal
CS_2	covalent	bonding between two nonmetals
P_4O_{10}	covalent	bonding between two nonmetals

6. $CaCl_4$ is not likely to exist. Calcium (Group 2A) forms a 2+ cation and chlorine (Group 7A) forms a 1- anion, making $CaCl_2$ the likely formula of the compound between these two elements.

8. The dot structure for BCl_3 is shown at left. The coordinate covalent bond between B and N is shown as a pair of electrons in the dot structure at right.

:Cl: :Cl:
:Cl – B – Cl: :Cl – B – Cl:
H – N – H
H

The bond between B-N and the three B-Cl bonds provide an octet configuration for Boron.

10. Which of the following are odd-electron species:

Specie	electrons	total	electron number
NO_2	5 + 2(6)	17	**odd**
SF_4	6 + 4(7)	34	even
SO_3	6 + 3(6)	24	even
O_2^-	2(6) + 1	13	**odd**

12. Bond order in acetylene and phosgene:

:Cl:
H – C≡C – H :Cl – C = O

The C-H bonds are bond order =1 (single bond); the C-C bond is bond order 3.

The Cl-C bonds (in phosgene) are bond order = 1; the C-O bond is bond order 2.

14. The N-O bond order in the nitrate ion:

$$\left[:\ddot{O} \overset{1}{-} \underset{\displaystyle |_{1}}{\overset{\displaystyle :\ddot{O}:}{N}} \overset{2}{=} \ddot{O}: \right]^{-}$$

The N-O bonds labeled (1) are of bond order 1, the N-O bond labeled (2) is of bond order 2. This is only one structure that we could draw for the nitrate ion. The others would show a similar electron distribution. We have **4 pairs** of electrons making 3 bonds so the "average" bond order is 4/3 or 1⅓.

16. Bond dissociation energy—The energy (enthalpy) required to break a bond in a molecule with the reactants and products (in the gas phase) under standard conditions. Bond-breaking reactions **always** require the input of energy, and so always have a + sign.

18. To estimate the enthalpy change for the formation of water (g) from hydrogen and oxygen one needs: O=O (double bond) energy ; H-H bond energy ; H-O bond energy

$$\begin{aligned} \Delta H\text{reaction} &= \Sigma\, E(\text{bonds broken}) - \Sigma\, E\ (\text{bonds made}) \\ &= [E\ (O{=}O)\ \text{bond} + 2\ E\ (H{-}H)\text{bond}] - 4[E\ (O{-}H)\ \text{bond}] \\ &= [498\ \text{kJ} + 2 \bullet 436\ \text{kJ}] - 4[463\ \text{kJ}] = -\ 482\ \text{kJ} \end{aligned}$$

20. Difference between electronegativities and electron affinities:

Electronegativity is the ability of an atom to attract electrons to itself (in a molecule). Electron affinity is the energy change when an anion of the element loses an electron (in the gas phase)—isolated atoms.

22. The principle of **electroneutrality** states that atoms in molecules or ions should have **formal charges as small as possible.** The resonance structure for carbon dioxide shown below can be eliminated since—using this structure the formal charges on the atoms are: (1+ 0 1-).

$$:O{\equiv}C{-}\ddot{\underset{..}{O}}:$$

The resonance structure showing 2 C=O double bonds has a formal charge of 0 on **all 3 atoms**— a much better "picture".

24. The **electron pair geometry** shows the *relative positions of all the valence electrons* around a central atom in a molecule or ion. The **molecular geometry** is concerned with the *relative positions of the **atoms*** in a molecule or ion. Water has 4 electron pairs around O (electron-pair geometry—tetrahedral), but only 2 atoms attached to O (molecular geometry—bent).

26. Four electron pairs form a **pyramidal molecule if** one of the electron pairs is a non–bonding electron pair. A **bent molecule** is obtained if **two** electron pairs are non-bonding pairs. In either case, the angle between two electron pairs and the central atom is approximately 109°.

Valence Electrons

28.

	Element	Group Number	Number of Valence Electrons
a.	N	5A	5
b.	B	3A	3
c.	Na	1A	1
d.	Mg	2A	2
e.	F	7A	7
f.	S	6A	6

The Octet Rule

30.

Group Number	Number of Bonds
1A	1
2A	2
3A	3
4A	4
5A	3 (or 4 in species such as NH_4^+)
6A	2 (or 3 as in H_3O^+)
7A	1
8A	0

Ionic Compounds

32. Compound with the most negative energy of ion pair formation ? least negative?
Using Coulomb's Law we see that the most negative IP energy will result from (a) increased charges on the ions and (b) decreased distance between the ions.

Compiling those data for the ions involved we get:

ion	charge	ionic radius (pm)
Na^+	1	116
Mg^{2+}	2	86
Cl^-	1	167
F^-	1	119
S^{2-}	2	170

Ignoring for the moment the charges on the electrons (since they are the same on all electrons, we can calculate terms for the numerators (# of + and - charges) and denominators (sum of ionic radii) for the three compounds:

$$NaCl \quad \frac{(1 \bullet 1)}{(116+167)} = \frac{1}{283} = \frac{1}{283}$$

$$MgF_2 \quad \frac{(2 \bullet 1)}{(86 + 119)} = \frac{2}{205} = \frac{1}{103}$$

$$MgS \quad \frac{(2 \bullet 2)}{(86 + 170)} = \frac{4}{256} = \frac{1}{64}$$

The third column represents the reduction of all numerators to unity. The result is that **MgS would have the largest negative energy of ion pair formation, and NaCl would have the least negative value.**

34. Arrange lattice energies from least negative to most negative:

Since CaO involves 2+ and 2- ions, the lattice energy for this compound is greater than the other compounds. Given that lattice energy is **inversely** related to the distance between the ions, the lattice energy for LiI is greater(more negative) than that for RbI ($Rb^+ > Li^+$). The small diameter of the fluoride ion (compared to iodide) indicates that the lattice energy for NaF would be more negative than for either LiI or RbI. The lattice energies are then:

least negative ----- RbI ----- LiI ----- NaF ----- CaO ----- **most negative**

36. Since melting a solid involves disassembling the crystal lattice of cations and anions, the greater the distance between cations and anions—the less the attraction between the cation and anion, and the easier it becomes to disassemble the lattice, and hence the lower the melting point. So if the anion-cation distance is decreased, **the melting point should increase.**

Lewis Electron Dot Structures

38. a. NF_3 : [1(5) + 3(7)] = 26 valence electrons

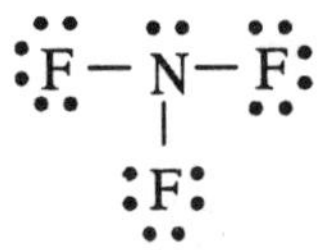

F – N – F
    |
    F

b. ClO_3^- : [1(7) + 3(6) + 1] = 26 valence electrons
↑
ion charge

[O – Cl – O
        |
        O] $^{-}$

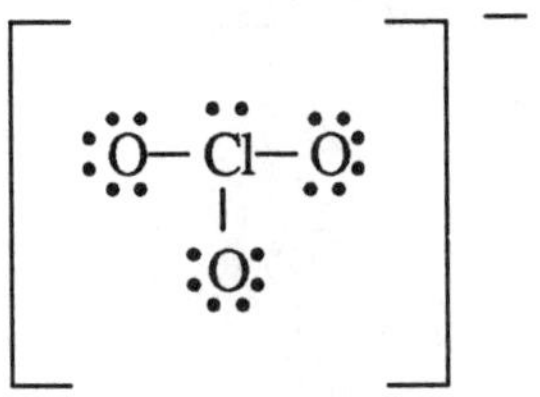

c. HOBr: [1(1) + 1(6) + 1(7)] = 14 valence electrons

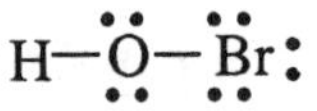

H–O–Br

d. SO_3^{2-} : [1(6) + 3(6) + 2] = 26 valence electrons
↑
ion charge

[O – S – O
       |
       O] $^{2-}$

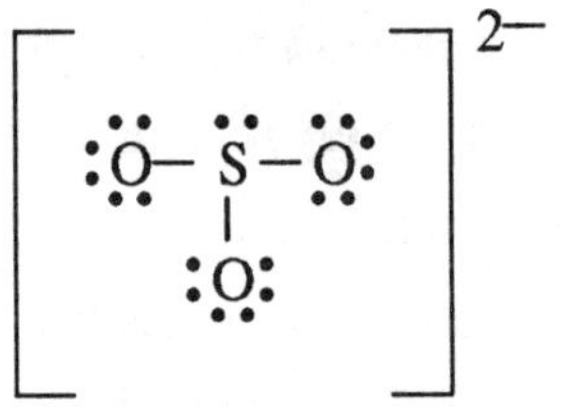

40. a. $CHClF_2$: [1(4) +1(1) + 1(7) + 2(7)] = 26 valence electrons

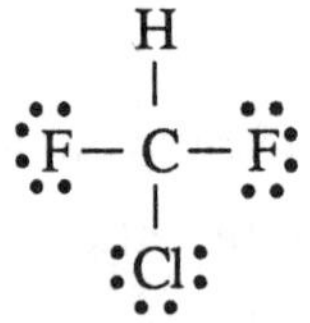

      H
      |
F – C – F
      |
     Cl

b. CH_3COOH: [3(1) + 2(4) + 2(6) + 1(1)] = 24 valence electrons

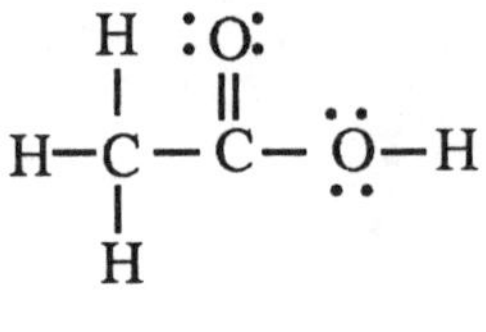

     H   O
     | ||
H–C – C – O–H
     |
     H

c. H_3CCN: [3(1) + 2(4) + 1(5)] = 16 valence electrons

H
|
H – C – C ≡ N:
|
H

d. H_3COH: [3(1) + 1(4) + 1(6) + 1(1)] = 14 valence electrons

H
|
H – C – Ö – H
|
H

42. Resonance structures for:

a. SO_2:

Ö = S̈ - Ö: ⟷ :Ö - S̈ = Ö

b. NO_2^-:

[Ö = N̈ – Ö:]⁻ ⟷ [:Ö – N̈ = Ö]⁻

c. SCN^-:

[:N̈ – C ≡ S:]⁻ ⟷ [N̈ = C = S̈]⁻ ⟷ [:N ≡ C – S̈:]⁻

44. a. BrF_3 : [1(7) + 3(7)] = 28 valence electrons

:F̈ – B̈r – F̈:
|
:F̈:

b. I_3^- : [3(7) + 1] = 22 valence electrons

[:Ï – Ï – Ï:]⁻

c. XeO_2F_2 : [1(8) + 2(6) + 2(7)] = 34 valence electrons

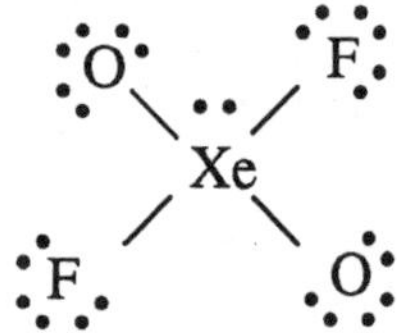

d. XeF_3^+ : [1(8) + 3(7)- 1] = 28 valence electrons

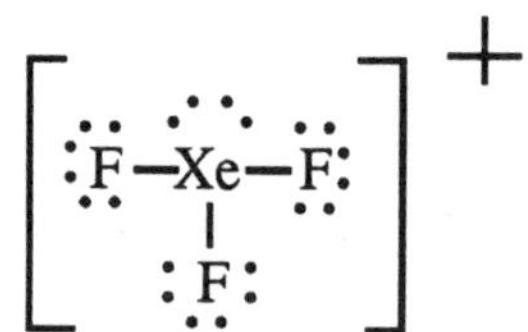

Bond Properties

46.

	Specie	Number of bonds	Bond Order : Bonded Atoms
a.	H_2CO	3	1 : CH 2: C = O
b.	SO_3^{2-}	3	1 : SO
c.	NO_2^+	2	2 : NO
d.	NOCl	2	1: N-Cl 2: N=O

48. In each case the shorter bond length should be between the atoms with smaller radii--if we assume that the bond orders are equal.

a. B-Cl	B is smaller than Ga	b. C-O	C is smaller than Sn
c. P-O	O is smaller than S	d. C=O	O is smaller than N

50. The CO bond in carbon monoxide is shorter. The CO bond in carbon monoxide is a triple bond, thus it requires more energy to break than the CO double bond in H_2CO.

52. The bond order for NO_2^+ is 2, for NO_2^- is 3/2 while the bond order for NO_3^- is 4/3. The Lewis dot structure for the NO_2^+ ion indicates that both NO bonds are double, while in the nitrate ion, any resonance structure (there are three) shows one double bond and two single bonds. The nitrite ion—in either resonance structure (there are two)—has one double and one single bond. Hence the **NO bonds in the nitrate ion will be longest** while those in the **NO_2^+ ion will be shortest**.

Bond Energies and Reaction Enthalpies

54.

$$H_3C\text{-}CH_2\text{-}\overset{H}{\overset{|}{C}}=\overset{H}{\overset{|}{C}}\text{-}H \quad + \quad H_2 \longrightarrow \quad H_3C\text{-}CH_2\text{-}\underset{H}{\underset{|}{\overset{H}{\overset{|}{C}}}}\text{-}\underset{H}{\underset{|}{\overset{H}{\overset{|}{C}}}}\text{-}H$$

Energy input: 1 mol C=C = 1 mol • 602 kJ/mol = 602 kJ

1 mol H-H = 1 mol • 436 kJ/mol = 436 kJ

Total input = 1038 kJ

Energy release: 1 mol C-C = 1 mol • 346 kJ/mol = 346 kJ

2 mol C-H = 2 mol • 413 kJ/mol = 826 kJ

Total released = 1172 kJ

Energy change: 1038 kJ - 1172 kJ = - 134 kJ

56. Heat of reaction for : $CO\ (g) + Cl_2\ (g) \rightarrow Cl_2CO\ (g)$

Energy input: 1 mol C≡O = 1 mol • 1072 kJ/mol = 1072 kJ

1 mol Cl-Cl = 1 mol • 242 kJ/mol = 242 kJ

Total input = 1314 kJ

Energy release: 2 mol C-Cl = 2 mol • 339 kJ/mol = 678 kJ

1 mol C=O = 1 mol • 732 kJ/mol = 732 kJ

Total released = 1410 kJ

Energy change: 1314 kJ - 1410 kJ = - 96 kJ

58. $OF_2\ (g) + H_2O\ (g) \rightarrow O_2\ (g) + 2\ HF\ (g)$ $\Delta H = -318$ kJ

Energy input : 2 mol O-F = 2 x (where x = O-F bond energy)

2 mol O-H = 2 mol • 463 kJ/mol = 926 kJ

Total input = (926 + 2x) kJ

Energy release: 1 mol O=O = 1 mol • 498 kJ/mol = 498 kJ

2 mol H-F = 2 mol • 565 kJ/mol = 1130 kJ

Total release = 1628 kJ

$$-318\text{ kJ} = 926\text{ kJ} + 2x - 1628\text{ kJ}$$
$$384\text{ kJ} = 2x$$
$$192\text{ kJ/mol} = \text{O-F bond energy}$$

Formal Charges

60. Formal charge on each atom in the following:

a. N_2H_4

Atom	Formal Charge
H	1 - 1/2(2) = 0
N	5 - 2 - 1/2(6) = 0

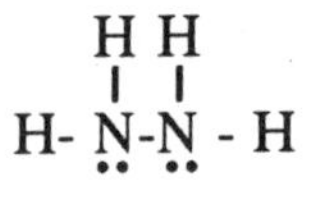

b. PO_4^{3-}

Atom	Formal Charge
P	5 - 1/2(8) = 1
O	6 - 6 - 1/2(2) = -1
	Sum = -3 (charge on ion)

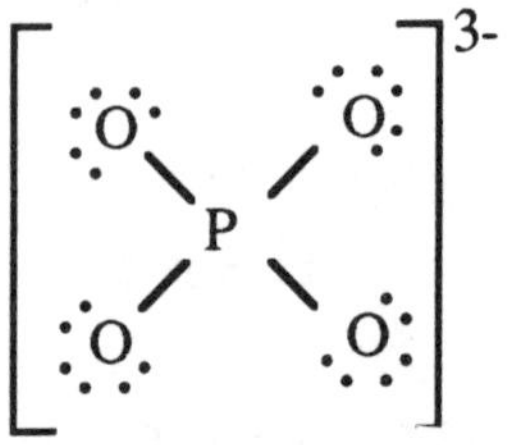

c. BH_4^-

Atom	Formal Charge
B	3 - 1/2(8) = -1
H	1 - 1/2(2) = 0
	Sum = -1 (charge on ion)

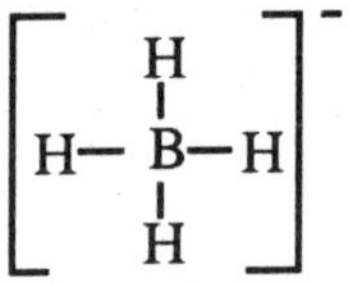

d. NH_2OH

Atom	Formal Charge
N	5 - 2- 1/2(6) = 0
H	1 - 1/2(2) = 0
O	6 - 4 - 1/2(4) = 0

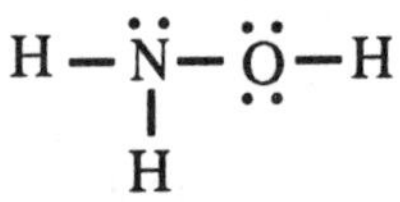

62. Formal charge on each atom in the following:

a. NO_2^+

Atom	Formal Charge
O	6 - 4- 1/2(4) = 0
N	5 - 0 - 1/2(8) = +1

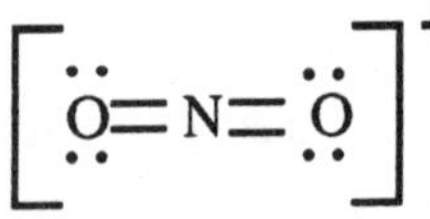

b. NO_2^-

Atom	Formal Charge
O1	6 - 4 - 1/2(4) = 0
O 2	6 - 6 - 1/2(2) = -1
N	5 - 2 - 1/2(6) = 0

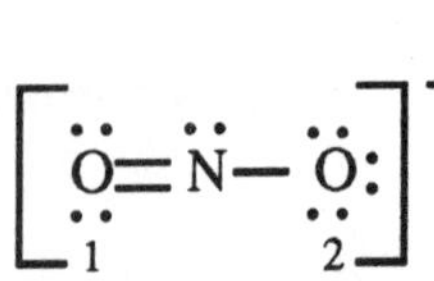

c. NF_3

Atom	Formal Charge
F	7 - 6 - 1/2(2) = 0
N	5 - 2 - 1/2(6) = 0

d. HNO_3

Atom	Formal Charge
O1	6 - 4 -1/2(4) = 0
O2	6 - 6 - 1/2(2) = -1
O3	6 - 4- 1/2(4) = 0
N	5 - 0 - 1/2(8) = 1
H	1 - 1/2(2) = 0

[Note: This is only 1 possible structure]

Electronegativity and Bond Polarity

64. Indicate the more polar bond (Arrow points toward the more negative atom in the dipole).

a. C-O > C-N
→ →

b. P-Cl > P-Br
→ →

c. B - O > B - S
→ →

d. B-F > B-I
→ →

66. For the bonds in acrolein the polarities are as follows:

	H-C	C-C	C=O
$\frac{\Delta\chi}{\Sigma\chi}$	0.09	0	0.16

a. The C-C bonds are nonpolar, the C-H bonds are slightly polar, and the C=O bond is polar.

b. The most polar bond in the molecule is the C=O bond, with the oxygen atom being the negative end of the dipole.

Charge Distribution in Molecules

68. Atom(s) on which the negative charge resides in:

a. BF_4^- Formal charges : B = -1; F = 0 but fluorine is MUCH more electronegative than B, so the negative charge (predicted to reside on B) resides on the fluorines--and is indeed distributed over the molecule.

b. BH_4^- Formal charges: B = -1: H = 0 Hydrogen is only slightly more electronegative than B (2.1 compared to 2.0), so the negative charge would reside on the H atoms (although the B-H bonds are **not very polar**).

c. OH^- Formal charges: O = -1 ; H = 0 Oxygen is much more electronegative than H so the negative charge resides on the oxygen atom.

d. $CH_3CO_2^-$

Formal charges : C = 0; O_1 = 0, O_2 = -1 Oxygen is more electronegative than C, so the charge would reside on the oxygens as opposed to the C. The picture shown here is a bit misleading, since **either oxygen** could have the double bond to the C, so **two resonance structures are available** with the negative charge distributed (delocalized) over both C-O bonds.

$$H_3C-\overset{\overset{:\ddot{O}:}{\|1}}{C}-\overset{2}{\ddot{\underset{..}{O}}}{}^-$$

70. a. Resonance structures of N_2O :

$$:N\equiv N-\ddot{\underset{..}{O}}: \longleftrightarrow \ddot{\underset{..}{N}}=N=\ddot{\underset{..}{O}} \longleftrightarrow :\ddot{\underset{..}{N}}-N\equiv O:$$

(atoms numbered 1 2 beneath N1 and N2 in each structure)

b. Formal Charges:

N_1	5 - 2 - 1/2(6) = 0	5 - 4 - 1/2(4) = -1	5 - 6 - 1/2(2) = -2
N_2	5 - 0 - 1/2(8) = +1	5 - 0 - 1/2(8) = +1	5 - 0 - 1/2(8) = +1
O	6 - 6 - 1/2(2) = -1	6 - 4 - 1/2(4) = 0	6 - 2 - 1/2(6) = +1

c. Of these three structures, the first is the most reasonable in that the most electronegative atom, O, bears a formal charge of -1.

72. Resonance structures for NO_2^- :

$$\left[\ddot{\underset{..}{O}}=\ddot{N}-\ddot{\underset{..}{O}}:\right]^- \longleftrightarrow \left[:\ddot{\underset{..}{O}}-\ddot{N}=\ddot{\underset{..}{O}}\right]^-$$

(oxygens numbered 1 and 2 in each structure)

Formal charges:

O_1 6 - 4- 1/2(4) = 0	O_1 6 - 6 - 1/2(2) = -1
N 5 - 2 - 1/2(6) = 0	N 5 - 2 - 1/2(6) = 0
O_2 6 - 6 - 1/2(2) = -1	O_2 6 - 4 - 1/2(4) = 0

74. The electron dot structure for the cyanide ion may be represented with the structure shown at right. The ion is symmetric with respect to the electron distribution—that is each atom has **one** lone pair and participates in a **triple bond**. Carbon has a formal charge of minus 1, while N has a formal charge of zero so we should write HCN— with the positive H ion attracted to the more negative "end" of the CN⁻ ion.

$[:C\equiv N:]^-$

Molecular Geometry

76. Using the Lewis structure describe the Electron-pair and molecular geometry:

a.

H–N(H)–Cl

Electron-pair: tetrahedral
Molecular : pyramidal

b.

:Cl–O–Cl:

Electron-pair: tetrahedral
Molecular: bent

c.

$[N=C=S]^-$

Electron-pair: linear
Molecular : linear

d.

H–O–F:

Electron-pair: tetrahedral
Molecular: bent

78. Using the Lewis structure describe the Electron-pair and molecular geometry:

a.

O=C=O

Electron-pair: linear
Molecular: linear

b.

$[\ddot{O}{=}\ddot{N}{-}\ddot{O}:]^-$

Electron-pair: trigonal planar
Molecular: bent

c.

$:\ddot{O}{-}\ddot{O}{=}\ddot{O}$

Electron-pair: trigonal planar
Molecular: bent

d.

$[:\ddot{O}{-}\ddot{Cl}{-}\ddot{O}:]^-$

Electron-pair: tetrahedral
Molecular: bent

All the species having at least one lone pair on the central atom have **bent molecular geometry**.

80. Using the Lewis structure describe the electron-pair and molecular geometry.
[Lone pairs on F have been omitted for clarity.]

a.

$[F{-}\ddot{Cl}{-}F]^-$

Electron-pair: trigonal bipyramidal
Molecular : linear

b.

F—Cl—F
|
F

Electron-pair: trigonal bipyramidal
Molecular: T-shaped

c.

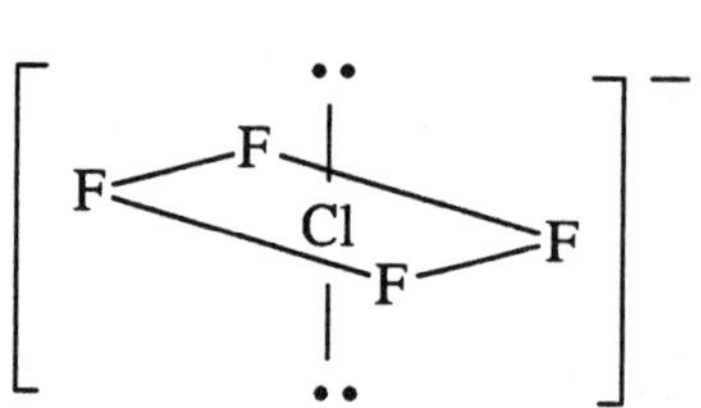

Electron-pair: octahedral
Molecular: square planar

d.

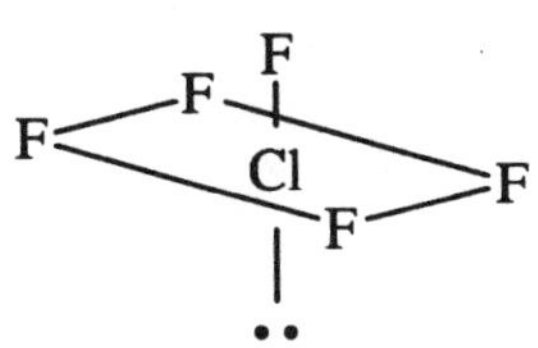

Electron-pair: octahedral
Molecular: square pyramidal

82. a. O-S-O angle in SO_2 : Slightly less than 120°; The lone pair of S should reduce the predicted 120° angle slightly.

b. F-B-F angle in BF_3 : 120°

c. (1) H-C-H angle in CH_3CN : 109°

(2) C-C ≡N angle in CH_3CN : 180°

d. Cl-C-Cl in Cl_2CO Slightly less than 120°; The two lone pairs of electrons on O will reduce the predicted 120° angle slightly.

84. Estimate the values of the angles indicated in the model of phenylalanine below:

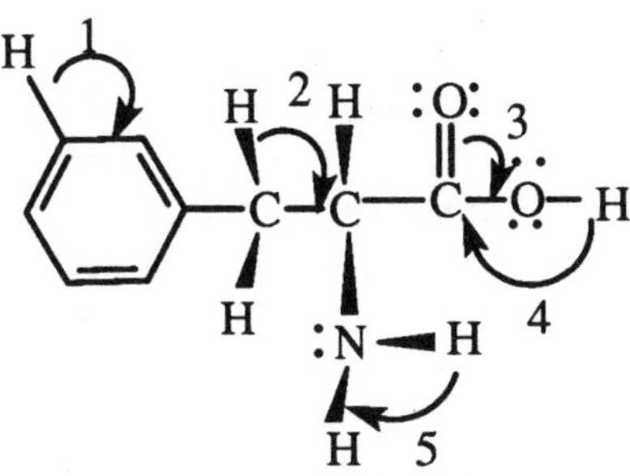

Angle 1:	H-C-C	120°	three groups around the C atom
Angle 2:	H-C-C	109°	four groups around the C atom
Angle 3:	O-C-O	120°	three groups around the C atom
Angle 4:	C-O-H	109°	four groups around the O atom
Angle 5:	H-N-H	109°	four groups around the N atom

Since the geometry around the C is tetrahedral, the C-C-C chain can not be linear.

86. Approximate values for the angles: (A = axial ; E= equatorial)

	Bond	Compound	Angles		
a.	F-Se-F	SeF_4	F(A)-Se-F(A)	180°	see SF_4 page 419
			F(A)-Se-F(E)	90°	
			F(E)-Se-F(E)	120°	
b.	O-S-F	OSF_4	O-S-F(A)	90°	
			O-S-F(E)	120°	
	F-S-F		F(A)-S-F(E)	90°	
			F(E)-S-F(E)	120°	
c.	F-Br-F	BrF_5	F(E)-Br-F(E)	90°	
			F(A)-Br-F(E)	90°	
d.	F-P-F	PF_6^-	F(A)-P-F(E)	90°	
			F(E)-P-F(E)	90°	
			F(A)-P-F(A)	180°	

88. NO_2^+ has two structural pairs around the N atom. We predict that the O-N-O bond angle would be approximately 180°. NO_2^- has three structural pairs (one lone pair). The geometry around this central atom would be trigonal planar with a bond angle of approximately 120°. The linear geometry around N in NO_2^+ would provide the larger bond angle.

Molecular Polarity

90. For the molecules: H_2O NH_3 CO_2 ClF CCl_4

i. Using the electronegativities to determine bond polarity:

	H_2O	NH_3	CO_2	ClF	CCl_4
$\frac{\Delta\chi}{\Sigma\chi}$	$\frac{1.4}{5.6}$	$\frac{0.9}{5.1}$	$\frac{1.0}{6.0}$	$\frac{10}{7.0}$	$\frac{0.5}{5.5}$

Reducing these fractions to a decimal form indicates that the H-O bonds in water are the most polar of these bonds.

[Note: $\Delta\chi$ represents the **difference** in electronegativities between the elements while $\Sigma\chi$ is the **sum** of the electronegativities of the elements.]

ii. The nonpolar compounds are:

CO_2 The O-C-O bond angle is 180°, thereby canceling the C-O dipoles.

CCl_4 The Cl-C-Cl bond angles are approximately 109°, with the Cl atoms directed at the corners of a tetrahedron. Such an arrangement results in a net dipole moment of zero.

iii. The F atom in ClF is more negatively charged.(Electronegativity of F = 4.0, Cl = 3.0)

92. Molecular polarity of the following: $BeCl_2$, HBF_2, CH_3Cl, SO_3

$BeCl_2$ and SO_3 are nonpolar. For HBF_2, the hydrogen and fluorine atoms are arranged at the corners of a triangle. The "negative end" of the molecule lies on the plane between the fluorine atoms, and the H atom is the "positive end." For CH_3Cl, the chlorine atom is the negative end and the H atoms form the positive end.

General Questions

94. Lewis structure(s) for the following: What are similarities and differences ?

a. CO_2

$$\left[\ddot{\underset{\cdot\cdot}{O}}=C=\ddot{\underset{\cdot\cdot}{O}}\right] \longleftrightarrow \left[:\ddot{\underset{\cdot\cdot}{O}}-C\equiv O:\right] \longleftrightarrow \left[:O\equiv C-\ddot{\underset{\cdot\cdot}{O}}:\right]$$

b. N_3^-

$$\left[\ddot{\underset{\cdot\cdot}{N}}=N=\ddot{\underset{\cdot\cdot}{N}}\right]^- \longleftrightarrow \left[:\ddot{\underset{\cdot\cdot}{N}}-N\equiv N:\right]^- \longleftrightarrow \left[:N\equiv N-\ddot{\underset{\cdot\cdot}{N}}:\right]^-$$

c. OCN^-

$$\left[\ddot{\underset{\cdot\cdot}{O}}=C=\ddot{\underset{\cdot\cdot}{N}}\right]^- \longleftrightarrow \left[:\ddot{\underset{\cdot\cdot}{O}}-C\equiv N:\right]^- \longleftrightarrow \left[:O\equiv C-\ddot{\underset{\cdot\cdot}{N}}:\right]^-$$

Each of these species has 16 electrons, and each is linear. Carbon dioxide is neutral while the azide ion and isocyanate ion are charged. The isocyanate ion is also polar, while carbon dioxide and azide are not polar.

96. a. Angle 1 = 120°; Angle 2 = 180°; Angle 3 = 120°

b. The C=C double bond is shorter than the C-C single bond.

c. The C=C double bond is stronger than the C-C single bond.

d. The C≡N bond is the most polar bond, with the N atom being the negative end of the bond dipole.

98. Using the Lewis structure describe the Electron-pair and Molecular geometry:

a.

:F̤̈–B–F̤̈:
|
:F̤̈:

Electron-pair: trigonal planar
Molecular : trigonal planar

b.

:F̈:
|
:F̤̈–C–F̤̈:
|
:F̤:

Electron-pair: tetrahedral
Molecular: tetrahedral

c.

:F̤̈–P̈–F̤̈:
|
:F̤̈:

Electron-pair: tetrahedral
Molecular: pyramidal

d.

:F̤̈–Ö̤–F̤̈:

Electron-pair: tetrahedral
Molecular: bent

e.

H–F̤̈:

Electron-pair: tetrahedral
Molecular: linear

For molecules that contain no lone pairs on the central atom (a, b) the electron-pair and molecular geometries are the same.

100. a. In XeF_2 the bonding pairs occupy the axial positions with the lone pairs located in the (preferred) equatorial plane.

b. In ClF_3 two of the three equatorial positions are occupied by the lone pairs of electrons on the Cl atom. (See question 80.b)

102. The decomposition of urea to hydrazine and carbon monoxide:

$$H_2N-C(=O)-NH_2 \longrightarrow H_2N-NH_2 + :C\equiv O:$$

Break	N-C bonds (2)	2 • 305 kJ/mol	
	C=O bond (1)	1 • 732 kJ/mol	bonds broken: 610 + 732 = 1342 kJ
Make	N-N bond (1)	1 • 163 kJ/mol	
	C≡O bond (1)	1 • 1072 kJ/mol	bonds made: 163 + 1072 = 1235 kJ

Change in energy: 1342 kJ - 1235 kJ = 107 kJ

Conceptual Questions

104. a. Resonance structures for CNO^- with formal charges.

$$[\ddot{C}=N=\ddot{O}]^- \longleftrightarrow [:\ddot{C}-N\equiv O:]^- \longleftrightarrow [:C\equiv N-\ddot{O}:]^-$$

Formal Charges:

-2 +1 0	-3 +1 +1	-1 +1 -1

b. The most reasonable structure is the one at the far right because it is the one in which the formal charges on the atoms are at a minimum, and oxygen has the negative formal charge.

c. The instability of the ion could be attributed to the fact that the least electronegative atom in the ion (C) bears a negative charge in all resonance structures.

106. The alternate structure for an amide is:

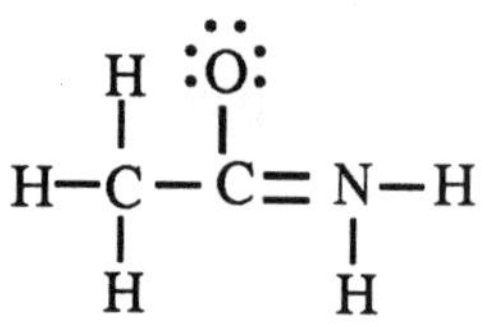

```
   H  :Ö:                        H  :O:
   |   |                         |   ||   ..
H—C — C= N—H                  H—C — C— N—H
   |       |                     |        |
   H       H                     H        H
```

and the "usual structure" is

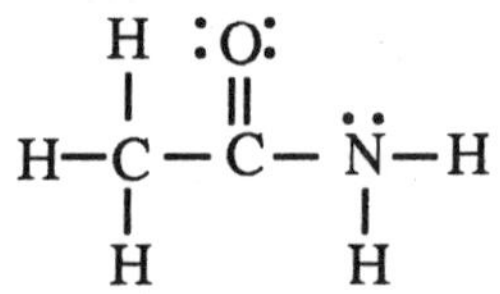

Calculate the formal charges on the O,C,N associated with the "amide" group.

"alternate structure" O: 6 - 6 - 1/2(2) = -1 and for the "usual": O: 6 - 4 - 1/2(4) = 0

C: 4 - 0 - 1/2(8) = 0 C: 4 - 0 - 1/2(8) = 0

N: 5 - 0 - 1/2 (8) = +1 N: 5 - 2 - 1/2 (6) = 0

Note that the "usual" structure has formal charges of 0 **on all atoms** while the alternate structure has a formal charge of +1 on N (which is more electronegative than C).

Challenging Question

108. a. Formal charges on S and O atoms:

Recall: Formal charge = Group # - nonbonding electrons - 1/2 (bonding electrons)

S: 6 - 4 - 1/2(4) = 0 and for O: 6 - 4 - 1/2(4) = 0

b. A model of 2-furylmethanethiol is shown below with most angles identified:

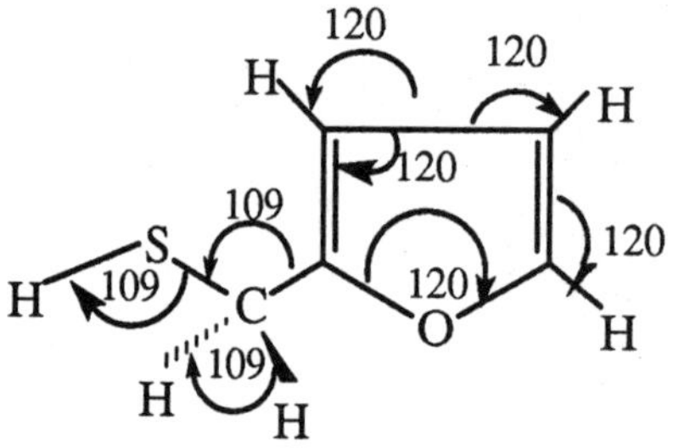

c. Identify shorter carbon-carbon bonds in the molecule.

The four carbons in the ring participating in C=C double bonds are shorter than e.g. the C-C single bond (shown on the side chain).

d. Most polar bond in the molecule:

Using electronegativity differences as a measure of polarity, the C-O bonds (shown at the bottom of the ring in this diagram) will be the most polar.

The electronegativity of C is 2.5 while that of O is 3.5 for a Δ of 1.0

e. The most electronegative atom in the molecule is **oxygen.**

f. While one normally expects oxygen to be sp^3 hybridized, the oxygen in the furan ring assumes an sp^2 hybridizattion (and a 120° bond angle) to complete the aromaticity of the ring.

g. The molecule is as a whole polar, owing to the S-H sidechain.

Summary Question

110. a. The production of ClF_3 from the elements may be represented:

$$Cl_2(g) + 3\,F_2(g) \rightarrow 2\,ClF_3(g)$$

b. Mass of ClF_3 expected:

Moles Cl_2: $0.71 \text{ g } Cl_2 \cdot \frac{1 \text{ mol } Cl_2}{70.9 \text{ g } Cl_2} = 0.010 \text{ mol } Cl_2$

Moles F_2: $1.00 \text{ g } F_2 \cdot \frac{1 \text{ mol } F_2}{38.00 \text{ g } F_2} = 0.0263 \text{ mol } F_2$

Ratios: Moles required: $\frac{3 \text{ mol } F_2}{1 \text{ mol } Cl_2}$

Moles available: $\frac{0.0263 \text{ mol } F_2}{0.010 \text{ mol } Cl_2}$

F_2 is the **limiting reagent.**

$$0.0263 \text{ mol } F_2 \cdot \frac{2 \text{ mol } ClF_3}{3 \text{ mol } F_2} \cdot \frac{92.45 \text{ g } ClF_3}{1 \text{ mol } ClF_3} = 1.62 \text{ g } ClF_3$$

c. The electron dot structure for ClF_3 is: [The three lone pairs on F's have been omitted for clarity.]

```
  ••
F—Cl—F
  |
  F
```

d. The electron-pair geometry for the molecule is trigonal bipyramidal. The molecular geometry is T-shape.

Chapter 10
Bonding and Molecular Structure: Orbital Hybridization and Molecular Orbitals

Review Questions

4. Hybrid orbital set used :

electron-pair geometry	hybrid orbital set used	number of orbitals needed
tetrahedral	sp^3	4
linear	sp	2
trigonal-planar	sp^2	3
octahedral	d^2sp^3 or sp^3d^2	6
trigonal-bipyramidal	dsp^3	5

6. Consider the species: BF_4^-, SiF_4, and SF_4:
   a. A molecule and a ion that is isoelectronic with BF_4^- :
      BF_4^- has 32 electrons (3 from B, 4 • 7 from the F atoms, and 1 from the negative charge. CF_4 is isolelectronic as is the ion NF_4^+
   b. SiF_4 and SF_4 have differing numbers of valence electrons and are **not** isoelectronic.
   c. The number of atoms attached to the central atom (B,Si, and S respectively) are the same--4 in all three cases, so the hybrid orbitals are **the same.**

8. Phosphorus and Sulfur are both atoms capable of having **more than 4 pairs** of valence electron pairs. PF_5 and SF_6 are molecules with 5 and 6 pairs of valence electrons respectively. PF_5 would utilize dsp^3 hybrid orbitals while SF_6 would use d^2sp^3 hybrids.

10. Four principles of Molecular Orbital Theory:
    a. Orbital conservation— The number of molecular orbitals in a molecule **is equal** to the number of atomic orbitals brought by all the atoms in the molecule.
    b. Energy released in bonding MOs— **Bonding MOs are lower in energy** that their parent orbitals while **antibonding MOs are higher in energy** than the parent orbitals.
    c. Pauli principle & Hund's rule obeyed—When electrons are placed in MOs, the placement is done so that **electrons are placed in the lowest energy orbitals** available, and when placed **in the same orbital the spins are paired.**

d. Similar atomic orbitals combine most effectively to form MOs—e.g. 1s orbitals combine with 1s orbitals--not 2s orbitals.

12. Connection between bond order, bond length, and bond energy:

molecule	bond order
C_2H_6	C-C bond =1
C_2H_4	C-C bond = 2
C_2H_2	C-C bond = 3

In the three compounds noted above, the bond order (of the C-C bonds) increases from 1 to 2 to 3. This change is accompanied by a **shortening of the bond**—that is the C-C bond in ethylene is shorter than the C-C bond in ethane and the C-C bond in acetylene is **shorter than** the C-C bond in ethylene. The shorter bonds are also more difficult to rupture, so the bond energy **increases** as the **length decreases**.

14. The term **localized** as it pertains to bonding theories refers to the idea that electron pairs are contained (or localized) in orbitals between two—and only two—atoms, while **delocalized** refers to the orbitals (in MO theory) which are thought of as being spread out (or delocalized) over the entire molecule.

Valence Bond Theory

16. NF_3 has **tetrahedral** electron-pair geometry and **pyramidal** molecular geometries.

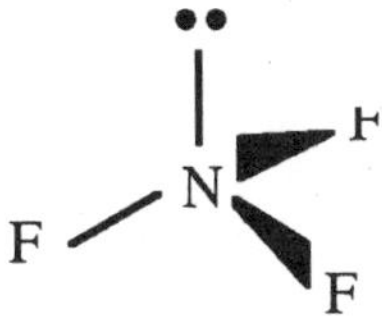

With four electron-pairs attached, the N is sp^3 hybridized. The lone pair occupies one of the four **sp^3** hybrid orbitals with the remaining three orbitals overlapping the **p** orbitals on F, forming the N-F sigma bonds.

18. The Lewis electron dot structure of $CHCl_3$:

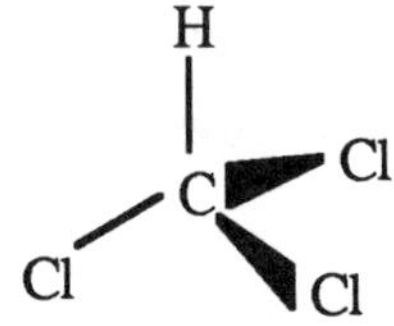

Electron pair geometry = tetrahedral Molecular geometry = tetrahedral. The H-C bonds are a result of the overlap of the hydrogen **s** orbital with **sp^3** hybrid orbitals on carbon. The Cl-C bonds are formed by the overlap of the **sp^3** hybrid orbitals on carbon with the **p** orbitals on chlorine.

20. Orbital sets used by the underlined atoms:

   a. $\underline{B}Br_3$: sp^2
   b. $\underline{C}O_2$: sp
   c. $\underline{C}HCl_3$: sp^3
   d. $H_2\underline{C}O$: sp^2

22. Hybrid orbital sets used by the underlined atoms:

   a. the C atoms and the O atom in dimethylether: $\underline{C}$: sp^3 ; $\underline{O}$: sp^3
   In the case of either the carbon OR oxygen atoms in dimethylether, each atom is bound to *four other groups*. This would require **four orbitals**
   b. The carbon atoms in propene: $\underline{C}H_3$: sp^3 ; $\underline{C}H$ and $\underline{C}H_2$: sp^2
   The methyl carbon is attached to four groups (three H and 1 C), and needs then four orbitals. The methylene and methine have bonds to four groups (2H and 2C in the case of CH_2 and four groups (1H and 3 C) in the case of CH-.
   c. The C atoms and the N atom in gycine: $\underline{N}$: sp^3; $\underline{C}H_2$: sp^3, $\underline{C}$=O: sp^2
   The N atom is attached to 4 groups (3 atoms and 1 lone pair), the CH_2 carbon has 4 groups attached (2 H atoms and 1N and 1 C atom). The carbonyl carbon is attached to only 3 groups (1C and 2 O atoms)

24. Hybrid orbital sets used by the underlined atoms:

	a. $\underline{Se}F_6$: sp^3d^2	b. $\underline{Se}F_4$: sp^3d	c. $\underline{I}Cl_2^-$: sp^3d	d. $\underline{Xe}F_4$ sp^3d^2
Electron Pair geometry:	octahedral	trigonal bipyramidal	trigonal bipyramidal	octahedral
Molecular geometry:	octahedral	see-saw	linear	square planar

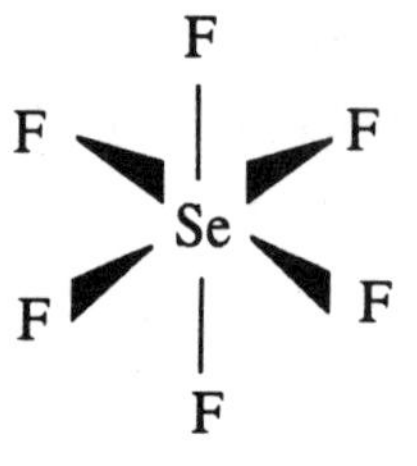

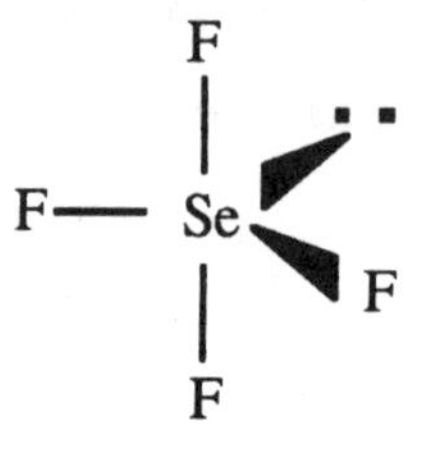

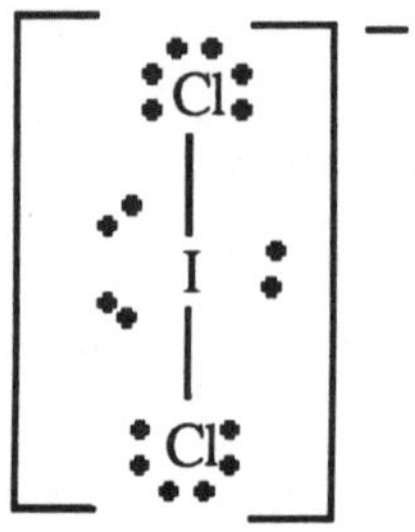

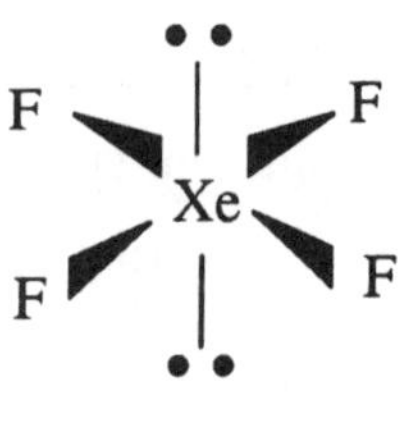

26. Lewis structures of HSO_3F and SO_3F^-:

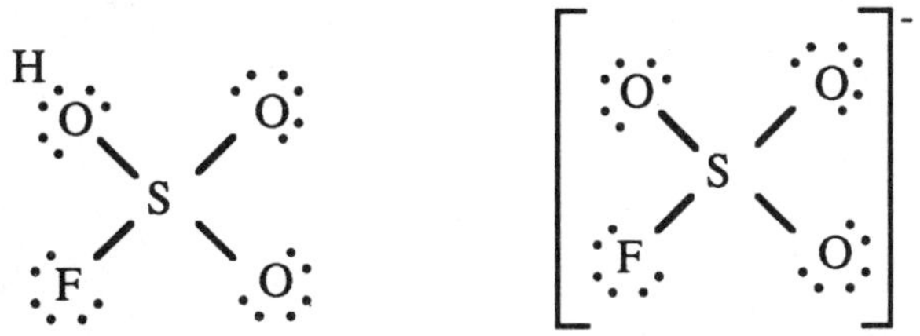

There are 32 valence electrons in each species. The molecular geometry around the sulfur atom in each specie is tetrahedral, with the hydribidization being sp^3.

28. For the molecule $COCl_2$: Hybridization of C = sp^2

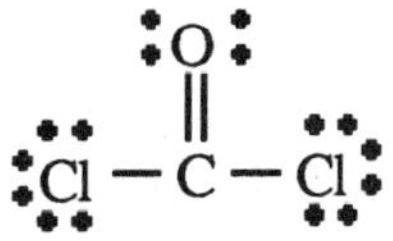

Bonding: 1 sigma bond between each chlorine and carbon (**sp** hybrid orbitals)

1 sigma bond between carbon and oxygen (**sp** hybrid orbitals)

1 pi bond between carbon and oxygen (**p** orbital)

Molecular Orbital Theory

30. Configuration for H_2^+: $(\sigma_{1s})^1$

Bond order for H_2^+: 1/2 (no. bonding e^- - no. antibonding e^-) = 1/2

The bond order for molecular hydrogen is <u>one</u> (1), and the H-H bond is stronger in the H_2 molecule than in the H_2^+ ion.

32. The molecular orbital diagram for C_2^{2-}, the acetylide ion:

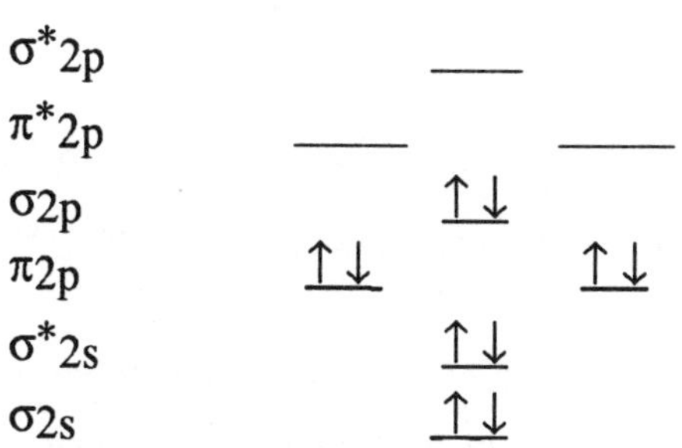

σ^*2p		___	
π^*2p	___		___
$\sigma2p$		↑↓	
$\pi2p$	↑↓		↑↓
σ^*2s		↑↓	
$\sigma2s$		↑↓	

There are 2 net pi bonds and 1 net sigma bond in the ion, giving a bond order of 3. On adding two electrons to C_2 (added to $\sigma 2p$) to obtain C_2^{2-}, the bond order increases by one. The ion is **diamagnetic**.

34. a. Using Figure 10.27 as a model for heteronuclear diatomic molecules, the electron configuration (showing only the outer level electrons) for CO is:

σ^*2p		___	
π^*2p	___		___
$\sigma 2p$		↑↓	
$\pi 2p$	↑↓		↑↓
σ^*2s		↑↓	
$\sigma 2s$		↑↓	

b. The HOMO is the $\sigma 2p$

c. There are no unpaired electrons, hence CO is diamagnetic.

d. There is one net sigma bond, and two net pi bonds for an overall bond order of 3.

General Questions

36. The hybrid orbitals used by sulfur

a.	SO_2	3 electron-pair groups	sp^2
b.	SO_3	3 electron-pair groups	sp^2
c.	SO_3^{2-}	4 electron-pair groups	sp^3
d.	SO_4^{2-}	4 electron-pair groups	sp^3

38. For the ion NO_3^- :

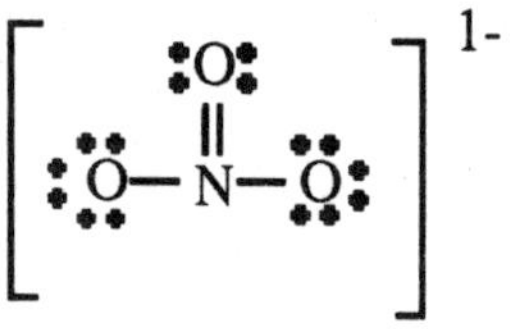

$$\left[\ddot{O}{=}N(-\ddot{\underset{..}{O}}:)_2 \right]^{1-}$$

Hybridization of N: sp^2

The N=O bond is formed by overlap of an **sp^2** hybrid orbital on both N and O to form a σ bond,while the unhybridized **p** orbitals on N and O overlap side-to-side to form a π bond.

40. Both carbon 1 and 2 are **sp^2** hybridized (3 electron-pair groups). Since angles A, B, and C are associated with sp^2 hybridized carbons, each of the three angles is approximately 120°.

42. a. The H-C-C bond angle designated **A is 120°**. The central carbon atom is attached to 3 groups—hence a bond angle of 120° maximizes the distance between the bonded atoms. The C-O-C bond angle designated as **B is 109°**. The central oxygen atom is attached to 4 groups—hence a bond angle of 109°.

The H-C-H bond angle designated as **C is 109°**. The central carbon atom is attached to 4 groups.
The C-C-O bond angle designated **D is 120°**. The central carbon atom is attached to 3 groups.

b.

Carbon #	Hybrid orbitals	groups attached to that carbon
1	sp^2	3
2	sp^2	3
3	sp^3	4

44. a.

Atom #	Hybrid orbitals	groups attached to that atom
1 (N)	sp^3	4
2 (C)	sp^3	4
3 (C)	sp^2	3
4 (N)	sp^2	3
5 (C)	sp^2	3

b.

Bond Angle	angle	
A	109°	C is attached to four groups
B	120°	C is attached to three groups
C	109°	N is attached to four groups. (Participation in the aromatic nature of the ring will make this angle closer to 120 °, and the ring will be planar.).

46. The valence bond theory for the peroxide ion (O_2^{2-}) shows a single bond connecting the two oxygen atoms (with three lone pairs on each oxygen). The bond order would be 1.
The MO theory would give $(\sigma_{2s})^2(\sigma^*_{2s})^2(\pi_{2p})^4(\sigma_{2p})^2(\pi^*_{2p})^4$. With **two more** electrons in bonding orbitals than in antibonding orbitals, the bond order predicted here would be **one as well.**

48. The Lewis dot structures are:

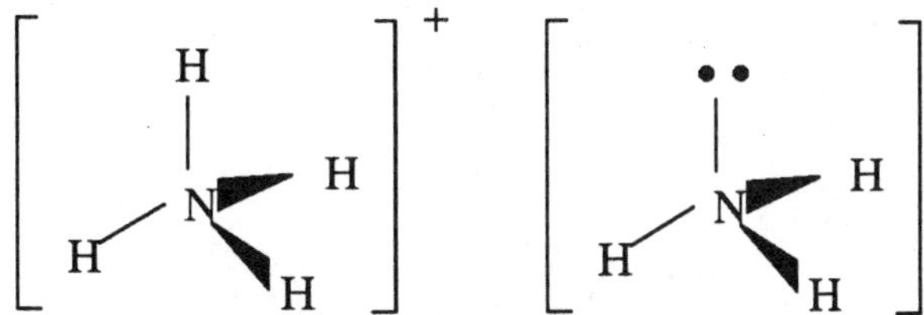

In both cases, the hybridization of the N atom is sp^3 since in both cases the N atom is attached to 4 groups. In the case of the ammonia molecule (right), the repulsion of the lone pair of electrons would reduce the H-N-H bond angle a bit below the predicted 109° , while in the case of the ammonium ion (left), the lone pair is now participating in a bond to an H atom, and the H-N-H bond angle is 109°.

50. The Lewis structures for IO_4^- and IO_5^{3-} are shown below:

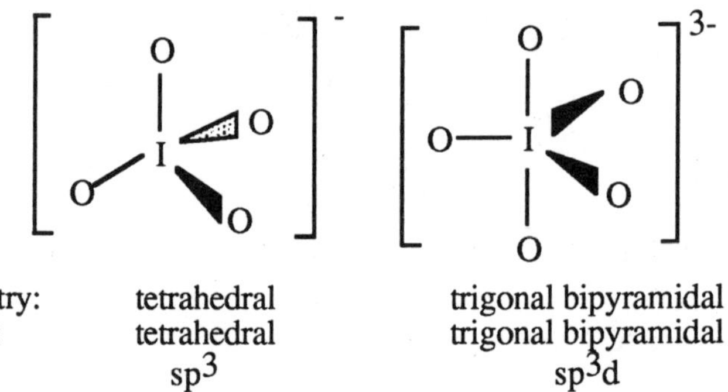

Electron-pair geometry:	tetrahedral	trigonal bipyramidal
Molecular geometry:	tetrahedral	trigonal bipyramidal
Hybridization of I:	sp^3	sp^3d

The lone pair of electrons on the oxygen atoms have been omitted for clarity. In **each case** there are three lone pairs on each oxygen atom.

52. molecule	MO configuration	magnetic character	bond order
Li_2	$(\sigma 2s)^2$	diamagnetic	1
Be_2	$(\sigma 2s)^2(\sigma^* 2s)^2$	diamagnetic	0
B_2	$(\sigma 2s)^2(\sigma^* 2s)^2(\pi 2p)^2$	paramagnetic	1
C_2	$(\sigma 2s)^2(\sigma^* 2s)^2(\pi 2p)^4$	diamagnetic	2
N_2	$(\sigma 2s)^2(\sigma^* 2s)^2(\pi 2p)^4(\sigma 2p)^2$	diamagnetic	3
O_2	$(\sigma 2s)^2(\sigma^* 2s)^2(\pi 2p)^4(\sigma 2p)^2(\pi^* 2p)^2$	paramagnetic	2
F_2	$(\sigma 2s)^2(\sigma^* 2s)^2(\pi 2p)^4(\sigma 2p)^2(\pi^* 2p)^4$	diamagnetic	1
Ne_2	$(\sigma 2s)^2(\sigma^* 2s)^2(\pi 2p)^4(\sigma 2p)^2(\pi^* 2p)^4(\sigma^* 2p)^2$	diamagnetic	0

54. a. The Lewis structure for the sulfamate ion:

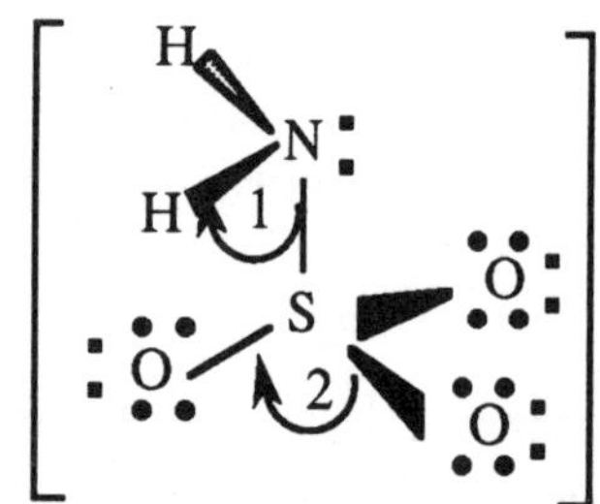

The S-N-H bond angles (1) and the O-S-O bond angles (2) are approximately 109 degrees.

b. In the reaction for $NH_2^- + SO_3 \rightarrow H_2NSO_3^-$, we expect the hybridization of N to be sp^3 **both before and after** the reaction. **After the reaction the hybridization of S is sp^3** while **before the reaction, the hybridization of S (in SO_3) is sp^2.**

Challenging Questions

56. Using the orbital diagram we predict the electron configuration to be (with 4 electrons from C and 5 from N) $(\sigma_{2s})^2(\sigma^*_{2s})^2(\pi_{2p})^4(\sigma_{2p})^1$
   a. The highest energy MO to which an electron is assigned is the σ_{2p}.
   b. Bond order = 1/2 (# bonding electrons - # non-bonding electrons)
      = 1/2 (7 - 2) or 5/2 or 2.5.
   c. There is a **net of 1/2** sigma bond —$(\sigma_{2p})^1$, and 2 net π bonds— $(\pi_{2p})^4$.
   d. The molecule has an unpaired electron so it is paramagnetic.

58. The Lewis structure for amphetamine is given below. All but one of the H atoms has been left off the C6 ring for clarity. Note that **no H** would be attached to the C atom that bears the side chain:

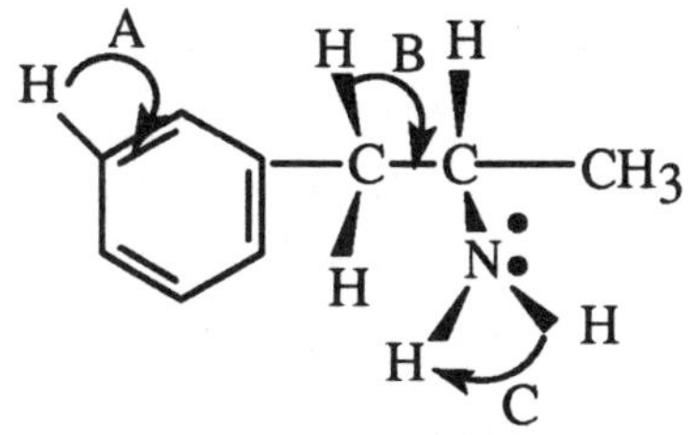

   a. The hybrid orbitals used by the C atoms in the ring are sp^2, while those of the sidechain are sp^3 (in the ring each C is attached to three groups—two C and 1 H—while on the sidechains, each C is attached to four groups). The hybridization used by N is also sp^3.
   b. Bond angle A:120 degrees; angle B: 109 degrees; angle C: 109 degrees. Angles B & C are those expected around a central atom with four groups attached, while angle A is the angle expected when 3 groups are attached to a central atom.
   c. Since each atom is attached to every other atom with a sigma bond, there are a total of 23 sigma bonds in the molecule. The C6 ring can be thought of as having 3 pi bonds—represented in the structure with the = .
   d. While the C6 ring is essentially non-polar (with C-C and C-H bonds) the sidechain (with the NH_2 group) would have a polar nature.

e. The proton would attach to the "most negative part" of the molecule—the lone pair of electrons on the N atom.

60. The essential part of the Lewis structure for the peptide linkage is shown below:

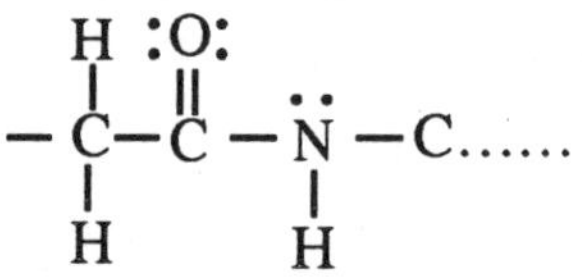

```
     H  :O:
     |   ||   ..
   — C — C — N — C.....
     |        |
     H        H
```

a. The carbonyl carbon (C=O) is attached to 3 groups and has a hybridization of sp^2. The N atom is connected to 4 groups (3 atoms, 1 lone pair) and has a hybridization of sp^3.

b. Another structure is feasible:

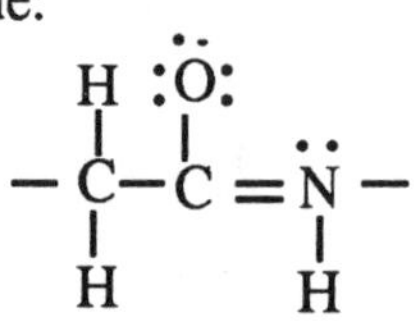

```
        ..
     H  :O:
     |   |    ..
   — C — C = N —
     |        |
     H        H
```

This would have formal charges on the carbonyl carbon of 0, on the oxygen of -1, and on the N of +1.

Since the "preferred" structure has a formal charge of 0 on **all three atoms**, the first structure is the more favorable.

c. If one views the "less preferred" resonance structure as a contributor, one can see that both the carbonyl carbon and the N are sp^2 hybridized. This leaves one "p" orbital on O, C, and N unhybridized, and capable of side-to-side overlap (also known as π type overlap) between the C and the O and between the C and the N, forming a planar region in the molecule.

Summary Question

62. The *enol* $\rightarrow$ *keto* transformation:

$$H_3C{-}C(OH){=}CH{-}C({=}O){-}CH_3 \rightleftharpoons H_3C{-}C({=}O){-}CH_2{-}C({=}O){-}CH_3$$

a. The keto and enol forms are **not resonance hybrids.** Resonance hybrids are in general structures that differ one from the other by the location of a multiple bond.

While the transformation does result in the "movement" of the C-C double bond, it also has H atoms that move.

b. Hybridization in the *enol* form

This C atom changes from sp^2 in the *enol* form to sp^3 in the *keto* form.

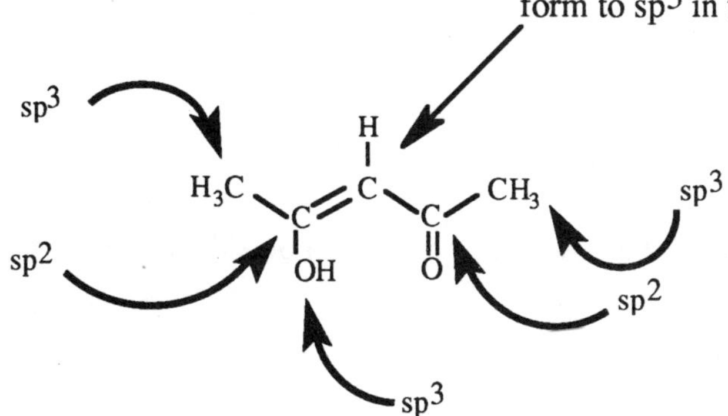

c. C atom geometries in *enol*. and *keto* forms

This C atom changes from a trigonal planar geometry to tetrahedral geometry in the *keto* form.

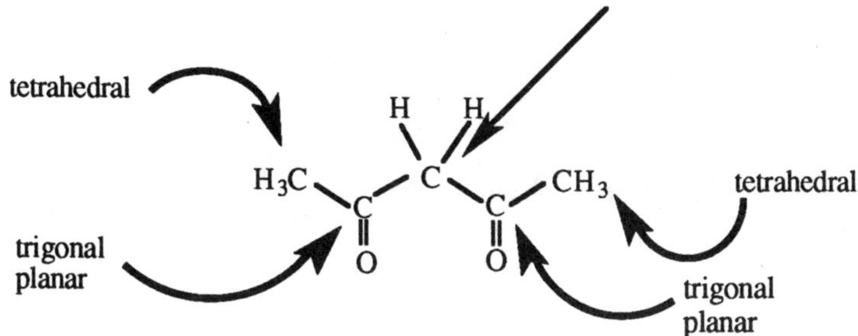

d. Resonance structures for the acac⁻ ion:

$$\left[CH_3-\underset{\underset{:\ddot{O}:}{|}}{C}=CH-\underset{\underset{:O:}{\|}}{C}-CH_3\right]^- \longleftrightarrow \left[CH_3-\underset{\underset{:O:}{\|}}{C}-\ddot{C}H-\underset{\underset{:O:}{\|}}{C}-CH_3\right]^- \longleftrightarrow \left[CH_3-\underset{\underset{:O:}{\|}}{C}-CH=\underset{\underset{:\ddot{O}:}{|}}{C}-CH_3\right]^-$$

e. $15.0 \text{ g Cr compound} \cdot \dfrac{1 \text{ mol compound}}{349.33 \text{ g}} = 4.29 \times 10^{-2} \text{ mol Cr compound}$

$4.29 \times 10^{-2} \text{ mol Cr compound} \cdot \dfrac{1 \text{ mol } CrCl_3}{1 \text{ mol Cr compound}} \cdot \dfrac{158.36 \text{ g}}{1 \text{ mol compound}} = 6.80 \text{ g } CrCl_3 \text{ required}$

$4.29 \times 10^{-2} \text{ mol Cr compound} \cdot \dfrac{3 \text{ mol acac}}{1 \text{ mol Cr compound}} \cdot \dfrac{100.12 \text{ g}}{\text{mol compound}} = 12.9 \text{ g acac required}$

$4.29 \times 10^{-2} \text{ mol Cr compound} \cdot 3 \text{ mol NaOH} \cdot \dfrac{40.0 \text{ g}}{\text{mole compound}} = 5.15 \text{ g NaOH required}$

Chapter 11
Bonding and Molecular Structure: Carbon - More Than Just Another Element

Alkanes

2. Alkanes have the formula C_nH_{2n+2}. The alkane with 12 carbon atoms therefore has the formula $C_{12}H_{26}$.

4. Structural isomers for C_8H_{18}:

$$\begin{array}{l} \qquad\ CH_3 \\ \qquad\ \ | \\ CH_3CCH_2CHCH_3 \\ \qquad\ \ | \qquad\ | \\ \qquad\ CH_3\ \ CH_3 \end{array}$$ 2,2,4-trimethylpentane

$$\begin{array}{l} \qquad\qquad CH_3 \\ \qquad\qquad\ | \\ CH_3CHCHCHCH_3 \\ \qquad\ | \qquad\ \ | \\ \qquad CH_3\ \ CH_3 \end{array}$$ 2,3,4-trimethylpentane

$$\begin{array}{l} \qquad\qquad\quad CH_3 \\ \qquad\qquad\qquad | \\ CH_3CH_2CHCHCH_3 \\ \qquad\qquad\ | \\ \qquad\qquad CH_2CH_3 \end{array}$$ 3-ethyl-2-methylpentane

$$\begin{array}{l} \qquad\qquad CH_3 \\ \qquad\qquad\ | \\ CH_3CH_2CCH_2CH_3 \\ \qquad\qquad\ | \\ \qquad\qquad CH_2CH_3 \end{array}$$ 3-ethyl-3-methylpentane

6. Name the following compound:

$$\begin{array}{l} \qquad\qquad\qquad\qquad\quad CH_3 \\ \qquad\qquad\qquad\qquad\qquad | \\ CH_3CHCH_2CH_2CHCH_3 \\ \qquad\ | \\ \qquad CH_2CH_3 \end{array}$$ 2,5-dimethylheptane

Note that the longest chain has **seven** carbons, so the "root" name is heptane. Numbering the carbons from right-to-left (in this diagram) places methyl groups on carbons 2 and 5. Numbering from left-to-right gives numbers of 3 and 6--so the lower sum is used.

An isomer of the compound is.

$CH_3CH_2CH_2CH_2CH_2CH_2CH_2CH_2CH_3$ nonane

This compound represent the "straight chain" isomer of the alkane with 9 carbons. There are 35 isomers of the alkane with formula C_9H_{20}.

8. Cycloheptane's structure :

The molecule has seven (hence the ..hept... root) carbons in the ring. Geometry tells us that each of the angles inside a closed figure (like the heptagon) will **add to 360 degrees.** Since there are 7 such angles (one at each C-C-C), the angles at any one carbon atom would be 360/7 or 51.4 degrees. For the ring to be flat the C-C-C bond angles would **need to be** about 51 degrees. Since these are sp^3 carbons (each has 2 carbons and 2 hydrogens attached) the C-C-C bond angle is approximately 109 degrees. So the ring **is not planar.**

Alkenes and Alkynes

10. cis- and trans- isomers of : 3-methyl-2-hexene

cis

$CH_3(H)C{=}C(CH_2CH_2CH_3)(CH_3)$

trans

$CH_3(H)C{=}C(CH_3)(CH_2CH_2CH_3)$

12. a. Structures and names for alkenes with formula C_5H_{10}.

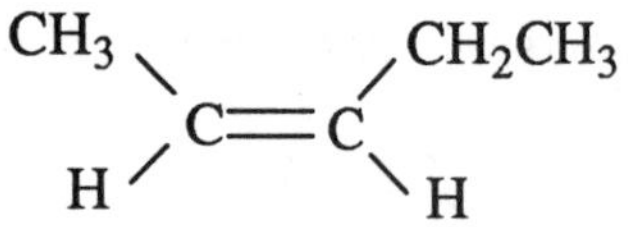

$CH_3(H)C{=}C(H)CH_2CH_3$

cis-2-pentene

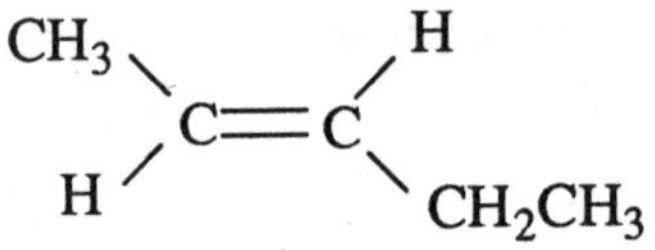

$CH_3(H)C{=}C(H)CH_2CH_3$

trans-2-pentene

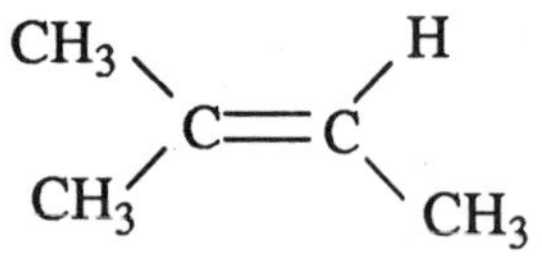

$(CH_3)_2C{=}C(H)CH_3$

2-methyl-2-butene

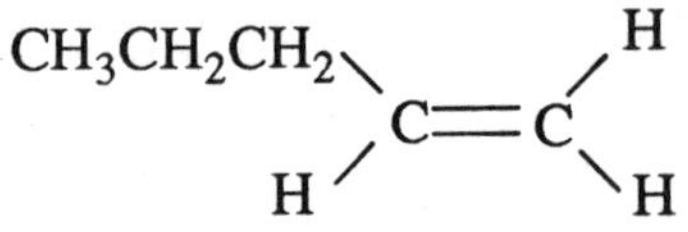

$CH_3CH_2CH_2(H)C{=}CH_2$

1-pentene

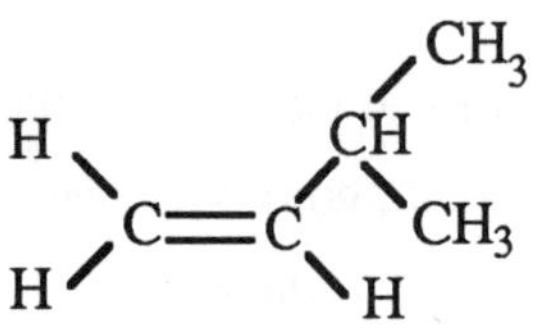

$H_2C{=}C(H)CH(CH_3)_2$

3-methyl-1-butene

$H_2C{=}C(CH_2CH_3)CH_3$

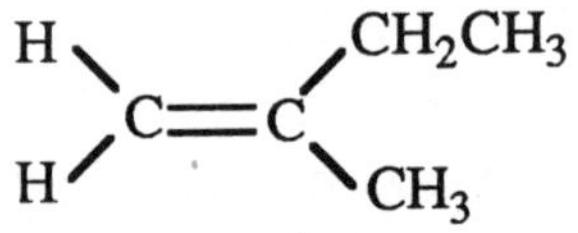

2-methyl-1-butene

b. The cycloalkane with the formula C_5H_{10}

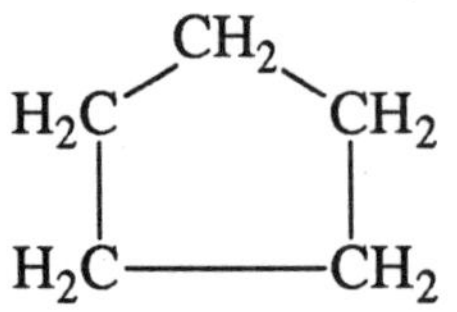

cyclo-$(CH_2)_5$: H_2C, CH_2, CH_2, CH_2, H_2C in a five-membered ring

14. Structure and names for products of:

a. $CH_3CH{=}CH_2$ + Br_2 → $CH_3CHBrCH_2Br$

1,2-dibromopropane

b. $CH_3CH_2CH{=}CHCH_3$ + H_2 → $CH_3CH_2CH_2CH_2CH_3$

pentane

16. The alkene which upon addition of HBr yields: $CH_3CH_2CH_2CH_2Br$

The addition of HBr is accomplished by adding (H) on one "side" of the double bond and (Br) on the other. If we can mentally "subtract" an H and Br from the formula above we get: $CH_3CH_2\mathbf{CH_2}CH_2\mathbf{Br} \rightarrow CH_3CH_2CH{=}CH_2$

Benzene and Aromatic Compounds

18. a. m-dichlorobenzene b. p-bromotoluene

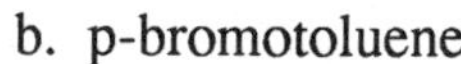

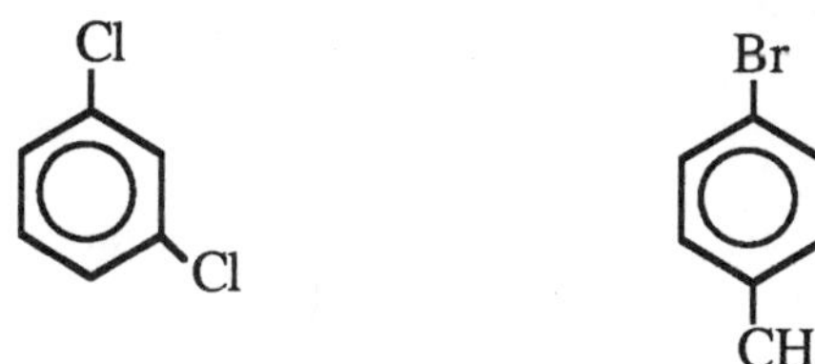

20. Ethylbenzene may be prepared from benzene in the following manner:

$$C_6H_6 \xrightarrow[CH_3CH_2Cl]{AlCl_3} C_6H_5CH_2CH_3$$

Alcohols and Amines

22. Systematic names of the alcohols:

a. 1-propanol	primary alcohol
b. 1-butanol	primary alcohol
c. 2-methyl-2-propanol	tertiary alcohol
d. 2-methyl-2-butanol	tertiary alcohol

24. Formulas and structures of the amines:

a. ethylamine $C_2H_5NH_2$

CH_3CH_2–N(H)–H

b. dipropylamine $(CH_3CH_2CH_2)_2NH$

$CH_3CH_2CH_2$–N(H)–$CH_2CH_2CH_3$

c. butyldimethylamine

(CH3CH2CH2CH2)(CH3)2N

$CH_3CH_2CH_2CH_2-N(CH_3)-CH_3$

26. Structural formulas for alcohols with the formula $C_4H_{10}O$:

1-butanol $CH_3CH_2CH_2CH_2OH$

2-butanol $CH_3CH_2CH(OH)CH_3$

2-methyl-1-propanol $CH_3CH(CH_3)CH_2OH$

2-methyl-2-propanol $CH_3C(OH)(CH_3)CH_3$

28. Complete the reactions of :

a. $2\ CH_3CH_2OH \xrightarrow{H_2SO_4,\ 180°C} CH_3CH_2OCH_2CH_3 + H_2O$

b. $CH_3CH_2CH_2OH + Na \rightarrow CH_3CH_2CH_2O^-Na^+ + 1/2\ H_2$

Compounds with a Carbonyl Group

30. Structural formulas for :

a. 2-pentanone

$CH_3-CH_2-CH_2-C(=O)-CH_3$

b. hexanal

$CH_3-CH_2-CH_2-CH_2-CH_2-C(=O)-H$

c. pentanoic acid

$CH_3CH_2CH_2CH_2CO_2H$ or $CH_3(CH_2)_3CO_2H$

32. Name the following compounds:

a. $CH_3CH_2\underset{\displaystyle CH_3}{\underset{|}{C}}HCH_2CO_2H$ 3-methylpentanoic acid

b. $CH_3CH_2\overset{\displaystyle O}{\overset{\|}{C}}OCH_3$ methyl propanoate

c. $CH_3\overset{\displaystyle O}{\overset{\|}{C}}OCH_2CH_2CH_2CH_3$ butyl ethanoate

d. CO_2H on a benzene ring, Br para — p-bromobenzoic acid

34. a. The product of oxidation is pentanoic acid:

$$CH_3CH_2CH_2CH_2\overset{\displaystyle O}{\overset{\|}{C}}-OH$$

b. The reduction product is 1-pentanol:

$$CH_3CH_2CH_2CH_2CH_2OH$$

c. The reduction yields 2-octanol:

$$CH_3CH_2CH_2CH_2CH_2CH_2\overset{\displaystyle OH}{\overset{|}{C}}HCH_3$$

d. The oxidation of 2-octanone with potassium permanganate gives <u>no reaction</u>.

36. $CH_3CH_2CH_2OH \xrightarrow{KMnO_4} CH_3CH_2CO_2H$

$$CH_3CH_2CO_2H + HOCH_2CH_2CH_3 \xrightarrow{H^+} CH_3CH_2\overset{\displaystyle O}{\overset{\|}{C}}OCH_2CH_2CH_3$$

38. The products of the hydrolysis of the ester are 1-butanol and sodium acetate:

$CH_3CH_2CH_2CH_2OH$ CH_3CO_2Na

Polymers

35. a. An equation for the formation of polyvinyl acetate from vinyl acetate:

$$n\ CH_2CHOCOCH_3 \longrightarrow [-CH_2CH(OCOCH_3)-]_n$$

b. The structure for polyvinylacetate:

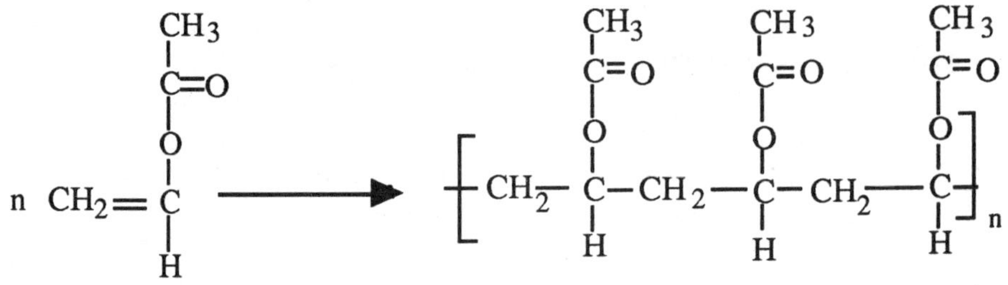

$$n\ CH_2{=}C(H)(O{-}C({=}O){-}CH_3) \longrightarrow [-CH_2-CH(OCOCH_3)-CH_2-CH(OCOCH_3)-CH_2-CH(OCOCH_3)-]_n$$

c. Prepare polyvinyl alcohol from polyvinyl acetate:

Polyvinyl acetate is an ester. Hydrolysis of the ester (structure shown above) with NaOH will produce the sodium salt, $NaC_2H_3O_2$, and polyvinyl alcohol. Acidification with a strong acid (e.g. HCl) will produce polyvinyl alcohol (structure below).

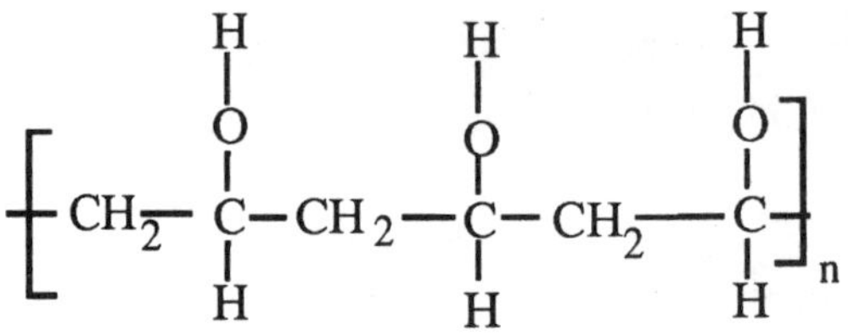

$$[-CH_2-CH(OH)-CH_2-CH(OH)-CH_2-CH(OH)-]_n$$

42. $Cl_2C{=}CH_2$ (1,1-dichloroethene) $\quad ClHC{=}CH_2$ (chloroethene)

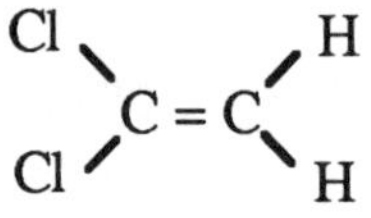

1,1-dichloroethene 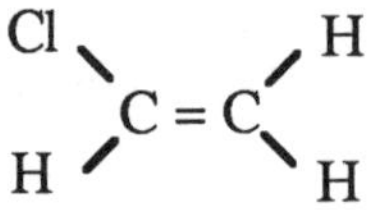 chloroethene

The reaction proceeds with the free radical addition of the copolymers:

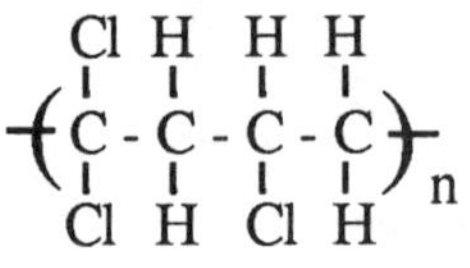

$$(-CCl_2-CH_2-CHCl-CH_2-)_n$$

General Questions

44. Structures for:

a. 2,2-dimethylpentane

$$\begin{array}{ccccccccccc} & & H & & CH_3 & & H & & H & & H & \\ & & | & & | & & | & & | & & | & \\ H & - & C & - & C & - & C & - & C & - & C & - & H \\ & & | & & | & & | & & | & & | & \\ & & H & & CH_3 & & H & & H & & H & \end{array}$$

b. 3,3-diethylpentane

$$\begin{array}{ccccccccccccc} & & & & & & CH_3 & & & & & & \\ & & & & & & | & & & & & & \\ & & H & & H & & CH_2 & & H & & H & & \\ & & | & & | & & | & & | & & | & & \\ H & - & C & - & C & - & C & - & C & - & C & - & H \\ & & | & & | & & | & & | & & | & & \\ & & H & & H & & CH_2 & & H & & H & & \\ & & & & & & | & & & & & & \\ & & & & & & CH_3 & & & & & & \end{array}$$

c. 2-methyl-3-ethylpentane

$$\begin{array}{ccccccccccccc} & & H & & CH_3 & & H & & H & & H & & \\ & & | & & | & & | & & | & & | & & \\ H & - & C & - & C & - & C & - & C & - & C & - & H \\ & & | & & | & & | & & | & & | & & \\ & & H & & H & & CH_2 & & H & & H & & \\ & & & & & & | & & & & & & \\ & & & & & & CH_3 & & & & & & \end{array}$$

d. 4-ethylheptane

$$\begin{array}{ccccccccccccc} & & H & & H & & H & & H & & H & & \\ & & | & & | & & | & & | & & | & & \\ H & - & C & - & C & - & C & - & C & - & C & - & CH_3 \\ & & | & & | & & | & & | & & | & & \\ & & H & & H & & CH_2 & & H & & H & & \\ & & & & & & | & & & & & & \\ & & & & & & CH_2 & & & & & & \\ & & & & & & | & & & & & & \\ & & & & & & CH_3 & & & & & & \end{array}$$

46. Structural isomers for $C_3H_6Cl_2$:

```
    H   H   H                  H   Cl  H
    |   |   |                  |   |   |
H — C — C — C — H          H — C — C — C — H
    |   |   |                  |   |   |
    H   Cl  Cl                 H   Cl  H
```

1,2-dichloropropane

2,2-dichloropropane

```
    Cl  H   H                  H   H   H
    |   |   |                  |   |   |
H — C — C — C — H          H — C — C — C — H
    |   |   |                  |   |   |
    Cl  H   H                  Cl  H   Cl
```

1,1-dichloropropane

1,3-dichloropropane

48. Structural isomers for trimethylbenzene:

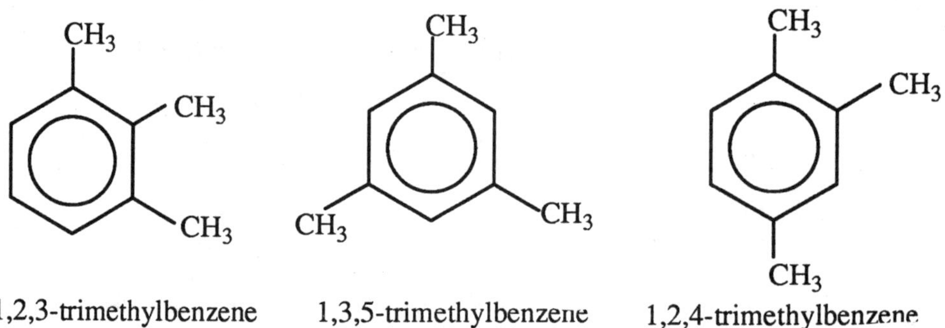

1,2,3-trimethylbenzene   1,3,5-trimethylbenzene   1,2,4-trimethylbenzene

50. The decarboxylase enzyme would remove the COOH functionality-- releasing CO_2 and appending the H in the former location of the COOH group.

```
                          H                                          H
                          |                                          |
H2N-CH2-CH2-CH2-CH2-C-NH2   →   H2N-CH2-CH2-CH2-CH2-C-NH2
                          |                                          |
                          C-OH                                       H
                          ||
                          O
```

lysine   cadaverine

52. Repeating unit of the polymer, Kevlar

$$HO_2C-C_6H_4-CO_2H + H_2N-C_6H_4-NH_2 \longrightarrow \left[-\overset{O}{\overset{\|}{C}}-C_6H_4-\overset{O}{\overset{\|}{C}}-\underset{H}{N}-C_6H_4-\underset{H}{N}- \right]_n$$

54. Structure for glyceryl trilaurate and saponification products:

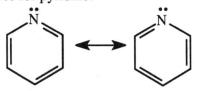

$$\begin{array}{l} CH_2\text{-O-}\overset{O}{\overset{\|}{C}}\text{ - }(CH_2)_{10}CH_3 \\ | \\ HC\text{- O - }\overset{O}{\overset{\|}{C}}\text{ - }(CH_2)_{10}CH_3 \\ | \\ CH_2\text{-O-}\overset{O}{\overset{\|}{C}}\text{ - }(CH_2)_{10}CH_3 \end{array} \xrightarrow{NaOH} \begin{array}{l} CH_2OH \\ | \\ CHOH \\ | \\ CH_2OH \end{array} + 3\,Na^{+}\,{}^{-}O_2C(CH_2)_{10}CH_3$$

glyceryl trilaurate glycerol sodium laurate

Conceptual Questions

56. Resonance structures for pyridine:

$$C_5H_5N \longleftrightarrow C_5H_5N$$

Both pyridine and benzene have the same number of electrons (30 valence). The molecular formulas also differ in that for pyridine one C-H unit has been replaced by a N atom.

58. To discriminate between the two isomers, react the two with elemental bromine. Cyclopentane will not react with bromine, while 1-pentene will react with bromine.

60. a. Energy is required to rotate the C—C bond because the pi bond present in the alkene prohibits free rotation of the C—C bond.

b. Energy is required for the rotation since as the rotation occurs the hydrogen atoms on C_2 and C_3 are brought closer together, and the electron clouds on the H atoms repel each other as they are brought to their closest approach.

Summary Question

62. a. The empirical formula of maleic acid:

Calculate the mass of C, H, and O respectively

$$0.190 \text{ g } CO_2 \cdot \frac{12.01 \text{ g C}}{44.01 \text{ g } CO_2} = 0.0518 \text{ g C}$$

$$0.0388 \text{ g } H_2O \cdot \frac{2.0158 \text{ g H}}{18.02 \text{ g } H_2O} = 0.00434 \text{ g H}$$

The mass of O is then 0.125 g - (0.0518 + 0.00434) = 0.0689 g O

The moles of atoms of C, H, and O are:

$$0.0518 \text{ g C} \cdot \frac{1 \text{ mol C}}{12.011 \text{ g C}} = 0.00431 \text{ mol C}$$

$$0.00434 \text{ g H} \cdot \frac{1 \text{ mol H}}{1.0079 \text{ g H}} = 0.00431 \text{ mol H}$$

$$0.0689 \text{ g O} \cdot \frac{1 \text{ mol O}}{16.00 \text{ g O}} = 0.00431 \text{ mol O}$$

Expressing the ratio of mole C: mole H: mole O, the empirical formula is then CHO.

b. 0.261 g of acid requires 34.60 mL of 0.130 M NaOH

Moles of NaOH = 0.130 mol/L • 0.03460 L = 0.00450 mol NaOH

Since maleic is a diprotic acid, the moles of maleic acid present is half the number of moles of NaOH, and the molar mass

$$\frac{0.261 \text{ g acid}}{0.00225 \text{ mol acid}} = 116 \text{ g/mol}$$

The empirical formula CHO would have a mass of 29.0 g

$$\frac{116 \text{ g maleic acid}}{1 \text{ mol maleic acid}} \cdot \frac{1 \text{ empirical formula}}{29.0 \text{ g maleic acid}} = 4 \frac{\text{empirical formulas}}{\text{mole maleic acid}} \text{ or } C_4H_4O_4$$

c. Lewis structure for $C_4H_4O_4$

```
       :O:     :O:
    ..  ||      ||  ..
   H-O-C-C=C-C-O-H
    ..     |  |     ..
           H  H
```

d. Hybridization used by the C atoms: Each carbon in the molecule has three electron-pair groups around it, making the C atoms sp^2 hybridized.

e. Bond angles: Each C atom would have three 120° angles separating the three groups attached.

Chapter 12
Gases and Their Properties

Numerical Questions

10. a. $440 \text{ mm Hg} \cdot \frac{1 \text{ atm}}{760 \text{ mm Hg}} = 0.58 \text{ atm}$ (2 sf in 440)

b. $440 \text{ mm Hg} \cdot \frac{1.013 \text{ bar}}{760 \text{ mm Hg}} = 0.59 \text{ bar}$

c. $440 \text{ mm Hg} \cdot \frac{101.325 \times 10^3 \text{ Pa}}{760 \text{ mm Hg}} = 5.9 \times 10^4 \text{ Pa}$

12. The levels in the U-tube manometer indicate a gas pressure of 56.3 mm Hg.

$56.3 \text{ mm Hg} \cdot \frac{1 \text{ atm}}{760 \text{ mm Hg}} = 0.0741 \text{ atm}$ (3 sf in 56.3)

$56.3 \text{ mm Hg} \cdot \frac{1 \text{ torr}}{1 \text{mm Hg}} = 56.3 \text{ torr}$

$56.3 \text{ mm Hg} \cdot \frac{101.325 \text{ kPa}}{760 \text{ mm Hg}} = 7.51 \text{ kPa}$

The Gas Laws

14. **Boyle's law** states that the pressure a gas exerts is inversely proportional to the volume it occupies, or for a given amount of gas--PV = constant. We can write this as:

$$P_1V_1 = P_2V_2$$

So (67.5 mm Hg)(500. mL) = (P_2)(125 mL)

and $\frac{(67.5 \text{ mm Hg})(500. \text{ mL})}{125 \text{ mL}} = 270. \text{ mm Hg}$

Note that the **volume decreased** by a factor of 4, and the **pressure increased** by a factor of 4.

16. **Charles' law** states that V α T (in Kelvin) or $\frac{V_1}{T_1} = \frac{V_2}{T_2}$

$\frac{3.5 \text{ L}}{295 \text{ K}} = \frac{V_2}{310 \text{ K}}$ and rearranging to solve for V_2 yields

$$V_2 = \frac{(3.5\ L)(310\ K)}{(295\ K)} = 3.7\ L$$

Note that with the **increase in T** there has been an **increase in volume.**

18. The reaction can be written: $2\ H_2 + O_2 \rightarrow 2\ H_2O$. Avogadro's Law states that, since the volume of a gas is **proportional to** the number of moles of gas, 3.6 L of H_2 would require 1.8 L of O_2, if both volumes are measured at the same temperature and pressure

20. Using the general gas law we can write: $\frac{P_1V_1}{T_1} = \frac{P_2V_2}{T_2}$

and for a fixed volume: $\frac{P_1}{T_1} = \frac{P_2}{T_2}$ or if we rearrange, we obtain $P_2 = P_1 \cdot \frac{T_2}{T_1}$

$$P_2 = 360\ \text{mm Hg} \cdot \frac{268.2\ K}{298.7\ K} = 320\ \text{mm Hg (2 significant figures)}$$

22. Using the general gas law: $\frac{P_1V_1}{T_1} = \frac{P_2V_2}{T_2}$ or $V_2 = \frac{V_1P_1T_2}{P_2T_1}$

$$\text{so } V_{\text{Flask B}} = 25.0\ \text{mL} \cdot \left(\frac{436.5\ \text{mm Hg}}{94.3\ \text{mm Hg}}\right) \cdot \left(\frac{297.7\ K}{293.7\ K}\right) = 117\ \text{mL}$$

Note that the **decrease in pressure** would lead us to expect a **larger volume for flask B**—a fact borne out by our calculation!

24. Using the general gas law: $\frac{P_1V_1}{T_1} = \frac{P_2V_2}{T_2}$ or $V_2 = \frac{V_1P_1T_2}{P_2T_1}$

$$\text{so } V_{(2\ \text{miles})} = 1.2 \times 10^7\ L \cdot \left(\frac{737\ \text{mm Hg}}{600.\ \text{mm Hg}}\right) \cdot \left(\frac{240.\ K}{289.2\ K}\right) = 1.2 \times 10^7\ L$$

Note that the volume of the balloon at 2 miles is (to 2 significant figures) exactly the volume of the balloon on the ground. Note that while we would anticipate a larger volume (owing to a decreased pressure), the much lower temperature (at 2 miles) offsets the expected volume increase.

The Ideal Gas Law

26. The pressure of 1.25 g of gaseous carbon dioxide may be calculated with the Ideal Gas Law:

$$1.25\ g\ CO_2 \cdot \frac{1\ \text{mol}\ CO_2}{44.01\ g\ CO_2} = 0.0284\ \text{mol}\ CO_2$$

Rearranging PV = nRT to solve for P, we obtain:

$$P = \frac{nRT}{V} = \frac{(0.0284\ \text{mol})(0.082057\ \frac{\text{L} \cdot \text{atm}}{\text{K} \cdot \text{mol}})(295.7\ \text{K})}{0.750\ \text{L}} = 0.919\ \text{atm}$$

28. The volume of the flask may be calculated by realizing that the gas will expand to fill the flask.

$$2.2\ \text{g}\ CO_2 \cdot \frac{1\ \text{mol}\ CO_2}{44.0\ \text{g}\ CO_2} = 0.050\ \text{mol}\ CO_2$$

$$P = 318\ \text{mm Hg} \cdot \frac{1\ \text{atm}}{760\ \text{mm Hg}} = 0.418\ \text{atm}$$

$$V = \frac{(0.050\ \text{mol})(0.082057\ \frac{\text{L} \cdot \text{atm}}{\text{K} \cdot \text{mol}})(295\ \text{K})}{0.418\ \text{atm}} = 2.9\ \text{L}$$

30. Rearranging PV = nRT to solve for n, we obtain: $n = \frac{PV}{RT}$

Converting 737 mm Hg to atmospheres, we obtain

$$737\ \text{mm Hg} \cdot \frac{1\ \text{atm}}{760\ \text{mm Hg}} = 0.970\ \text{atm}$$

$$n = \frac{(0.970\ \text{atm})(1.2 \times 10^7\ \text{L})}{(0.082057\ \frac{\text{L} \cdot \text{atm}}{\text{K} \cdot \text{mol}})(298.2\ \text{K})} = 4.8 \times 10^5\ \text{moles of He}$$

and since each mole of He has a mass of 4.00 g,

$$4.8 \times 10^5\ \text{mol He} \cdot \frac{4.00\ \text{g He}}{1\ \text{mol He}} = 1.9 \times 10^6\ \text{g He}$$

32. Rearranging PV = nRT to solve for n, we obtain: $n = \frac{PV}{RT}$

Converting 715. mm Hg to atmospheres, we obtain

$$715.\ \text{mm Hg} \cdot \frac{1\ \text{atm}}{760\ \text{mm Hg}} = 0.941\ \text{atm}$$

$$n = \frac{(0.941\ \text{atm})(0.452\ \text{L})}{(0.082057\ \frac{\text{L} \cdot \text{atm}}{\text{K} \cdot \text{mol}})(296.2\ \text{K})} = 0.0175\ \text{moles of the unknown gas.}$$

Since this number of moles of the gas has a mass of 1.007 g, we can calculate the molar mass:

$$\frac{1.007\ \text{g of the unknown gas}}{0.0175\ \text{moles of the unknown gas}} = 57.5\ \text{g/mol}$$

34. Calculate the empirical formula:

$$\% F = 100.0\% - (11.79\% C + 69.57\% Cl) = 18.64\% F$$

The moles of each element:

$$18.64 \text{ g F} \cdot \frac{1 \text{ mol F}}{18.998 \text{ g F}} = 0.9812 \text{ mol F}$$

$$11.79 \text{ g C} \cdot \frac{1 \text{ mol C}}{12.011 \text{ g C}} = 0.9816 \text{ mol C}$$

$$69.57 \text{ g Cl} \cdot \frac{1 \text{ mol Cl}}{35.453 \text{ g Cl}} = 1.962 \text{ mol Cl}$$

If we express these # of moles in a ratio (by dividing each of the moles by the smallest #— 0.9812) we get an empirical formula of $FCCl_2$.

Calculate the molar mass: (Express the pressure in units of atmospheres)

$$21.3 \text{ mm Hg} \cdot \frac{1 \text{ atm}}{760. \text{ mm Hg}} = 0.0280 \text{ atm}$$

$$n = \frac{PV}{RT} = \frac{(0.0280 \text{ atm})(0.458 \text{ L})}{(0.082057 \frac{\text{L} \cdot \text{atm}}{\text{K} \cdot \text{mol}})(298 \text{ K})} = 5.25 \times 10^{-4} \text{ mol}$$

Since 0.107 g corresponds to 5.25×10^{-4} mol, the molar mass is:

$$\frac{0.107 \text{ g}}{5.25 \times 10^{-4} \text{ mol}} = 204 \text{ g/mol}$$

With an empirical formula of CCl_2F (Empirical formula weight = 102), the molecular formula must be $C_2Cl_4F_2$.

36. Write the ideal gas law as: Molar Mass $= \frac{dRT}{P}$ where d = density in grams per liter.

Solving for d, we obtain: $\frac{(\text{Molar Mass}) \cdot P}{R \cdot T} = d$

The average molar mass for air is approximately 29 g/mol.

$$\frac{(29 \text{ g/mol})(0.20 \text{ mm Hg} \cdot \frac{1 \text{ atm}}{760 \text{ mm Hg}})}{(0.082057 \frac{\text{L} \cdot \text{atm}}{\text{K} \cdot \text{mol}})(250 \text{ K})} = 3.7 \times 10^{-4} \text{ g/L} = d$$

38. Molar mass = $\dfrac{(0.355 \text{ g/L})(0.082057 \frac{\text{L} \cdot \text{atm}}{\text{K} \cdot \text{mol}})(290.\text{ K})}{(189 \text{ mm Hg} \cdot \frac{1 \text{ atm}}{760 \text{ mm Hg}})}$ = 34.0 g/mol

40. Solve the Ideal Gas Law for volume to obtain

(1) $V_1 = \dfrac{n_1RT_1}{P_1}$

We are interested in the mass (and therefore moles) of oxygen needed to fill this balloon to the same volume, measured at the same pressure but a different temperature. We could express this as:

(2) $V_1 = \dfrac{n_2RT_2}{P_1}$

In equations (1) and (2) we have the common term V_1 so we may write:

(3) $\dfrac{n_1RT_1}{P_1} = \dfrac{n_2RT_2}{P_1}$

Since both R and P_1 are constant in these conditions, equation (3) may be simplified:

$$n_1T_1 = n_2T_2$$

Calculating the moles of oxygen represented by 12.0 g gives:

$$n_1 = 12.0 \text{ g } O_2 \cdot \frac{1 \text{ mol } O_2}{32.00 \text{ g } O_2} = 0.375 \text{ mol } O_2$$

Substituting for T_1 (300. K) and T_2 (354 K) and solving for n_2 we obtain:

$$\frac{(0.375 \text{ mol } O_2) \cdot (300.\text{K})}{354 \text{ K}} = 0.318 \text{ mol } O_2 \text{ or } 10.2 \text{ g } O_2$$

42. To calculate the temperature at which P = 7.25 atm, rearrange the Ideal Gas Law.

$$\frac{PV}{nR} = T = \frac{(7.25 \text{ atm})(1.52 \text{ L})}{(0.406 \text{ mol})(0.082057 \frac{\text{L}\cdot\text{atm}}{\text{K}\cdot\text{mol}})} = 331 \text{ K or } 58 \text{ °C}$$

Gas Laws and Stoichiometry

44. Determine the amount of H_2 generated when 2.2 g Fe reacts:

$$2.2 \text{ g Fe} \cdot \frac{1 \text{ mol Fe}}{55.85 \text{ g Fe}} \cdot \frac{1 \text{ mol } H_2}{1 \text{ mol Fe}} = 0.039 \text{ mol } H_2$$

Note that the latter factor is achieved by examining the balanced equation!

The pressure of this amount of H_2 is :

$$P = \frac{nRT}{V} = \frac{(0.039 \text{ mol } H_2)(62.4 \frac{\text{L} \cdot \text{torr}}{\text{K} \cdot \text{mol}})(298 \text{ K})}{10.0 \text{ L}}$$

$$= 73 \text{ torr or } 73 \text{ mm Hg} \qquad (2 \text{ sf})$$

46. Calculate the moles of N_2 needed:

$$n = \frac{PV}{RT} = \frac{(1.3 \text{ atm})(25.0 \text{ L})}{(0.082057 \frac{\text{L} \cdot \text{atm}}{\text{K} \cdot \text{mol}})(298 \text{ K})} = 1.33 \text{ mol } N_2$$

The mass of NaN_3 needed to produce this is obtained from the stoichiometry of the equation:

$$1.33 \text{ mol } N_2 \cdot \frac{2 \text{ mol } NaN_3}{3 \text{ mol } N_2} \cdot \frac{65.0 \text{ g } NaN_3}{1 \text{ mol } NaN_3} = 58 \text{ g } NaN_3 \quad (2 \text{ sf})$$

48. $N_2H_4 (g) + O_2 (g) \rightarrow N_2 (g) + 2 H_2O (g)$

$$1.00 \text{ kg } N_2H_4 \cdot \frac{1.0 \times 10^3 \text{ g } N_2H_4}{1.0 \text{ kg } N_2H_4} \cdot \frac{1 \text{ mol } N_2H_4}{32.0 \text{ g } N_2H_4} \cdot \frac{1 \text{ mol } O_2}{1 \text{ mol } N_2H_4}$$

$$= 3.13 \times 10^1 \text{ mole } O_2$$

$$P(O_2) = \frac{n(O_2) \cdot R \cdot T}{V} = \frac{(3.13 \times 10^1 \text{mol})(0.082057 \frac{\text{L} \cdot \text{atm}}{\text{K} \cdot \text{mol}})(296 \text{ K})}{450 \text{ L}}$$

$$P(O_2) = 1.69 \text{ atm or } 1.7 \text{ atm to } 2 \text{ sf}$$

50. This problems has two parts: 1. How many moles of Ni are present?

2. How many moles of CO are present?

1. # moles of Ni present:

$$0.450 \text{ g Ni} \cdot \frac{1 \text{ mol Ni}}{58.693 \text{g Ni}} = 7.67 \times 10^{-3} \text{ mol Ni}$$

and since 1 mol $Ni(CO)_4$ is formed for **each mol of Ni**, one can form

7.67×10^{-3} mol $Ni(CO)_4$

2. # moles of CO present:

Using the Ideal Gas Law:

$$n = \frac{\frac{418}{760} \text{ atm} \cdot (1.50\text{L})}{(0.082057 \frac{\text{L}\cdot\text{atom}}{\text{K}\cdot\text{mol}})(298\text{K})} = 0.0337 \text{ mol CO}$$

which would be capable of forming

$$0.0337 \text{ mol CO} \cdot \frac{1 \text{ mol Ni(CO)}_4}{4 \text{ mol CO}} = 8.43 \times 10^{-3} \text{ mol Ni(CO)}_4$$

Since the amount of nickel limits the maximum amount of $Ni(CO)_4$ that can be formed, the maximum mass of $Ni(CO)_4$ is then:

$$7.67 \times 10^{-3} \text{ mol Ni(CO)}_4 \cdot \frac{170.7 \text{ g Ni(CO)}_4}{1 \text{ mol Ni(CO)}_4} = 1.31 \text{ g Ni(CO)}_4$$

Gas Mixtures

52. We know that the total pressure will be equal to the sum of the pressure of each gas (also called the *partial pressure* of each gas).

$$1.0 \text{ g H}_2 \cdot \frac{1 \text{ mol H}_2}{2.02 \text{ g H}_2} = 0.50 \text{ mol H}_2 \text{ and } 8.0 \text{ g Ar} \cdot \frac{1 \text{ mol Ar}}{39.9 \text{ g Ar}} = 0.20 \text{ mol Ar}$$

We can calculate the **total pressure** using the Ideal Gas law:

$$P = \frac{n \cdot R \cdot T}{V} = \frac{(0.70 \text{ mol})(0.082057 \frac{\text{L} \cdot \text{atm}}{\text{K} \cdot \text{mol}})(300 \text{ K})}{3.0 \text{ L}} = 5.7 \text{ atm}$$

The pressure of **each gas** can be calculated by multiplying the total pressure (5.7 atm) by the mole fraction of the gas.

$$\text{Pressure of H}_2 = \frac{0.50 \text{ mol H}_2}{0.70 \text{ mol H}_2 + \text{Ar}} \cdot 5.7 = 4.1 \text{ atm and the}$$

$$\text{Pressure of Ar} = \frac{0.20 \text{ mol Ar}}{0.70 \text{ mol H}_2 + \text{Ar}} \cdot 5.7 = 1.6 \text{ atm}$$

Note that the total pressure is indeed (4.1+1.6) or 5.7 atm

54. $P_{total} = P_{halothane} + P_{oxygen}$ = 170 mm Hg + 570 mm Hg = 740 mm Hg

a. Since we know that the pressure a gas exerts is **proportional** to the # of moles of gas present we can calculate the ratio of moles by using their partial pressures:

$$\frac{\text{moles of halothane}}{\text{moles of oxygen}} = \frac{170 \text{ mm Hg}}{570 \text{ mm Hg}} = 0.30$$

b. $$160 \text{ g oxygen} \cdot \frac{1 \text{ mol oxygen}}{32.0 \text{ g oxygen}} \cdot \frac{0.30 \text{ mol halothane}}{1 \text{ mol oxygen}} \cdot \frac{197 \text{ g halothane}}{1 \text{ mol halothane}} =$$

2.9×10^2 g halothane (2 sf)

56. Let's calculate the # of moles of ammonium dichromate (and from that the # of moles of gaseous products expected:

$$5.0 \text{ g } (NH_4)_2Cr_2O_7 \cdot \frac{1 \text{ mol } (NH_4)_2Cr_2O_7}{252 \text{ g } (NH_4)_2Cr_2O_7} = 0.020 \text{ mol } (NH_4)_2Cr_2O_7 \quad (2 \text{ sf})$$

The balanced equation tells us that 1 mol of $(NH_4)_2Cr_2O_7$ produces 1 mol of N_2 and 4 mol of H_2O so we anticipate (1 • 0.020) mol N_2 and (4 • 0.020)mol H_2O.

[If you multiply the # of mole of $(NH_4)_2Cr_2O_7$ (0.0198 x 5), one gets 0.099 mol of gas (total)]

The total pressure would then be

$$P = \frac{n \cdot R \cdot T}{V} = \frac{(0.099 \text{ mol})(0.082057 \frac{L \cdot atm}{K \cdot mol})(296 \text{ K})}{3.0 \text{ L}} = 0.80 \text{ atm}$$

Since the total pressure is the sum of the pressure of $N_2 + H_2O$ (g), and 1/5 of the total moles of gas is N_2, the pressure of N_2 is 1/5 • 0.80 atm or 0.16 atm, and that of water is (0.80 - 0.16) 0.64 atm.

58. Dalton's Law tells us that in a mixture of gases the total pressure is the sum of the pressures of the individual gases (partial pressures)

$$P_T = P_{N_2} + P_{H_2O} \quad \text{so}$$

$$P_{N_2} = P_T - P_{H_2O} = 747 \text{ mm Hg} - 15.5 \text{ mm Hg}$$

$$= 732 \text{ mm Hg}$$

Kinetic Molecular Theory

60. a. Kinetic energy depends only on the temperature so the average kinetic energies of these two gases are equal.

b. Since the kinetic energies are equal, we can state:

$$KE(H_2) = KE(CO_2)$$

$$1/2\ m(H_2) \cdot \overline{V}^2(H_2) = 1/2\ m(CO_2) \cdot \overline{V}^2(CO_2)$$

Where m = mass of a molecule and $\overline{V}$ = average velocity of a molecule

So $$m(H_2) \cdot \overline{V}^2(H_2) = m(CO_2) \cdot \overline{V}^2(CO_2)$$

and $$\frac{\overline{V}^2(H_2)}{\overline{V}^2(CO_2)} = \frac{m(CO_2)}{m(H_2)}$$

Now the molar mass of H_2 = 2.0 g and the molar mass of CO_2 = 44 g

$$\frac{\overline{V}_{H_2}}{\overline{V}_{CO_2}} = \sqrt{\frac{m_{CO_2}}{m_{H_2}}} = \sqrt{\frac{44}{2.0}} = 4.7$$

The hydrogen molecules have an average velocity which is 4.7 times the average velocity of the CO_2 molecules.

c. Since the temperatures and the volumes are equal for these two gas samples, the pressure is proportional to the amount of gas present.

$$V_A = \frac{n_A R T_A}{P_A} \quad \text{and} \quad V_B = \frac{n_B R T_B}{P_B}$$ now $T_A = T_B$ and $V_A = V_B$ so

$$\frac{n_A R}{P_A} = \frac{n_B R}{P_B} \quad \text{or} \quad \frac{n_A}{P_A} = \frac{n_B}{P_B}$$

Since the pressure in Flask B (2 atm) is twice that of Flask A (1 atm), there are two times as many moles (and molecules) of gas in Flask B (CO_2) as there are in Flask A (H_2).

d. Since Flask B contains twice as many moles of CO_2 as Flask A contains of H_2, the ratio of masses of gas present are:

$$\frac{\text{Mass (Flask B)}}{\text{Mass (Flask A)}} = \frac{(2 \text{ mole } CO_2)(44 \text{ g } CO_2/\text{mol } CO_2)}{(1 \text{ mol } H_2)(2 \text{ g } H_2/\text{mol} H_2)} = \frac{44}{1}$$

Note that any number of moles of CO_2 and H_2 (in the ratio of 2 : 1) would provide the same answer.

62. Since two gases at the same temperature have the same kinetic energy

$$KE_{O_2} = KE_{CO_2}$$

and since the average $KE = 1/2\, m\bar{u}^2$

where $\bar{u}$ is the average speed of a molecule, we can write.

$$1/2\, M_{O_2}\bar{U}^2_{O_2} = 1/2\, M_{CO_2}\bar{U}^2_{CO_2} \quad \text{or} \quad M_{O_2}\bar{U}^2_{O_2} = M_{CO_2}\bar{U}^2_{CO_2}$$

$$\text{and} \quad \frac{M_{O_2}}{M_{CO_2}} = \frac{\bar{U}^2_{CO_2}}{\bar{U}^2_{O_2}}$$

and solving for the average velocity of CO_2 :

$$\bar{U}^2_{CO_2} = \frac{M_{O_2}}{M_{CO_2}} \cdot \bar{U}^2_{O_2}$$

Taking the square root of both sides

$$\bar{U}_{CO_2} = \sqrt{\frac{M_{O_2}}{M_{CO_2}}} \cdot \bar{U}_{O_2} = \sqrt{\frac{32.0 \text{ g } O_2 / \text{mol } O_2}{44.0 \text{ g } CO_2 / \text{mol } CO_2}} \cdot 4.28 \times 10^4 \text{cm/s}$$

$$= 3.65 \times 10^4 \text{ cm/s}$$

64. The species will have average molecular speeds which are inversely proportional to their molar masses.

Slowest						Fastest	
CH_2F_2	<	Ar	<	N_2	<	CH_4	
54		40		28		16	(to integral values)

Diffusion and Effusion

66. Compare the rates of effusion of the gases Argon and Helium:

$$\frac{\text{Rate of effusion of He}}{\text{Rate of effusion of Ar}} = \sqrt{\frac{\text{M of Ar}}{\text{M of He}}} = \sqrt{\frac{40}{4}}$$

The square root of 10 (40/4) is 3.2 (to 2 significant figures). This tells us that **helium effuses approximately 3.2 times faster than argon.**

68. Determine the molar mass of a gas which effuses at a rate 1/3 that of He:

$$\frac{\text{Rate of effusion of He}}{\text{Rate of effusion of unknown}} = \sqrt{\frac{\text{M of unknown}}{\text{M of He}}}$$

$$\frac{3}{1} = \sqrt{\frac{\text{M of unknown}}{4.0 \text{ g/mol}}}$$

Squaring both sides gives: $9 = \frac{M}{4.0}$ or M = 36 g/mol

Non-Ideal Gases

70. According to the Ideal Gas Law, the pressure would be:

$$P = \frac{n \cdot R \cdot T}{V} = \frac{(8.00 \text{ mol})(0.082057 \frac{L \cdot atm}{K \cdot mol})(300 \text{ K})}{4.00 \text{ L}} = 49.2 \text{ atm}$$

The van der Waal's equation is: $\left[P + a\left(\frac{n}{V}\right)^2\right][V - bn] = nRT$.

Substituting we get:

$$\left[P + 6.49\frac{atm \bullet L^2}{mol^2}\left(\frac{8.00 \text{ mol}}{4.00 \text{ L}}\right)^2\right]\left[4.00L - 0.0562\frac{L}{mol} \bullet 8.00mol\right] = 8.00mol \bullet 0.082057\frac{atm \bullet L}{K \bullet mol} \bullet 300K$$

Simplifying: $[P + 25.96\text{atm}][4.00L - 0.45L] = 196.94\text{atm} \bullet L$ and

$$P = \frac{196.94 \text{ atm} \bullet L}{3.55L} - 25.96 \text{ atm} = 29.5 \text{ atm}$$

General Questions

72.

	atm	mm Hg	kPa	bar
Standard atmosphere:	1	1 atm • $\frac{760.\text{ mm Hg}}{1\text{ atm}}$ = 760. mm Hg	1 atm • $\frac{101.325\text{ kPa}}{1\text{ atm}}$ = 101.325 kPa	1 atm • $\frac{1.013\text{ bar}}{1\text{ atm}}$ = 1.013 bar
Partial pressure of N_2 in the atmosphere	593 mm Hg • $\frac{1\text{ atm}}{760\text{ mm Hg}}$ = 0.780 atm	**593**	0.780 atm • $\frac{101.3\text{ kPa}}{1\text{ atm}}$ = 79.1 kPa	0.780 atm • $\frac{1.013\text{ bar}}{1\text{ atm}}$ =0.791 bar
Tank of compressed H_2	133 bar • $\frac{1\text{ atm}}{1.013\text{ bar}}$ = 131 atm	131 atm • $\frac{760.\text{ mm Hg}}{1\text{ atm}}$ = 9.98 x 10^4 mm Hg	131 atm • $\frac{101.3\text{ kPa}}{1\text{ atm}}$ = 1.33 x 10^4 kPa	**133**
Atmospheric pressure at top of Mt. Everest	33.7 kPa • $\frac{1\text{ atm}}{101.3\text{ kPa}}$ = 0.333 atm	0.333 atm • $\frac{760\text{ mm Hg}}{1\text{ atm}}$ = 253 mm Hg	**33.7**	0.333 atm • $\frac{1.013\text{ bar}}{1\text{ atm}}$ = 0.337 bar

74. Rewriting the ideal gas law we obtain:

$$P \cdot V = n \cdot R \cdot T$$

$$P \cdot V = \frac{\text{mass}}{\text{Molar mass}} \cdot R \cdot T$$

Rearranging this equation gives: $\text{Molar mass} \cdot P = \frac{\text{mass}}{V} \cdot R \cdot T$

and noting that $D = \frac{\text{mass}}{V}$ we write: $\text{Molar mass} = \frac{D \cdot R \cdot T}{P}$

Converting 331 mm Hg to atm yields:

$$331\text{ mm Hg} \cdot \frac{1\text{ atm}}{760\text{ mm Hg}} = 0.436\text{ atm}$$

$$\text{Molar mass} = \frac{0.855\,\frac{\text{g}}{\text{L}} \cdot 0.082057\,\frac{\text{L} \cdot \text{atm}}{\text{K} \cdot \text{mol}} \cdot 273.2\text{ K}}{0.436\text{ atm}} = 44.0\,\frac{\text{g}}{\text{mol}}$$

76. Since P and the amount of gas are fixed, the ideal gas law in these situations can be written

$$\frac{P_1V_1}{T_1} = \frac{P_2V_2}{T_2} \quad \text{and} \quad \text{since } P1 = P_2 \quad \frac{V_1}{T_1} = \frac{V2}{T2}$$

$$\frac{25.5 \text{ mL}}{363 \text{ K}} = \frac{21.5 \text{ mL}}{T_2} \quad \text{and} \quad T_2 = \frac{(21.5 \text{ mL})(363 \text{ K})}{25.5 \text{ mL}}$$

T_2 = 306 K or 33 °C

78. To calculate molecules of water in 1 cm^3, let's calculate the number of moles of water in 1 cm^3 (1×10^{-3} L) :

$$n = \frac{PV}{RT} = \frac{(23.8 \text{ torr})(1 \times 10^{-3} \text{ L})}{(62.4 \frac{\text{L} \cdot \text{torr}}{\text{K} \cdot \text{mol}})(298 \text{ K})} = 1.28 \times 10^{-6} \text{ mol } H_2O$$

and multiplying by Avogadro's number:

$$(1.28 \times 10^{-6} \text{ mol})(\frac{6.022 \times 10^{23} \text{ molecules } H_2O}{1 \text{ mol}}) = 7.71 \times 10^{17} \text{ molecules}$$

80. The amount of N_2 can be calculated.

First, calculate the pressure exerted by the nitrogen:

P(Total) = P(H_2O) + P(N_2)

736.0 mm Hg = 18.7 mm Hg + P(N_2)

717.3 mm Hg = P(N_2) = 0.944 atm

Now calculate the # of moles of N_2:

$$n_{N_2} = \frac{(0.944 \text{ atm})(0.295 \text{ L})}{(0.082057 \frac{\text{L} \cdot \text{atm}}{\text{K} \cdot \text{mol}})(294.2 \text{ K})} = 1.15 \times 10^{-2} \text{ mol } N_2$$

According to the equation in which sodium nitrite reacts with sulfamic acid, one mole of $NaNO_2$ produces one mole of N_2.

$$1.15 \times 10^{-2} \text{ mol } N_2 \cdot \frac{1 \text{ mol } NaNO_2}{1 \text{ mol } N_2} \cdot \frac{69.00 \text{ g } NaNO_2}{1 \text{ mol } NaNO_2} = 0.796 \text{ g } NaNO_2$$

$$\text{Weight percentage of } NaNO_2 = \frac{0.796 \text{ g } NaNO_2}{1.232 \text{ g sample}} \times 100 = 64.6\% \text{ } NaNO_2$$

82. The partial pressure of each gas can be calculated from

$$P_2 = P_1 \cdot \frac{V_1}{V_2} \quad \text{where } V_2 \text{ in each case is 5.0 L}$$

$$\text{Partial pressure of He} = 145 \text{ mm Hg} \cdot \frac{3.0 \text{ L}}{5.0 \text{ L}} = 87 \text{ mm Hg}$$

$$\text{Partial pressure of Ar} = 355 \text{ mm Hg} \cdot \frac{2.0 \text{ L}}{5.0 \text{ L}} = 140 \text{ mm Hg}$$

$$P_{total} = P_{He} + P_{Ar} = 87 \text{ mm Hg} + 140 \text{ mm Hg} = 230 \text{ mm Hg}$$

84. Use the Ideal Gas Law to calculate the molar mass of the new gas.

(Remembering that $n = \frac{\text{mass}}{M}$)

$$PV = \frac{\text{mass}}{M} RT \qquad \text{or} \qquad M = \frac{\text{mass RT}}{PV}$$

$$M = \frac{(0.150 \text{ g})(62.4 \frac{\text{L} \cdot \text{torr}}{\text{K} \cdot \text{mol}})(294 \text{ K})}{(17.2 \text{ torr})(1.850 \text{ L})} = 86.5 \text{ g/mol}$$

Since we know the compound **must** have at least one atom of Cl, F, and O, let's subtract the mass of 1 mol of each of these atoms from the molar mass of 86.5.

$$86.5 \text{ g cpd} - (35.5 \text{ g Cl} + 16.0 \text{ g O} + 19.0 \text{ g F}) = 16 \text{ g}$$

So the compound must have an additional oxygen atom — ClO_2F.

86. The number of moles of He in the balloon can be calculated with the Ideal Gas Law. First calculate the P of He in the balloon:

gauge = total - barometric or

gauge + barometric = total pressure = 22 mm Hg + 755 mm Hg = 777 mm Hg or 777 torr

$$n = \frac{PV}{RT} = \frac{(777 \text{ torr})(0.305 \text{ L})}{(62.4 \frac{\text{L} \cdot \text{torr}}{\text{K} \cdot \text{mol}})(298 \text{ K})} = 0.0127 \text{ mol He}$$

Conceptual Questions

88. In a 1.0-L flask containing 10.0 g each of O_2 and CO_2 at 25 °C.

a. The gas with the greater partial pressure:

Partial pressure is a relative measure of the number of moles of gas present. The molar mass of oxygen is approximately 32 g/mol while that of CO_2 is approximately 44 g/mol. There will be a greater number of moles of oxygen in the flask—hence the **partial pressure of O_2 will be greater.**

b. The gas with the greater average speed:

The kinetic energy of each gas is given as: KE = $1/2\ mu^2$, where u is the average speed of the molecules. Since the average KE of both gases are the same (They are at the same T), The lighter of the gases **(O_2)will have a greater average speed.**

c. The gas with the greater average kinetic energy:

The KE depends on temperature, and since both gases are at the same T, the **average KE of the two gases is the same.**

90. In two cylinders of equal volume, one containing CO and the other acetylene, C_2H_2:

a. In which cylinder is the pressure greater at 25 °C—if each contains 1kg of the gas?

The molar mass of CO is approximately 28 g/mol while that of C_2H_2 is approximately 26 g/mol. One kg of CO would contain approximately 35.7 mol of gas while one kg of acetylene would contain approximately 38.5 mol of the gas. Given equal volumes, and the fact that (according to Avogadro's law) the pressure is proportional to the number of moles of gas present, the **pressure in the cylinder of C_2H_2 would be greater.**

b. Which cylinder contains the greater number of molecules?

Given the statement in part a above, the **cylinder containing C_2H_2 would contain the greater number of molecules.**

92. Which of the following samples is a gas?

a. Material expands 10% when a sample of gas originally at 100 atm is suddenly allowed to exist at one atmosphere pressure: This sample is **not a gas**, since a gas would expand in volume by 100-fold (to exist at 1 atm).

b. A 1.0-mL sample of material weighs 8.2 g: This sample is **not a gas** since the density of the sample (at 8.2 g/mL) is too great.

c. Material is transparent and pale green in color: **Insufficient information** to tell. Liquids could also be pale green and transparent.

d. One cubic meter of material contains as many molecules as an equal volume of air at the same temperature and pressure: This material **is a gas**, since one cubic meter of a liquid or solid would contain a greater number of molecules than a cubic meter of air.

94. Four tires contain four different gases:

a. Since the four tires have the same volume and pressure at the same temperature, **each contains the same number of molecules** as the other three.

b. The relative mass of an atom of the unknown gas compared to an atom of helium. Since the tires have the same number of molecules, and the unknown gas has a mass of 160. g while that of He is 16.0 g, the **atoms of unknown gas are (160/16.0) or 10 times heavier than atoms of helium.**

c. The molecules of each of the four gases **all have the same kinetic energy** since kinetic energy depends on temperature. However since kinctic energy is a function of the mass of the gas (= 1/2 mu^2), the **molecules of the lightest gas—helium—will have the greatest average speed.**

Challenging Questions

96. Masses of CO and CO_2 and partial pressures of CO, CO_2 and O_2 in a 550 mL tank at 24°C with a total pressure of 1.56 atm. The amount of O_2 present is easily calculated:

$0.0870 \text{ g } O_2 \cdot \frac{1 \text{ mol } O_2}{32.00 \text{g } O_2} = 2.72 \times 10^{-3} \text{ mol } O_2$ and from this, the pressure of O_2:

$$P = \frac{(2.72 \times 10^{-3} \text{ mol})(0.082057 \frac{\text{L} \cdot \text{atm}}{\text{K} \cdot \text{mol}})(297 \text{ K})}{0.550 \text{ L}} = 0.121 \text{ atm } O_2$$

The P (CO_2) = Ptotal - [P(CO)+ P(O_2)] = 1.56 atm - 1.34 atm = 0.22 atm

and since we know the partial pressure of oxygen, the partial pressure of CO is:
P (CO) = Ptotal - P(CO_2) - P(O_2) = 1.56 atm - 0.22 atm - 0.121 atm = 1.22 atm
The numbers of moles of each of the gases can be calculated:

$$n(CO) = \frac{PV}{RT} = \frac{(1.22 \text{ atm})(0.550 \text{ L})}{(0.082057 \frac{\text{L} \cdot \text{atm}}{\text{K} \cdot \text{mol}})(297 \text{ K})} = 2.75 \times 10^{-2} \text{ mol CO}$$

and since 1 mol of CO has a mass of 28.01 g:

$$2.75 \times 10^{-2} \text{ mol CO} \cdot \frac{28.01 \text{ g CO}}{1 \text{ mol CO}} = 0.771 \text{ g CO}$$

$$n(CO_2) = \frac{PV}{RT} = \frac{(0.22 \text{ atm})(0.550 \text{ L})}{(0.082057 \frac{\text{L} \cdot \text{atm}}{\text{K} \cdot \text{mol}})(297 \text{ K})} = 4.96 \times 10^{-3} \text{ mol } CO_2$$

and since 1 mol of CO_2 has a mass of 44.01 g:

$$4.96 \times 10^{-3} \text{ mol } CO_2 \cdot \frac{44.01 \text{ g } CO_2}{1 \text{ mol } CO_2} = 0.22 \text{ g } CO_2$$

98. The formula for a compound of $Fe_x(CO)_y$:

$$Fe_x(CO)_y + O_2 \rightarrow Fe_2O_3 + CO_2$$

Given 1.50 L of gas at 25 °C and 44.0 mm Hg, we can calculate the # of moles of CO_2:

$$n(CO_2) = \frac{PV}{RT} = \frac{(44.9 \text{ torr})(1.50 \text{ L})}{(62.4 \frac{\text{L} \cdot \text{torr}}{\text{K} \cdot \text{mol}})(298 \text{ K})} = 3.62 \times 10^{-3} \text{ mol } CO_2 .$$

The equation tells us that for each CO (in the iron compound) we get 1 mol of CO_2 when the iron carbonyl compound decomposes. So with 3.62×10^{-3} mol of carbon dioxide (produced) we have 3.62×10^{-3} mol of CO (in the compound). CO has a molar mass of 28.01 g, so the mass associated with 3.62×10^{-3} mol of CO is:

$$3.62 \times 10^{-3} \text{ mol} \cdot \frac{28.01 \text{ g CO}}{1 \text{ mol CO}} = 0.101 \text{ g CO}$$

The mass of the sample is 0.142 g, so the mass of iron in that sample is (0.142- 0.101) or 0.041 g Fe. The number of moles of Fe that correspond to this mass of Fe is:

$$0.041 \text{ g Fe} \cdot \frac{1 \text{ mol Fe}}{55.85 \text{ g Fe}} = 7.26 \times 10^{-4} \text{ mol Fe.}$$

Dividing the # moles of CO by the # moles of Fe we obtain:

$$\frac{3.62 \times 10^{-3} \text{ mol CO}}{7.26 \times 10^{-4} \text{ mol Fe}} = 5 ,$$

so the apparent formula for the compound is $Fe(CO)_5$.

Adding the atomic masses for 1 Fe, 5 C, and 5 O, we obtain a molar mass of 195.9 g/mol. So 195.9 g/mol • 7.26×10^{-4} mol = 0.142 g (the mass of the sample). The formula of the compound is then $Fe(CO)_5$.

100. The correct stoichiometric ratio of SiH_4 to O_2 is 1 mol SiH_4: 2 mol O_2. Let's **assume** that we have exactly 1 mole of silane and 2 moles of oxygen. The number of moles we pick won't matter—but the relative amounts will! Dalton's Law tells us that the total pressure is the sum of the partial pressures, so we can write:

$$P_{total} = P_{silane} + P_{oxygen}$$

Additionally we know that the pressure a gas exerts is related to its mole fraction so:

$$P_{silane} = \frac{\text{moles silane}}{\text{moles silane + moles oxygen}} \cdot P_{total} = \frac{1 \text{ mol}}{3 \text{ mol}} \cdot 120 \text{ mm Hg} = 40 \text{ mm Hg}$$

$$P_{oxygen} = \frac{\text{moles oxygen}}{\text{moles silane + moles oxygen}} \cdot P_{total} = \frac{2 \text{ mol}}{3 \text{ mol}} \cdot 120 \text{ mm Hg} = 80 \text{ mm Hg}$$

The total pressure in the flask can be calculated by noting that since the reactants are present in the stoichiometric amount, the contents of the flask after reaction is **only silicon dioxide—a solid—and gaseous water**. The balanced equation indicates that we will obtain 1 mol of water for each mole of oxygen used. Hence the **pressure in the flask will be the same as the pressure of oxygen prior to the reaction—80 mm Hg.**

Summary Questions

102. a. Valence electrons for ClO_2 : 1(7) + 2(6) = 19 electrons

b. Electron dot structure for ClO_2^- : (20 electrons)

$$\left[:\ddot{O}: - \ddot{\underset{..}{Cl}} - :\ddot{O}: \right]^-$$

c. The hybridization for Cl is **sp^3** (4 electron pairs attached). The ion has a **bent shape**.

d. The molecule ozone has a "central" oxygen atom with 3 electron groups attached. (1 lone pair; 1 double bond; 1 single bond — with the double bond being delocalized between the three O atoms.) The predicted geometry would give a bond angle of about 120 ° (measured is approximately 117 °). The bond angle for ClO_2^- would have a smaller angle (approximately 109°).

e. To determine the mass of ClO_2, determine the limiting reagent (if there is one).
Moles of ClO_2:

$$15.6 \text{ g } NaClO_2 \cdot \frac{1 \text{ mol } NaClO_2}{54.99 \text{ g } NaClO_2} = 0.284 \text{ mol } NaClO_2$$

Moles of Cl_2:

$$n = \frac{(1050 \text{ torr})(1.45 \text{ L})}{(62.4 \frac{\text{L} \cdot \text{torr}}{\text{K} \cdot \text{mol}})(295 \text{ K})} = 0.0827 \text{ mol } Cl_2$$

Note (from the balanced equation) that each mol of Cl_2 requires 2 mol of $NaClO_2$ (which we have in excess). So Cl_2 will limit the amounts of product obtainable.

The mass of ClO_2 obtainable is:

$$0.0827 \text{ mol } Cl_2 \cdot \frac{2 \text{ mol } ClO_2}{1 \text{ mol } Cl_2} = 0.165 \text{ mol } ClO_2$$

$$0.165 \text{ mol } ClO_2 \cdot \frac{67.45 \text{ g } ClO_2}{1 \text{ mol } ClO_2} = 11.2 \text{ g } ClO_2$$

The pressure the ClO_2 will exert is:

$$P = \frac{nRT}{V} = \frac{(0.165 \text{ mol } ClO_2)(62.4 \frac{L \cdot torr}{K \cdot mol})(298 \text{ K})}{(1.25 \text{ L})}$$

$$= 2460 \text{ torr or } 2460 \text{ mm Hg (or } 3.24 \text{ atm)}$$

Chapter 13
Bonding and Molecular Structure: Intermolecular Forces, Liquids, and Solids

Intermolecular Forces

16.

change	intermolecular
a. melt ice	hydrogen bonds (dipole-dipole)
b. melt solid I_2	induced dipole-induced dipole
c. remove water from hydrated salt	ion-dipole
d. convert $NH_3(l)$ to $NH_3(g)$	hydrogen bonds (dipole-dipole)

18. Since I_2 is a non-polar molecule, the forces that must be overcome in the solid are induced dipole-induced dipole forces. The attractive forces between molecules of CH_3OH are hydrogen bonds. The forces between CH_3OH and I_2 molecules will be dipole-induced dipole.

20. To convert species from a liquid to a gas one must overcome intermolecular forces.

species	intermolecular
a. liquid O_2	induced dipole-induced dipole
b. mercury	induced dipole-induced dipole
c. methyl iodide	dipole-dipole
d. ethanol	hydrogen bonding and dipole-dipole

22. Increasing strength of intermolecular forces:

$Ne < CH_4 < CO < CCl_4$

Neon and methane are nonpolar species and possess only induced dipole-induced dipole interactions. Neon has a smaller molar mass than CH_4, and therefore weaker London (dispersion) forces. Carbon monoxide is a polar molecule. Molecules of CO would be attracted to each other by dipole-dipole interactions, but the CO molecule is not a very strong dipole. The CCl_4 molecule is a non-polar molecule, but very heavy (when compared to the other three). Hence the greater London forces that accompany larger molecules would result in the strongest attractions of this set of molecules.

The lower molecular weight molecules with weaker interparticle forces should be gases at 25 °C and 1 atmosphere: Ne, CH_4, CO.

24. Member of each pair with the higher boiling point:
   a. O_2 would have a higher boiling point than N_2 owing to its greater molar mass.
   b. SO_2 would boil higher since SO_2 is a polar molecule while CO_2 is non-polar.
   c. HF would boil higher since the strong hydrogen bonds exist in HF but not in HI.
   d. GeH_4 would boil higher since—while both molecules are non-polar, germane has the greater molar mass—and therefore stronger London forces.

26. Compounds which are capable of forming hydrogen bonds with water are those containing polar O-H bonds and lone pairs of electrons on N,O, or F.

a. CH_3-O-CH_3	no; no "polar H's" and the C-O bond is not very polar
b. CH_4	no
c. HF	yes: lone pairs of electrons on F and a "polar hydrogen".
d. CH_3CO_2H	yes: lone pairs of electrons on O atoms, and a "polar hydrogen" attached to one of the oxygen atoms
e. Br_2	no
f. CH_3OH	yes: "polar H" and lone pairs of electrons on O

28. a. LiCl would be more strongly hydrated than CsCl, since the smaller Li^+ would be more strongly attracted to water than Cs^+.

b. $Mg(NO_3)_2$ will be more likely hydrated since Mg^{2+} will be more strongly attracted to water than Na^+ (+2 > +1).

c. $NiCl_2$ — for the same reason as Mg^{2+} in part b.

Liquids

30. Heat required: $125 \text{ mL} \cdot \frac{0.7849 \text{ g}}{1 \text{ mL}} \cdot \frac{1 \text{ mol}}{46.07 \text{ g}} \cdot \frac{42.32 \text{ kJ}}{1 \text{ mol}} = 90.1 \text{ kJ}$

32. Using Figure 13.21:

   a. The equilibrium vapor pressure of water at 60 °C is approximately 150 mm Hg. Appendix G lists this value as 149.4 mm Hg.

   b. Water has a vapor pressure of 600 mm Hg at 93 °C.

   c. At 70 °C the vapor pressure of water is approximately 225 mm Hg while that of ethanol is approximately 520 mm Hg.

34. The vapor pressure of $(C_2H_5)_2O$ at 30. °C is **590 mm Hg.**

   Calculate the amount of $(C_2H_5)_2O$ to furnish this vapor pressure at 30.°C.

$$n = \frac{PV}{RT} = \frac{590\text{ mm} \cdot \frac{1\text{ atm}}{760\text{ mm}} \cdot 0.100\text{ L}}{0.082057\frac{\text{L} \cdot \text{atm}}{\text{K} \cdot \text{mol}} \cdot 303\text{ K}} = 3.1 \times 10^{-3}\text{ mol}$$

The total mass of $(C_2H_5)_2O$ [FW = 74.1 g] needed to create this pressure is about 0.23 g. Since there is adequate ether to provide this pressure, we anticipate the pressure in the flask to be approximately 590 mm Hg.

As the flask is cooled from 30.°C to 0 °C where the vapor pressure of ether is about 160 mm Hg, **some of the gaseous ether will condense** to form liquid ether.

36. a. From the figure, we can read the vapor pressure of CS_2 as approximately 620 mm Hg and for nitromethane as approximately 80 mm Hg.

   b. The principle intermolecular forces for CS_2 (a non-polar molecule) are **induced dipole-induced dipole**; for nitromethane they are **dipole-dipole.**

   c. The normal boiling point from the figure for CS_2 is 46 °C and for CH_3NO_2 100 °C.

   d. The temperature at which the vapor pressure of CS_2 is 600 mm Hg is about 39 °C.

   e. The vapor pressure of CH_3NO_2 is 60 mm Hg at approximately 34 °C.

38. CO can be liquefied at or below its critical temperature. The Tc for CO is 132.9 K (or approximately -140 °C), so CO **cannot** be liquefied **at or above room temperature.**

Metallic and Ionic Solids

40. This compound would have the formula AB since each black square (A) has one corresponding white square (B).

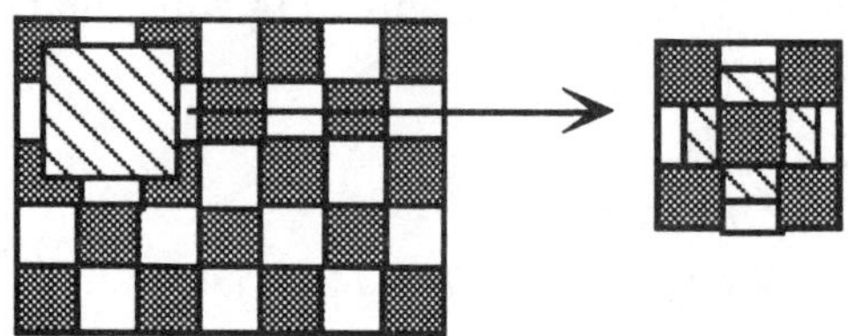

The diagonally crossed square is the "outline" of the unit cell. Note that it contains 1 solid square and quarters of four solid squares.

The diagonally crossed square also contains quarters of four "white" squares, making the overall ratio 4 "A" and 4 "B" squares, or a unit cell of AB.

42. Iridium (molar mass 192.22 g/mol) has a face-centered cubic unit cell, and a density of 22.56 g/cm^3 . What is the radius of an iridium atom ?

We need to find the dimensions of the unit cell (specifically the edge length), and the number of Ir atoms in the cell!

1. Let's determine the mass of the unit cell:

$$\frac{192.22 \text{ g Ir}}{1 \text{ mol Ir}} \cdot \frac{1 \text{ mol Ir}}{6.0221 \times 10^{23} \text{ Ir atoms}} \cdot \frac{4 \text{ Ir atoms}}{1 \text{ unit cell}} = \frac{1.2768 \times 10^{-21} \text{ g Ir}}{\text{unit cell}}$$

Regarding the last factor in this calculation, recall that ALL face-centered cubic unit cells have four atoms:

6 atoms (one in each fact) • 1/2 = 3 in the cell

8 atoms (one at each corner) • 1/8 = 1 in the cell

4 atoms in the cell

2. Now determine the volume of the unit cell, using the density of Ir:

$$\frac{1.2768 \times 10^{-21} \text{ g Ir}}{\text{unit cell}} \cdot \frac{1 \text{ cm}^3}{22.56 \text{ g Ir}} = 5.659 \times 10^{-23} \text{ cm}^3\text{/unit cell}$$

3. Now the length of one edge of the cell is the cube root of the volume:

$$V = l \cdot l \cdot l \quad \text{so } l = \sqrt[3]{V} = (5.659 \times 10^{-23} \text{ cm}^3)^{1/3} = 3.839 \times 10^{-8} \text{ cm}$$

Recall that the diagonal of the face of the unit cell = 4 atomic radii , and that from geometric considerations the diagonal $= \sqrt{2} \cdot$ edge

Let's solve for the face diagonal:

$$\text{diagonal} = \sqrt{2} \cdot \text{edge} = \sqrt{2} \cdot 3.839 \times 10^{-8}\ \text{cm} = 5.430 \times 10^{-8}\ \text{cm}$$

and since the diagonal contains 4 radii:

$$4 \text{ atomic radii} = 5.430 \times 10^{-8}\ \text{cm and}$$

$$\text{atomic radius} = \frac{5.430 \times 10^{-8}\ \text{cm}}{4} = 1.356 \times 10^{-8}\ \text{cm or } 135.6 \text{ picometers.}$$

44. Let's assume that copper has a face-centered cubic unit cell, and calculate the density from this assumption. Comparing it with the stated density (8.95 g/cm^3) should give us an idea about the correctness of our assumption.

The fcc unit cell has 4 Cu atoms/cell. The mass of **one** Cu atom is:

$$\frac{63.546 \text{ g Cu}}{1 \text{ mol Cu}} \cdot \frac{1 \text{ mol Cu}}{6.0221 \times 10^{23} \text{ Cu atoms}} = \frac{1.0552 \times 10^{-22} \text{ g}}{\text{Cu atom}}$$

$$\frac{4 \text{ Cu atoms}}{1 \text{ unit cell}} \cdot \frac{1.0552 \times 10^{-22} \text{ g}}{\text{Cu atom}} = \frac{4.2209 \times 10^{-22} \text{ g}}{\text{unit cell}}$$

The diagonal of the face of the unit cell = 4 atomic radii $= \sqrt{2}$ x edge

Solving for the edge gives:

$$\text{edge} = \frac{4 \text{ atomic radii}}{\sqrt{2}} = \frac{4 \cdot 127.8 \text{ pm}}{1.414} = 361.5 \text{ pm} = 3.615 \times 10^{-8}\ \text{cm}$$

The volume of this cell is then the edge^3 = $(3.615 \times 10^{-8}\ \text{cm})^3 = 4.723 \times 10^{-23}\ \text{cm}^3$

$$\text{Density} = \frac{M}{V} = \frac{4.2209 \times 10^{-22} \text{ g}}{4.723 \times 10^{-23} \text{ cm}^3} = 8.937 \text{ g/cm}^3$$

This is in good agreement with the measured density, hence **Cu is a face-centered cubic unit cell.**

46. Like question 44, let's assume a cell configuration, calculate the density and compare!

Assume that TlCl crystallizes in a simple cubic cell.

$$\frac{239.823 \text{ g TlCl}}{1 \text{ mol TlCl}} \cdot \frac{1 \text{ mol TlCl}}{6.0221 \times 10^{23} \text{ TlCl units}} \cdot \frac{1 \text{ TlCl unit}}{1 \text{ unit cell}} = \frac{3.98238 \times 10^{-22} \text{ g}}{\text{unit cell}}$$

The volume of this cell : edge^3 = $(3.85 \times 10^{-8}\ \text{cm})^3 = 5.71 \times 10^{-23}\ \text{cm}^3$

$$\text{Density} = \frac{M}{V} = \frac{3.98238 \times 10^{-22} \text{ g}}{5.71 \times 10^{-23} \text{ cm}^3} = 6.97 \text{ g/cm}^3$$

Given the good agreement between our calculated density and the reported density, TlCl solid is **a simple cubic unit cell.**

48. To determine the perovskite formula, determine the number of each atom belonging **uniquely** to the unit cell shown. The Ca atom is wholly contained within the unit cell. There are Ti atoms at each of the eight corners. Since each of these atoms belong to eight unit cells, the portion of each Ti atom belonging to the pictured unit cell is 1/8 so 8 Ti atoms x 1/8 = 1 Ti atom. The O atoms on an edge belong to 4 unit cells, so the fraction contained within the pictured cell is 1/4. There are twelve such O atoms, leading to 12 x 1/4 = 3 O atoms— and a formula of $CaTiO_3$.

50. Note that each corner is occupied by a Zn atom. Also each face is occupied by a Zn atom. The tetrahedral holes (four of them) are occupied by S atoms. The net formula would be

8 Zn (corner) x 1/8 = 1
6 Zn (faces) x 1/2 = 3
4 Zn atoms per unit cell
4 S (tetrahedral holes) • 1 = 4 ⇒ ZnS

Molecular and Network Solids

52. For the unit cell of diamond:
   a. The unit cell has 8 corner atoms (1/8 in the cell), 6 face atoms (1/2 in the cell), and 4 atoms wholly within the cell, for a total of **8 carbon atoms.**
   b. Diamond uses a fcc unit cell (The structure shown also has 4 atoms occupying holes in the lattice.
   c. The volume of the unit cell may be calculated if we first calculate the masses of the atoms involved:

$$\frac{12.01 \text{ g C atom}}{1 \text{ mol C atom}} \cdot \frac{1 \text{ mol C atom}}{6.022 \times 10^{23} \text{ C atom}} \cdot \frac{8 \text{ atoms}}{1 \text{ unit cell}} = 1.60 \times 10^{-22} \frac{\text{g C atom}}{\text{unit cell}}$$

Now we can calculate the volume—using the density:

$$D = \frac{M}{V} \quad \text{and } V = \frac{M}{D} = \frac{1.60 \times 10^{-22} \text{ g/unit cell}}{3.51 \text{ g/cm}^3} = 4.55 \times 10^{-23} \text{ cm}^3\text{/unit cell}$$

Since the length of an edge (l) cubed is the volume we can write:

$$\text{Volume} = l^3 = 4.55 \times 10^{-23}\ \text{cm}^3$$

$$l = 3.57 \times 10^{-8}\ \text{cm}$$

$$\text{or } 3.57 \times 10^{-8}\ \text{cm} \cdot \frac{1.00 \times 10^{12}\ \text{pm}}{1 \times 10^{2}\ \text{cm}} = 357\ \text{pm}$$

Phase Changes

54. a. The positive slope of the solid/liquid equilibrium line means the liquid CO_2 is **less dense** than solid CO_2.

b. At 5 atm and 0 °C, CO_2 is in the **gaseous phase**.

c. The phase diagram for CO_2 shows the critical pressure for CO_2 to be 73 atm, and the critical temperature to be +31 °C.

56. The heat required is a summation of three "steps":

1. heat the liquid(at -50.0 °C) to its boiling point (-33.3 °C)
2. "boil" the liquid—converting it to a gas and
3. warm the gas from -33.3 °C to 0.0 °C

Let's do them one at a time:

1. To heat the liquid (at -50.0 °C) to its boiling point (-33.3 °C):

$$q_{liquid} = 1.2 \times 10^4\ \text{g} \cdot 4.7 \frac{\text{J}}{\text{g} \cdot \text{K}} \cdot (239.9\ \text{K} - 223.2\ \text{K}) = 9.4 \times 10^5\ \text{J}$$

2. To boil the liquid:

$$23.33 \times 10^3 \frac{\text{J}}{\text{mol}} \cdot \frac{1\ \text{mol NH}_3}{17.0\ \text{g NH}_3} \cdot 1.2 \times 10^4\ \text{g} = 1.6 \times 10^7\ \text{J}$$

3. To heat the gas from -33.3 °C to 0.0 °C:

$$q_{gas} = 1.2 \times 10^4\ \text{g} \cdot 2.2 \frac{\text{J}}{\text{g} \cdot \text{K}} \cdot (273.2\ \text{K} - 239.9\ \text{K}) = 8.8 \times 10^5\ \text{J}$$

The total heat required is then:

$$9.4 \times 10^5\ \text{J} + 1.6 \times 10^7\ \text{J} + 8.8 \times 10^5\ \text{J} = 1.8 \times 10^7\ \text{J or } 1.8 \times 10^4\ \text{kJ}$$

Physical Properties of Solids

58. The heat evolved when 15.5 g of benzene freezes at 5.5 °C :

$$15.5\ \text{g benzene} \cdot \frac{1\ \text{mol benzene}}{78.1\ \text{g benzene}} \cdot \frac{9.95\ \text{kJ}}{1\ \text{mol benzene}} = -1.97\ \text{kJ}$$

Note once again the negative sign indicates that heat is evolved.

The quantity of heat needed to remelt this 15.5 g sample of benzene would be +1.97 kJ.

General Questions

60. Increasing strength of intermolecular forces :

Ar < CO_2 < CH_3OH

Argon and CO_2 are nonpolar species and possess only induced dipole-induced dipole (London) interactions. Ar has a smaller mass than CO_2, so London forces are expected to be less. The polar molecular CH_3OH is capable of forming the stronger hydrogen-bonds.

62.

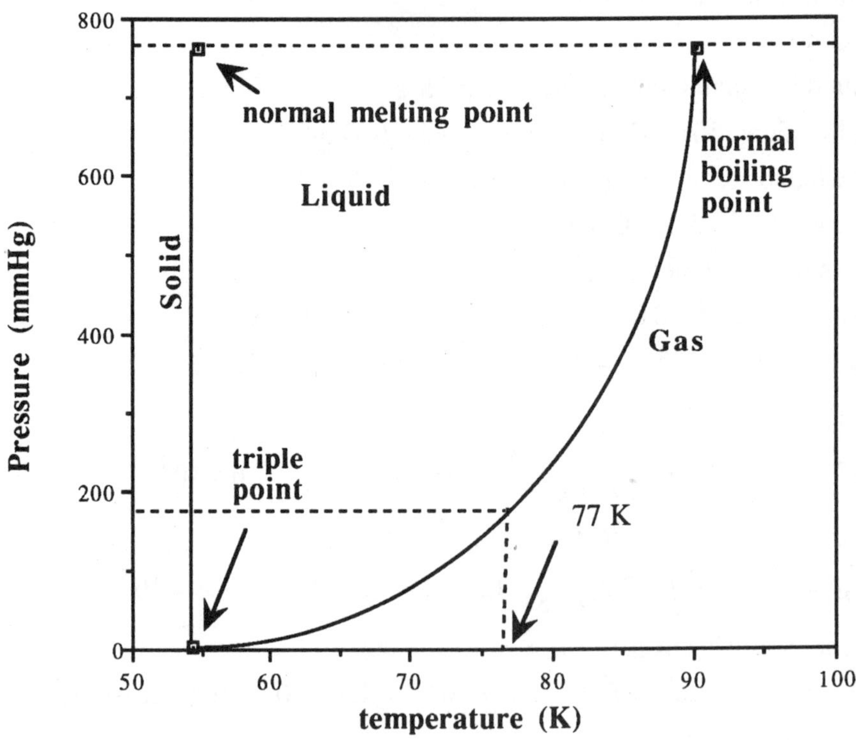

The estimated vapor pressure at 77 K is between 150-200 mm Hg. The very slight positive slope of the solid/liquid equilibrium line indicates the **solid is more dense than the liquid.**

64. Acetone readily absorbs water owing to **hydrogen bonding** between the C = O oxygen atom and the O—H bonds of water.

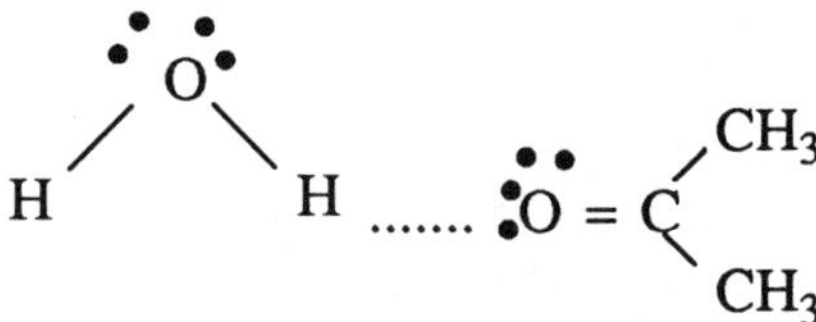

66. Volume of room = 3.0×10^2 cm • 2.5×10^2 cm • 2.5×10^2 cm = 1.9×10^7 cm^3

Convert volume to L:

$$1.9 \times 10^7\ cm^3 \cdot \frac{1\ L}{1.0 \times 10^3\ cm^3} = 1.9 \times 10^4\ L$$

To produce a pressure of 59 mm Hg, calculate the amount of ethanol required.

$$P = 59\ mm \cdot \frac{1\ atm}{760\ mm} = 7.8 \times 10^{-2}\ atm$$

$$V = 1.9 \times 10^4\ L \qquad \text{and } T = 25 + 273 = 298\ K$$

$$\frac{PV}{RT} = \frac{7.8 \times 10^{-2}\ atm \cdot 1.9 \times 10^4\ L}{0.082057 \frac{L \cdot atm}{K \cdot mol} \cdot 298\ K} = 6.0 \times 10^1\ \text{mol ethanol}$$

$$6.0 \times 10^1\ \text{mol ethanol} \cdot \frac{46.1\ \text{g ethanol}}{1\ \text{mol ethanol}} = 2.8 \times 10^3\ \text{g ethanol}$$

This mass of ethanol would occupy a volume of:

$$2.8 \times 10^3\ g\ C_2H_5OH \cdot \frac{1\ cm^3}{0.785\ g} = 3.5 \times 10^3\ cm^3$$

As only 1.0 L of C_2H_5OH (1.0×10^3 cm^3) was introduced into the room, **all the ethanol would evaporate**.

68. The **viscosity of ethylene glycol would be predicted to be greater** than that of ethanol since the glycol possesses two O-H groups per molecule while ethanol possesses one. Two OH groups/molecule would provide more hydrogen bonding!

70. For ammonia cooling:

$$q_{liquid} = 1.00 \text{ mol NH}_3 \cdot \frac{17.0 \text{ g NH}_3}{1.0 \text{ mol NH}_3} \cdot 4.70 \frac{\text{J}}{\text{g} \cdot \text{K}} \cdot (229.9 \text{ K} - 239.9 \text{ K}) = -800. \text{ J}$$

Note that the - sign indicates that heat is being lost by the ammonia to its surroundings.

For water cooling:

$$q_{liquid} = 1.00 \text{ mol H}_2\text{O} \cdot \frac{18.02 \text{ g H}_2\text{O}}{1.0 \text{ mol H}_2\text{O}} \cdot 4.184 \frac{\text{J}}{\text{g} \cdot \text{K}} \cdot (-10 \text{ K}) = -754 \text{ J}$$

Note that **ammonia releases more heat** than water. The larger specific heat capacity of liquid NH_3 compared to that of liquid H_2O would tell us that information without *doing the calculation.*

72. Using the vapor pressure curves:

a. The vapor pressure of ethanol at 60 °C is: 350 mm Hg (to the limits of this reader's ability to read the graph).

b. The stronger intermolecular forces in the liquid state are those of: Ethanol has a lower vapor pressure than carbon disulfide at every temperature—and hence stronger intermolecular forces. This is quite expected since ethanol has hydrogen bonding as the intermolecular force while CS_2 has only induced dipole forces.

c. The temperature at which heptane has a vapor pressure of 500 mm Hg is: 84 °C

d. The approximate normal boiling points of the three substances are:

Carbon disulfide	bp = 46°C (literature value 46.5 °C)
Ethanol	bp = 78°C (literature value 78.5 °C)
Heptane	bp = 99°C (literature value 98.4 °C)

e. At a pressure of 400 mm Hg and 70 °C, the state of the three substances is:

Carbon disulfide	state = gas
Ethanol	state = gas
Heptane	state = liquid

74. Tungsten in the unit cell

a. Type of unit cell: with one atom at the center, and atoms at each of the eight corners, this is a **body-centered-cubic** cell.

b. Number of tungsten atoms per unit cell: 2 (1-center + 8 (1/8)-for the corner atoms)

c. Radius of a tungsten atom:

According to Example 13.7, the

diagonal distance across the cube = $\sqrt{3} \cdot$ cell edge

Across the diagonal there are 2 radii and 1 diameters (4 radii)of atoms in contact so

$$4 \bullet \text{radii} = \sqrt{3} \bullet \text{cell edge or rearranging: radii} = \frac{\sqrt{3} \bullet \text{cell edge}}{4}$$

$$\text{radii} = \frac{1.732 \bullet 315.5\text{pm}}{4} = 136.6 \text{ pm or } 1.366 \times 10^{-8} \text{ cm}$$

Conceptual Questions

76. 1-propanol has the possibility of hydrogen bonding (with the -OH groups) between molecules. Methyl ethyl ether, on the other hand, has **no** polar -OH groups, eliminating the possibility of hydrogen bonding. The reduced intermolecular forces between molecules of methyl ethyl ether result in a **much lower** boiling point for the ether compared to the alcohol.

78. $CaCl_2$ cannot have the NaCl structure. As shown in Figure 13.24 of your text, the cubic structure possesses 4 net lattice ions (occupied by anions) per face-centered lattice and 4 octahedral holes (occupied by cations). This is suitable for salts of a 1:1 composition.

80. Assume that the anions (PO_4^{3-} in this case) assume the lattice positions in the face-centered cubic unit cell. This arrangement gives rise to 13 octahedral sites [1 in the center, and 12 that are on the edges of the unit cell (with 1/4 of each site in the unit cell)] . There are then 4 (1 + 12(1/4)) such sites with sodium ions. The anion sites (8 corners -with 1/8 in the unit cell and 6 faces--with 1/2 in the unit cell) total 4. So the stoichiometry for such a ionic solid would be MX **not M_3X.**

82. The can collapses as a result of the condensation of the gas in the can—which has filled the heated can—to a liquid. The distances between the particles of liquid are **much less** than the distances between the particles of gas. The resulting decrease in pressure inside the can causes the greater pressure outside the can to crush the can.

Challenging Questions

84. Calculate the number of atoms represented by the vapor:

The pressure of Hg is 0.00169 mm Hg $\bullet \frac{1\text{atm}}{760 \text{ mm Hg}} = 2.22 \times 10^{-6}$ atm

The number of moles/L is: $\frac{P}{RT} = \frac{n}{V} = \frac{2.22 \times 10^{-6} \text{ atm}}{(0.082057 \frac{L \cdot atm}{K \cdot mol})(297 \text{ K})} = 9.12 \times 10^{-8}$ mol/L

Converting this to atoms/m^3:

$$9.12 \times 10^{-8} \frac{mol}{L} \cdot \frac{1000 \text{ L}}{1 \text{ m}^3} \cdot \frac{6.022 \times 10^{23} \text{ atoms}}{1 \text{ mol}} = 5.5 \times 10^{19} \frac{\text{atoms}}{\text{m}^3}$$

Note that the information that the air was saturated with mercury vapor obviates the need to calculate the volume of the room.

86. For a simple cubic unit cell, each corner is occupied by an atom or ion. Each of these is contained within EIGHT unit cells contributing 1/8 to each. Within one unit cell, therefore, there is ($8 \times \frac{1}{8}$) 1 atom or ion. The volume occupied by that one net atom would be equal to 4/3 πr^3 with r representing the radius of the spherical atom or ion. The volume of the unit cell may be calculated by noting that the length of one side of the cell (an edge) corresponds to two radii (2r)—since the spheres touch. The volume of this cube is therefore $(2r)^3$. The empty space within the cell is therefore:

$(2r)^3 - 4/3\, \pi\, r^3$ and the fraction of space unoccupied is:

$$\frac{(8 - 4/3\ \pi\ )r^3}{8r^3} = \frac{8 - 4/3\ \pi}{8} = 0.476 \text{ or approximately } 48\ \%$$

Summary Question

88. a. The electron dot structure for SO_2:

(i) The OSO angle is approximately 120°.

(Slightly less due to the lone pair on S).

(ii) The electron-pair geometry is trigonal planar and the molecular geometry is bent.

b. The forces binding SO_2 molecules to each other are **dipole-dipole** forces since SO_2 molecules are polar.

c. Listed in order of increasing intermolecular forces:

$CH_4 < NH_3 < SO_2 < H_2O$

For H_2O hydrogen bonding is possible. SO_2 is a relatively heavy molecule that will exhibit dipole-dipole forces. The lighter NH_3 molecule will have the hydrogen bonding forces that would be absent in CH_4.

d. Enthalpy change for SO_2 (g) → SO_3 (g):

ΔH_{rxn} = [ΔH_f SO_3 (g)] - [ΔH_f SO_2 (g) + ΔH_f O_2 (g)]

= (- 395.72 kJ/mol) - (- 296.83 kJ/mol)

= - 98.89 kJ/mol

Enthalpy change for H_2SO_4 formation:

ΔH_{rxn} = [ΔH_f H_2SO_4 (aq)] - [ΔH_f SO_3 (g) + ΔH_f H_2O (ℓ)]

= - 909.27 kJ/mol - [- 395.72 kJ/mol + (- 285.83 kJ/mol)]

= - 227.72 kJ/mol

Chapter 14
Solutions and Their Behavior

NUMERICAL QUESTIONS

Concentration Units

10. For a solution containing 2.56 g of $C_4H_6O_5$ in 500.0 g of water, the molarity is:

$$2.56 \text{ g } C_4H_6O_5 \cdot \frac{1 \text{ mol } C_4H_6O_5}{134.1 \text{ g } C_4H_6O_5} \cdot \frac{1}{0.500 \text{ L}} = 0.0382 \text{ M}$$

The molality is:

$$2.56 \text{ g } C_4H_6O_5 \cdot \frac{1 \text{ mol } C_4H_6O_5}{134.1 \text{ g } C_4H_6O_5} \cdot \frac{1}{0.500 \text{ kg}} = 0.0382 \text{ molal}$$

The mole fraction of malic acid is:

$$500.0 \text{ g water} \cdot \frac{1 \text{ mol water}}{18.015 \text{ g water}} = 27.75 \text{ mol } H_2O$$

and the moles of malic acid calculated above are: 0.0191 mol $C_4H_6O_5$

$$\text{giving } X_{\text{malic acid}} = \frac{0.0191 \text{ mol } C_4H_6O_5}{0.0191 \text{ mol } C_4H_6O_5 + 27.75 \text{ mol } H_2O} = 6.87 \times 10^{-4}$$

and a X_{water} of (1 - 0.000687) or 0.9993.

The weight percent of malic acid is:

$$\frac{2.56 \text{ g } C_4H_6O_5}{2.56 \text{ g } C_4H_6O_5 + 500.0 \text{ g } H_2O} \cdot 100 = 0.509 \text{ \% malic acid}$$

12. Complete the following transformations for

NaI:

Weight percent:

$$\frac{0.15 \text{ mol NaI}}{1 \text{ kg solvent}} \cdot \frac{149.9 \text{ g NaI}}{1 \text{ mol NaI}} = \frac{22.5 \text{ g NaI}}{1 \text{ kg solvent}}$$

$$\frac{22.5 \text{ g NaI}}{1000 \text{ g solvent} + 22.5 \text{ g NaI}} \cdot 100 = 2.2 \text{ \% NaI}$$

Mole fraction:

1000 g H_2O = 55.51 mol H_2O

$$X_{NaI} = \frac{0.15 \text{ mol NaI}}{55.51 \text{ mol } H_2O + 0.15 \text{ mol NaI}} = 2.7 \times 10^{-3}$$

C_2H_5OH:

Molality:

$$\frac{5.0 \text{ g } C_2H_5OH}{100 \text{ g solution}} \bullet \frac{1 \text{ mol } C_2H_5OH}{46.07 \text{ g } C_2H_5OH} \bullet \frac{100 \text{ g solution}}{95 \text{ g solvent}} \bullet \frac{1000 \text{ g solvent}}{1 \text{ kg solvent}}$$

$$= 1.1 \text{ molal}$$

Mole fraction:

$$\frac{5.0 \text{ g } C_2H_5OH}{1} \bullet \frac{1 \text{ mol } C_2H_5OH}{46.07 \text{ g } C_2H_5OH} = 0.11 \text{ mol } C_2H_5OH$$

and for water : $\frac{95 \text{ g } H_2O}{1} \bullet \frac{1 \text{ mol } H_2O}{18.02 \text{ g } H_2O} = 5.27 \text{ mol } H_2O$

$$X_{C_2H_5OH} = \frac{0.11 \text{ mol } C_2H_5OH}{5.27 \text{ mol } H_2O + 0.11 \text{ mol } C_2H_5OH} = 0.020$$

$C_{12}H_{22}O_{11}$:

Weight percent:

$$\frac{0.15 \text{ mol } C_{12}H_{22}O_{11}}{1 \text{ kg solvent}} \bullet \frac{342.3 \text{ g } C_{12}H_{22}O_{11}}{1 \text{ mol } C_{12}H_{22}O_{11}} = \frac{51.3 \text{ g } C_{12}H_{22}O_{11}}{1 \text{ kg solvent}}$$

$$\frac{51.3 \text{ g } C_{12}H_{22}O_{11}}{1000 \text{ g } H_2O + 51.3 \text{ g } C_{12}H_{22}O_{11}} \times 100 = 4.9\ \%\ C_{12}H_{22}O_{11}$$

Mole fraction:

$$X_{C_{12}H_{22}O_{11}} = \frac{0.15 \text{ mol } C_{12}H_{22}O_{11}}{55.51 \text{ mol } H_2O + 0.15 \text{ mol } C_{12}H_{22}O_{11}} = 2.7 \times 10^{-3}$$

14. To prepare a solution that is 0.200 m Na_2CO_3:

$$\frac{0.200 \text{ mol } Na_2CO_3}{1 \text{ kg } H_2O} \bullet \frac{0.125 \text{ kg } H_2O}{1} \bullet \frac{106.0 \text{ g } Na_2CO_3}{1 \text{ mol } Na_2CO_3} = 2.65 \text{ g } Na_2CO_3$$

Note that the first two fractions above determine the # of mol of Na_2CO_3 to be 0.025 mol.

The mole fraction of Na_2CO_3 in the resulting solution:

$$\frac{125 \text{ g } H_2O}{1} \cdot \frac{1 \text{ mol } H_2O}{18.02 \text{ g } H_2O} = 6.94 \text{ mol } H_2O$$

$$X_{Na_2CO_3} = \frac{0.025 \text{ mol } Na_2CO_3}{0.025 \text{ mol } Na_2CO_3 + 6.94 \text{ mol } H_2O} = 3.59 \times 10^{-3}$$

16. To calculate the number of mol of $C_3H_5(OH)_3$:

$$0.093 = \frac{x \text{ mol } C_3H_5(OH)_3}{x \text{ mol } C_3H_5(OH)_3 + (425 \text{ g } H_2O \cdot \frac{1 \text{ mol } H_2O}{18.02 \text{ g } H_2O})}$$

$$0.093 = \frac{x \text{ mol } C_3H_5(OH)_3}{x \text{ mol } C_3H_5(OH)_3 + 23.58 \text{ mol } H_2O}$$

$0.093(x + 23.58) = x$ and solving for x we get 2.4 mol $C_3H_5(OH)_3$

Grams of glycerol needed: 2.4 mol $C_3H_5(OH)_3 \cdot \frac{92.1 \text{ g}}{1 \text{ mol}} = 220$ g $C_3H_5(OH)_3$

The molality of the solution is $\frac{2.4 \text{ mol } C_3H_5(OH)_3}{0.425 \text{ kg } H_2O} = 5.7$ m

18. For the compound K_2CO_3:

$$0.0125 \text{ molal solution} = \frac{x \text{ mol } K_2CO_3}{0.125 \text{ kg } H_2O} \text{ and}$$

$x = 1.56 \times 10^{-3}$ mol K_2CO_3 or

$$1.56 \times 10^{-3} \text{ mol } K_2CO_3 \cdot \frac{138.2 \text{ g } K_2CO_3}{1 \text{ mol}} = 0.216 \text{ g } K_2CO_3$$

Since 125. g H_2O corresponds to 6.94 mol H_2O,

$$\text{the } X_{K_2CO_3} = \frac{1.56 \times 10^{-3} \text{ mol } K_2CO_3}{1.56 \times 10^{-3} \text{ mol } K_2CO_3 + 6.94 \text{ mol } H_2O} = 2.25 \times 10^{-4}$$

For the compound C_2H_5OH:

13.5 g C_2H_5OH = 0.293 mol and 175. g H_2O = 9.71 mol

The molality of the solution is: $\frac{0.293 \text{ mol}}{0.175 \text{ kg}} = 1.67$ molal

The $X_{ethanol}$ of the solution is: $\frac{0.293}{0.293 + 9.71} = 0.0293$

For the compound $NaNO_3$:

The $X_{water} = 1 - 0.0934 = 0.9066$

and 555 g water = $555 \text{ g } H_2O \cdot \frac{1 \text{ mol } H_2O}{18.02 \text{ g } H_2O} = 30.8 \text{ mol } H_2O$

We can write: $\frac{\text{mol } H_2O}{\text{mol } H_2O + \text{mol } NaNO_3} = 0.9066$ and substituting

$\frac{30.8}{30.8 + x} = 0.9066$ so x = 3.17 and 3.17 mol of $NaNO_3$ = 270. g $NaNO_3$

For the molality:

$$\frac{3.17 \text{ mol of } NaNO_3}{0.555 \text{ kg } H_2O} = 5.72 \text{ molal}$$

20. The molality of a 95.0 % solution of H_2SO_4:

We need the # of moles of both solvent (water) and solute (sulfuric acid). A 95% solution of the acid contains 5 % of water.

Moles of solute : $95.0 \text{ g } H_2SO_4 \cdot \frac{1 \text{ mol } H_2SO_4}{98.1 \text{ g } H_2SO_4} = 0.968 \text{ mol } H_2SO_4$

Mass of solvent: $5.0 \text{ g } H_2O \cdot \frac{1 \text{ kg } H_2O}{1000 \text{ g } H_2O} = 0.0050 \text{ kg } H_2O$

Molality of solution = $\frac{0.968 \text{ mol } H_2SO_4}{0.0050 \text{ kg } H_2O} = 2.00 \times 10^2 \text{ m}$

The molarity of the H_2SO_4:

We need to calculate the volume of the solution.

So 100. g of solution $\cdot \frac{1000 \text{ mL}}{1840 \text{ g solution}} \cdot \frac{1\text{L}}{1000 \text{ mL}} = 0.0543 \text{ L solution}$

Molarity $= \frac{0.968 \text{ mol } H_2SO_4}{0.0543 \text{ L solution}} = 17.8 \text{ M}$

22. a. Mole fraction of NaOH:

$$\frac{10.7 \text{ mol NaOH}}{1 \text{ kg solvent}} \cdot \frac{1\text{kg solvent}}{1000 \text{ g solvent}} \cdot \frac{18.02 \text{ g solvent}}{1 \text{ mol solvent}} = \frac{10.7\text{mol NaOH}}{55.5 \text{ mol } H_2O}$$

$$X_{NaOH} = \frac{10.7 \text{ mol NaOH}}{55.5 \text{ mol } H_2O + 10.7 \text{ mol NaOH}} = 0.162$$

b. Weight percentage of NaOH:

$$\frac{10.7 \text{ mol NaOH}}{1000 \text{ g solvent}} \cdot \frac{40.0 \text{ g NaOH}}{1 \text{ mol NaOH}} = \frac{428 \text{ g NaOH}}{1000 \text{ g solvent}}$$

The mass of solution would be (428 g + 1000. g) 1428 g.

$$\frac{428 \text{ g NaOH}}{1428 \text{ g solution}} \cdot 100 = 30.0\% \text{ NaOH}$$

c. Molarity of the solution:

$$\frac{10.7 \text{ mol NaOH}}{1428 \text{ g solution}} \cdot \frac{1.33 \text{ g NaOH}}{1 \text{ cm}^3 \text{ solution}} \cdot \frac{1000 \text{ cm}^3}{1 \text{ L solution}} = 9.97 \text{ M NaOH}$$

24. The molality of the $Ca(NO_3)_2$ solution:

$$\frac{2.00 \text{ g } Ca(NO_3)_2}{0.75 \text{ kg solvent}} \cdot \frac{1 \text{ mol } Ca(NO_3)_2}{164.1 \text{ g } Ca(NO_3)_2} = 0.016 \text{ molal } Ca(NO_3)_2$$

One mol $Ca(NO_3)_2$ provides 3 mol of ions (1 Ca^{2+} and 2 NO_3).

The total molality would be (3 • 0.016) or 0.048 molal.

26. The concentration of ppm expressed in grams is:

$$0.18 \text{ ppm} = \frac{0.18 \text{ g solute}}{1.0 \times 10^6 \text{ g solvent}} = \frac{0.18 \text{ g solute}}{1.0 \times 10^3 \text{ kg solvent}} \text{ or } \frac{0.00018 \text{ g solute}}{1 \text{ kg water}}$$

$$\frac{0.00018 \text{ g } Li^+}{1 \text{ kg water}} \cdot \frac{1 \text{ mol } Li^+}{6.939 \text{ g } Li^+} = 2.6 \times 10^{-5} \text{ molal } Li^+$$

The Solution Process

28. Pairs of liquids that will be miscible:

a. $H_2O/CH_3CH_2CH_2CH_3$

Will **not** be miscible. Water is a polar substance, while butane is nonpolar.

b. C_6H_6/CCl_4

Will **be** miscible. Both liquids are nonpolar and are expected to be miscible.

c. H_2O/CH_3CO_2H

Will **be** miscible. Both substances can hydrogen bond, and we know that they mix—since a 5% aqueous solution of acetic acid is sold as "vinegar"

30. The enthalpy of solution for LiCl:

The process can be represented as LiCl (s) → LiCl (aq)

The $\Delta H_{reaction} = \Sigma\Delta H_f$ (product) - $\Sigma\Delta H_f$ (reactant)

$= (-445.6 \text{ kJ/mol})(1 \text{ mol}) - (-408.6 \text{ kJ/mol})(1 \text{ mol}) = -37.0 \text{ kJ}$

The similar calculation for NaCl is + 3.9 kJ. Note that the enthalpy of solution for NaCl is **endothermic** while that for LiCl is **exothermic.**

32. Raising the temperature of the solution will increase the solubility of NaCl in water. Hence to increase the amount of dissolved NaCl in solution one must **(c) raise the temperature of the solution and add some NaCl.**

Henry's Law

34. Molarity of $O_2 = k \cdot P_{O_2}$

$$= (1.66 \times 10^{-6} \frac{M}{\text{mm Hg}}) \cdot 40 \text{ mm Hg} = 7 \times 10^{-5} \text{ M}$$

$$\text{and } 7 \times 10^{-5} \frac{\text{mol}}{\text{L}} \cdot \frac{32.0 \text{ g } O_2}{1 \text{ mol } O_2} = 2.1 \times 10^{-3} \frac{\text{g } O_2}{\text{L}} \quad \text{(about 2 mg } O_2\text{)}$$

36. Solubility $= k \cdot P_{CO_2}$

$$0.0506 \text{ M} = (4.48 \times 10^{-5} \frac{M}{\text{mm Hg}}) \cdot P_{CO_2}$$

$$1130 \text{ mm Hg} = P_{CO_2} \text{ or } 1130 \text{ mm Hg} \cdot \frac{1 \text{ atm}}{760 \text{ mm Hg}} = 1.49 \text{ atm}$$

Vapor Pressure Changes

38. The vapor pressure can be calculated with Raoult's Law if we know the mole fraction of solvent.

$$\text{mol of urea} = 9.00 \text{ g } CO(NH_2)_2 \cdot \frac{1 \text{ mol } CO(NH_2)_2}{60.06 \text{ g } CO(NH_2)_2} = 0.150 \text{ mol } CO(NH_2)_2$$

$$\text{mol of water} = 10.0 \text{ g } H_2O \cdot \frac{1 \text{ mol } H_2O}{18.02 \text{ g } H_2O} = 0.555 \text{ mol } H_2O$$

$$X_{water} = \frac{0.555 \text{ mol } H_2O}{0.555 \text{ mol } H_2O + 0.150 \text{ mol } CO(NH_2)_2} = 0.787$$

The vapor pressure of water at 24 °C = 22.4 mm Hg so the vapor pressure of the solution is: $P^\circ_{water} \cdot X_{water} = 22.4 \text{ mm Hg} \cdot 0.787 = 17.6 \text{ mm Hg}$

40. Using Raoult's Law, we know that the vapor pressure of pure water (P°) multiplied by the mole fraction(X) of the solvent gives the vapor pressure of the solvent above the solution (P).

$$P_{water} = X_{water} P^\circ_{water}$$

The vapor pressure of pure water at 90 °C is 525.8 mm Hg (from Appendix G). Since the P_{water} is given as 457 mm Hg, the mole fraction of the water is:

$$\frac{457 \text{ mm Hg}}{525.8 \text{ mm Hg}} = 0.869$$

The 2.00 kg of water correspond to a mf of 0.869. This mass of water corresponds to:

$$2.00 \times 10^3 \text{ g } H_2O \cdot \frac{1 \text{ mol } H_2O}{18.02 \text{ g } H_2O} = 111 \text{ mol water.}$$

Representing moles of ethylene glycol as x we can write:

$$X_{H_2O} = \frac{\text{mol } H_2O}{\text{mol } H_2O + \text{mol } C_2H_4(OH)_2} = \frac{111}{111 + x} = 0.870$$

$$\frac{111}{0.870} = 111 + x \ ; \ 16.7 = x \text{ (mol of ethylene glycol)}$$

$$16.7 \text{ mol } C_2H_4(OH)_2 \cdot \frac{62.07 \text{ g } C_2H_4(OH)_2}{1 \text{ mol } C_2H_4(OH)_2} = 1.04 \times 10^3 \text{ g } C_2H_4(OH)_2$$

Boiling Point Elevation

42. From Table 14.3 we see that benzene normally boils at a temperature of 80.10 °C. If the solution boils at a temperature of 84.2 °C, the change in temperature is (84.2 - 80.10 °C) or 4.1 °C. Let's calculate the Δt, using the equation $\Delta t = K_{bp} \cdot m_{solute}$:

The molality of the solution is $\frac{0.200 \text{ mol}}{0.125 \text{ kg solvent}}$ or 1.60 m

The K_{bp} for benzene (from Table 14.3) is +2.53 °C/m

So $\Delta t = K_{bp} \cdot m_{solute} = +2.53 \text{ °C/m} \cdot 1.60 \text{ m} = +4.1 \text{ °C}$.

44. Calculate the molality of $C_{12}H_{10}$ in the solution.

$$0.515 \text{ g } C_{12}H_{10} \cdot \frac{1 \text{ mol } C_{12}H_{10}}{154.2 \text{ g } C_{12}H_{10}} = 3.34 \times 10^{-3} \text{ mol } C_{12}H_{10}$$

and the molality is : $\dfrac{3.34 \times 10^{-3} \text{ mol acenaphthalene}}{0.0150 \text{ kg } CHCl_3} = 0.223 \text{ molal}$

the boiling point elevation is:

$$\Delta t = m \cdot K_{bp} = 0.223 \text{ molal} \cdot \frac{+3.63 \text{ °C}}{\text{molal}} = 0.808 \text{ °C}$$

and the boiling point will be 61.70 + 0.808 = 62.51 °C

46. The change in the temperature of the boiling point is (80.51 - 80.10)°C or 0.41 °C.
Using the equation $\Delta t = m \cdot K_{bp}$; 0.41 °C = m • +2.53 °C/m, and the molality is:

$$\frac{0.41 \text{ °C}}{+2.53 \text{ °C/m}} = m = 0.16 \text{ molal}$$

The solution contains 50.0 g of solvent (or 0.0500 kg solvent). We can calculate the # of moles of phenanthrene:

$0.16 \text{ molal} = \dfrac{x \text{ mol } C_{14}H_{10}}{0.0500 \text{ kg}}$ or 8.1×10^{-3} mol $C_{14}H_{10}$, and since 1 mol of $C_{14}H_{10}$ has a mass of 178 g,

$$8.1 \times 10^{-3} \text{ mol } C_{14}H_{10} \cdot \frac{178 \text{ g } C_{14}H_{10}}{1 \text{ mol } C_{14}H_{10}} = 1.4 \text{ g } C_{14}H_{10}$$

48. The **solution with the highest boiling point** will have the **greatest number of particles** in solution.

If we assume total dissociation of the solutes given the molality of particles for the solution will be:

$0.10 \text{ m KCl} \rightarrow 0.10 \text{ m } K^{+} + 0.10 \text{ m } Cl^{-} = 0.20 \text{ m}$

0.10 m sugar $\rightarrow$ (covalently bonded specie) $= 0.10$ m

$0.080 \text{ m } MgCl_2 \rightarrow 0.080 \text{ m } Mg^{2+} + 0.16 \text{ m } Cl^{-} = 0.24 \text{ m}$

In order of increasing boiling point: 0.10 m sugar < 0.10 m KCl < 0.080 m $MgCl_2$

50. The change in the temperature of the boiling point is (80.26 - 80.10)°C or 0.16 °C.

Using the equation $\Delta t = m \bullet K_{bp}$; 0.16 °C = m • +2.53 °C/m, and the molality is:

$$\frac{0.16\ °C}{+2.53\ °C/m} = m = 0.063 \text{ molal}$$

The solution contains 11.12 g of solvent (or 0.01112 kg solvent). We can calculate the # of moles of the orange compound:

$$0.063 \text{ molal} = \frac{x \text{ mol compound}}{0.01112 \text{ kg solvent}} \text{ or } 7.0 \times 10^{-4} \text{ mol compound.}$$

This number of moles of compound has a mass of 0.255 g, so 1 mol of compound is:

$$\frac{0.255 \text{ g compound}}{7.0 \times 10^{-4} \text{ mol}} = 360 \text{ g/mol.}$$

The empirical formula, $C_{10}H_8Fe$, has a mass of 184 g, so the # of "empirical formula units" in one molecular formula is : $\frac{360 \text{ g/mol}}{184 \text{ g/empirical formula}} = 2$ mol/empirical formulas or a **molecular formula of $C_{20}H_{16}Fe_2$.**

52. The change in the temperature of the boiling point is (61.82 - 61.70)°C or 0.12 °C.

Using the equation $\Delta t = m \bullet K_{bp}$; 0.12 °C = m • +3.63 °C/m, and the molality is:

$$\frac{0.12\ °C}{+3.63\ °C/m} = m = 0.033 \text{ molal}$$

The solution contains 25.0 g of solvent (or 0.0250 kg solvent). We can calculate the # of moles of benzyl acetate:

$$0.033 \text{ molal} = \frac{x \text{ mol compound}}{0.0250 \text{ kg solvent}} \text{ or } 8.3 \times 10^{-4} \text{ mol compound.}$$

This number of moles of benzyl acetate has a mass of 0.125 g, so 1 mol of benzyl acetate is: $\frac{0.125 \text{ g compound}}{8.3 \times 10^{-4} \text{ mol}} = 150$ g/mol. (2 sf)

Freezing Point Depression

54. The solution freezes 16.0 °C lower than pure water.

a. We can calculate the molality of the ethanol:

$$\Delta t = m \bullet K_{fp}$$

$-16.0\ °C = m\ (-1.86\ °C/molal)$

$8.60 =$ molality of the alcohol

b. If the molality is 8.60 then there are 8.60 moles of C_2H_5OH ($8.60 \cdot 46.07$ g/mol = 396 g) in 1000 g of H_2O.

The weight percent of alcohol is $\frac{396\ g}{1396\ g} \cdot 100 = 28.5\ \%$ ethanol

56. The number of moles of LiF is : $52.5\ g\ LiF \cdot \frac{1\ mol\ LiF}{25.94\ g\ LiF} = 2.02\ mol\ LiF$

So $\Delta t_{fp} = \frac{2.02\ mol\ LiF}{0.306\ kg\ H_2O} \cdot -1.86\ °C/molal \cdot 2 = -24.6\ °C$

The anticipated freezing point is then 24.6 °C lower than pure water (0.0°C) or -24.6 °C

58. Determine the molality of the solution

$-0.040\ °C = m \cdot -1.86\ °C/molal = 0.215$ molal (or 0.22 to 2 sf)

$$\text{and } 0.22\ \text{molal} = \frac{\frac{0.180\ g\ solute}{MM}}{0.0500\ kg\ water}$$

$MM = 167$ or 170 (2 sf)

60. The change in the temperature of the freezing point is (69.40 - 70.03)°C or -0.63 °C.

Using the equation $\Delta t = m \cdot K_{fp}$; $-0.63\ °C = m \cdot -8.00\ °C/m$, and the molality is:

$\frac{-0.63\ °C}{-8.00\ °C/m} = m = 0.079$ molal (2 sf)

The solution contains 10.0 g of biphenyl (or 0.0100 kg solvent). We can calculate the # of moles of naphthalene:

$0.079\ molal = \frac{x\ mol\ naphthalene}{0.0100\ kg\ solvent}$ or 7.9×10^{-4} mol compound.

This number of moles of naphthalene has a mass of 0.100 g, so 1 mol of naphthalene is:

$\frac{0.100\ g\ naphthalene}{7.9 \times 10^{-4}\ mol} = 130$ g/mol (2 sf)

62. Solutions given in order of increasing melting point:

The solution with the greatest **number** of particles will have the lowest melting point.

The total molality of solutions is then:

a. 0.1 m sugar	• 1 particle/formula unit =	0.1 m	
b. 0.1 m NaCl	• 2 particles/formula unit =	0.2 m	[Na^+, Cl^-]
c. 0.08 m $CaCl_2$	• 3 particles/formula unit =	0.24 m	[Ca^{2+}, 2 Cl^-]
d. 0.04 m Na_2SO_4	• 3 particles/formula unit =	0.12 m	[2 Na^+, SO_4^{2-}]

The melting points would increase in the order: $CaCl_2$ < NaCl < Na_2SO_4 < sugar

Osmosis

64. Assume we have 100 g of this solution, the number of moles of phenylalanine is

$$3.00 \text{ g phenylalanine} \cdot \frac{1 \text{ mol phenylalanine}}{165.2 \text{ g phenylalanine}} = 0.0182 \text{ mol phenylalanine}$$

The molality of the solution is $\frac{0.0182 \text{ mol phenylalanine}}{0.09700 \text{ kg water}} = 0.187 \text{ molal}$

a. The freezing point :

$$\Delta t = 0.187 \text{ molal} \cdot -1.86 \text{ °C/molal} = -0.348 \text{ °C}$$

The new freezing point is 0.0 - 0.348 °C = -0.348 °C.

b. The boiling point of the solution

$$\Delta t = m K_{bp} = 0.187 \text{ molal} \cdot + 0.5121\text{°C/molal} = +0.0959 \text{ °C}$$

The new boiling point is then 100.00 + 0.0959 = +100.10 °C

c. The osmotic pressure of the solution:

If we assume that the **Molarity** of the solution is equal to the **molality**, then the osmotic pressure should be

$$\Pi = (0.187 \text{ mol/L})(0.0821 \frac{\text{L} \cdot \text{atm}}{\text{K} \cdot \text{mol}})(298 \text{ K}) = 4.58 \text{ atm}$$

The freezing point will be most easily measured.

66. $3.1 \text{ mm Hg} \cdot \frac{1 \text{ atm}}{760 \text{ mm Hg}} = (\text{M})(0.08205 \frac{\text{L} \cdot \text{atm}}{\text{K} \cdot \text{mol}})(298 \text{ K})$

1.67×10^{-4} = Molarity or 1.7×10^{-4} (2 sf)

$$1.7 \times 10^{-4} \frac{\text{mol bovine insulin}}{\text{L}} = \frac{\frac{1.00 \text{ g bovine insulin}}{\text{MM}}}{1 \text{ L}}$$

$$\text{Solving for MM} = 6.0 \times 10^3 \text{ g/mol}$$

Colloids

68. a. $BaCl_2(aq) + Na_2SO_4(aq) \rightarrow BaSO_4(s) + 2\ NaCl(aq)$

b. The $BaSO_4$ formed is of a colloidal size — not large enough to precipitate fully.

c. The particles of $BaSO_4$ grow with time, owing to a gradual loss of charge and become large enough to have gravity affect them —and settle to the bottom.

General Questions

70. a. The increased boiling point is calculated : $\Delta t = m \cdot K_{bp} \cdot i$
Since the solvent (and therefore K_{bp}) is constant for both solutions, we should look at the **molality** and **van't Hoff factors.** Sugar will remain as a molecular entity (i = 1), but Na_2SO_4 will dissociate into ions (2 Na^+ and 1 SO_4^{2-} : i = 3). The product, m • i, will be (0.10)(3) for Na_2SO_4 and (0.15)(1) for sugar. So the **Na_2SO_4 solution will have the higher boiling point.**

b. Lowering of vapor pressure is proportional to the mole fraction (the relative number of particles) of solute. The relationship can be written.

$$P_{water} = X_{water} P^{\circ}_{water}$$

This means that the vapor pressure of the solution (P_{water}) will be higher if X_{water} (the mf of water) is higher. Note that the mf of NH_4NO_3 (0.30 molal) will be greater than the mf of Na_2SO_4 (0.15 molal). This means that X_{water} will be greater for the Na_2SO_4 solution. **The vapor pressure of water will be higher for the 0.15 molal Na_2SO_4 solution.**

72. For DMG, $(CH_3CNOH)_2$, the MM is 116.1 g/mol

$$\text{So 53.0 g is : } 53.0\text{g} \cdot \frac{1 \text{ mol DMG}}{116.1 \text{ g DMG}} = 0.456 \text{ mol DMG}$$

$$525 \text{ g of } C_2H_5OH \text{ is : } 525 \text{ g} \cdot \frac{1 \text{ mol } C_2H_5OH}{46.07 \text{ g } C_2H_5OH} = 11.4 \text{ mol } C_2H_5OH$$

a. the X_{DMG}: $\frac{0.456 \text{ mol}}{(11.4 + 0.456) \text{ mol}} = 0.0385$

b. The molality of the solution: $\frac{0.456 \text{ mol DMG}}{0.525 \text{ kg}} = 0.869$ molal DMG

c. $P_{alcohol} = P°_{alcohol} \cdot X_{alcohol}$

$= (760. \text{ mm Hg})(1 - 0.0385) = 730.7$ mm Hg (731 to 3 sf)

d. The boiling point of the solution:

$\Delta t = m \cdot K_{bp} \cdot i = (0.869)(+1.22\ °C/\text{molal})(1)$

$= 1.06\ °C$

The new boiling point is 78.4 °C + 1.06 °C = 79.46 °C or 79.5 °C

74. Consider the aqueous solutions

(i) 0.20 m $HOCH_2CH_2OH$ (iii) 0.12 m KBr

(ii) 0.10 m $CaCl_2$ (iv) 0.12 m Na_2SO_4

(a) The solution with the highest boiling point and (b) that with the lowest freezing point would be that with the **highest particle concentration**. From the Δt equation, ($\Delta t = m \cdot K_{bp} \cdot i$), the product ($m \cdot i$) would give the particle concentration.

(c) The solution with the highest water vapor pressure would have the **lowest particle concentration**.

solution concentration	particle concentration
(i) 0.20 m $HOCH_2CH_2OH$	0.20 • 1
(ii) 0.10 m $CaCl_2$	0.10 • 3
(iii) 0.12 m KBr	0.12 • 2
(iv) 0.12 m Na_2SO_4	0.12 • 3

Solution (iv) has the highest boiling and lowest freezing point while solution (i) has the lowest particle concentration.

76. The change in temperature of the freezing point is: $\Delta t = m \cdot K_{fp} \cdot i$

Calculate the molality:

$35.0 \text{ g } CaCl_2 \cdot \frac{1 \text{ mol } CaCl_2}{111.0 \text{ g } CaCl_2} = 0.315 \text{ mol } CaCl_2$ in 0.150 kg water.

$m = \frac{0.315 \text{ mol } CaCl_2}{0.150 \text{ kg}} = 2.10 \text{ molal } CaCl_2$

$\Delta t = m \cdot K_{fp} \cdot i = (2.10 \text{ molal} \cdot -1.86\ °C/\text{molal} \cdot 2.7) = -10.6\ °C.$

The freezing point of the solution is 0.0 °C - 10.6 °C = -11 °C (2 sf)

78. The change in the temperature of the boiling point is 61.82 °C - 61.70 °C = 0.12 °C

$$\Delta t = m \cdot K_{bp} \text{ so } m = \frac{\Delta t}{K_{bp}} = \frac{0.12\ °C}{3.63\ °C/m} = 0.033\ m$$

The number of moles of compound can be calculated:

$$0.033\ m = \frac{x \text{ mol compound}}{0.0250 \text{ kg}}\ ;\ x = 8.3 \times 10^{-4} \text{ mol compound}$$

This # of moles of compound corresponds to 0.135 g of compound, so the molar mass is:

$$\frac{0.135 \text{ g compound}}{8.3 \times 10^{-4} \text{ mol compound}} = 160 \text{ g/mol}$$

The empirical formula, C_5H_6O, has a formula mass of approximately 82, so there are **2 empirical formula units in a molecule** — with the formula of $C_{10}H_{12}O_2$.

80. At 80 °C, 1092 g of ammonium formate will dissolve in 200 g of water (546 g/100 g water). At 0 °C, only 204 g of ammonium formate will dissolve in this mass of water (102 g/100 g water), so (1092 - 204 g) **888 g of ammonium formate** will precipitate.

82. The vapor pressure of water at 18 °C is 15.5 torr. So the desired humidity is 55 % of 15.5 torr or 8.53 torr.

$$8.53 \text{ torr} = 15.5 \text{ torr} \cdot X_{H_2O}$$

$$0.55 = X_{H_2O} \qquad \text{and } 0.45 = X_{glycerol}$$

Assume that there is a total of 1.00 mol of water plus glycerol. Then there would be 0.55 mol of water and 0.45 mol of glycerol.

The masses of this number of moles would be:

$$0.55 \text{ mol } H_2O \cdot \frac{18.0 \text{ g } H_2O}{1 \text{ mol } H_2O} = 9.9 \text{ g } H_2O$$

$$0.45 \text{ mol glycerol} \cdot \frac{92.1 \text{ g glycerol}}{1 \text{ mol glycerol}} = 41 \text{ g glycerol} \qquad (2 \text{ sf})$$

The % of glycerol is then: $\frac{41 \text{ g glycerol}}{50.9 \text{ g total}} = 81\ \%$ (2 sf)

Conceptual Questions

84. a. The acetic acid reacts with the principal component of the egg shell—a carbonate—producing carbon dioxide gas and water. The membrane does not react with the acid, remaining intact.

b. The egg—minus its shell—when placed in water has a higher solute concentration **inside the egg membrane**. The resulting osmotic pressure is an attempt to reduce the solute concentration—and the net transport of water from **outside to inside** the membrane results in a swelling of the egg.

c. When the egg is placed in a solution with a high solute concentration, the situation is the reverse of that in (b). The higher solute concentration is **outside the membrane** and the net transport of water from **inside the membrane to outside the membrane occurs**—with the concomitant shriveling of the egg.

86. The apparent molecular weight of acetic acid in benzene, determined by the depression of benzene's freezing point.

$$\Delta t = m \cdot K_{fp} \cdot i$$

$$(3.37\ ^\circ\text{C} - 5.50\ ^\circ\text{C}) = m(-5.12\ ^\circ\text{C/molal})\ i$$

$$\frac{-2.13\ ^\circ\text{C}}{-5.12\ ^\circ\text{C/molal}} = m \cdot i$$

$0.416 \text{ molal} = m \cdot i$ (momentarily assume $i = 1$)

and the apparent molecular weight is:

$$0.416 \text{ molal} = \frac{\frac{5.00 \text{ g acetic acid}}{\text{MM}}}{0.100 \text{ kg}}$$

$$120 \text{ g/mol} = \text{MM}$$

The apparent molecular weight of acetic acid in water

$$\Delta t = m \cdot K_{fp} \cdot i$$

$$(-1.49\ ^\circ\text{C} - 0.00\ ^\circ\text{C}) = m(-1.86\ ^\circ\text{C/molal})\ i$$

$$\frac{-1.49\ ^\circ\text{C}}{-1.86\ ^\circ\text{C/molal}} = m \cdot i$$

$0.801 \text{ molal} = m \cdot i$ (once again, momentarily $i = 1$)

and the apparent molecular weight is:

$$0.801 \text{ molal} = \frac{\frac{5.00 \text{ g acetic acid}}{\text{MM}}}{0.100 \text{ kg}}$$

$$62.4 \text{ g/mol} = \text{MM}$$

The accepted value for acetic acid's molecular weight is approximately 60 g/mol. Hence the value for i isn't much larger than 1, indicating that the degree of dissociation of acetic acid molecules in water is not great—a finding consistent with the designation of acetic acid as a weak acid. The apparently doubled molecular weight of acetic acid in benzene indicates that the acid must exist primarily as a dimer.

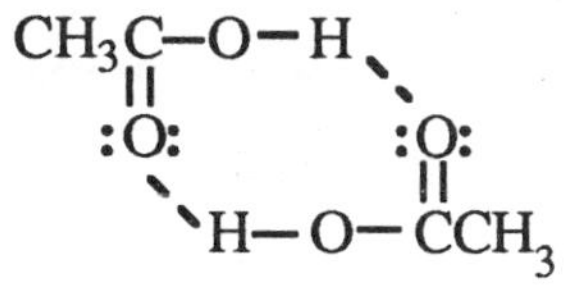

```
CH3C—O—H
   ||      \
  :O:       :O:
     \       ||
      H—O—CCH3
```

88. Note that all 3 alcohols—methanol, ethanol, and octanol—contain a polar "end"—the OH, and a non-polar "end"—the carbon chain. As the carbon chain lengthens, the non-polar carbon chain reduces the solubility of the alcohol in water.

90. Relationship between lattice energies and solubilities:

The lattice energy is defined as the energy associated with the formation of a solid ion pair from the gaseous cation and gaseous anion. Solution is a result of two processes:

1. The solid lattice separating into the gaseous ions (the **negative** of the lattice energy) and
2. The hydration of the gaseous ions into the aqueous solution.

In general the smaller the lattice energy the greater the solubility! It is noted that solubility is favored when the lattice energy is *smaller than or roughly equal to* the energy released upon hydration of the ions ($\Delta H_{hydration}$)

Challenging Problems

92. The enthalpies of solution for Li_2SO_4 and K_2SO_4:

The process is MX (s) $\rightarrow$ MX (aq)

Using the data for **Li_2SO_4**:

$\Delta H_{solution} = \Delta H_f\text{ (aq)} - \Delta H_f\text{ (s)} = (-1464.4\text{ kJ/mol}) - (-1436.4\text{ kJ/mol}) = -28.0\text{ kJ/mol}$

Using the data for **K_2SO_4**:

$\Delta H_{solution} = \Delta H_f\text{ (aq)} - \Delta H_f\text{ (s)} = (-1414.0\text{ kJ/mol}) - (-1437.7\text{ kJ/mol}) = 23.7\text{ kJ/mol}$

Note that for Li_2SO_4 *the process is* ***exothermic*** *while for* K_2SO_4 *the process is endothermic.*

Similar data for LiCl and KCl:

For LiCl: $\Delta H_{f\,(aq)} - \Delta H_{f\,(s)}$ = (-445.6 kJ/mol) - (-408.6 kJ/mol) = - 37.0 kJ/mol and

for KCl: $\Delta H_{f\,(aq)} - \Delta H_{f\,(s)}$ = (-419.5 kJ/mol) - (-436.7 kJ/mol) = 17.2 kJ/mol

Note the similarities of the chloride salts, with the lithium salt being ***exothermic*** *while the potassium salt is* ***endothermic****.*

94. Graham's Law says that the pressure of a mixture of gases (benzene and toluene) is the sum of the partial pressures. So, using Raoult's Law $P_{benzene} = X_{benzene} \cdot P^{\circ}_{benzene}$ and similarly for toluene. The total pressure is then:

$$P_{total} = P_{benzene} + P_{toluene}$$

$$= (\frac{2 \text{ mol benzene}}{3 \text{ mol}} \cdot 75 \text{ mm Hg}) + (\frac{1 \text{ mol toluene}}{3 \text{ mol}} \cdot 22 \text{ mm Hg}) = 57\text{mm Hg}$$

What is the mole fraction of each component in the liquid and in the vapor?

The **mf of the components in the liquid** are: $X_{benzene} = 2/3$ and $X_{toluene} = 1/3$

The **mf of the components in the vapor** are proportional to their pressures in the vapor state.

The $X_{benzene}$ is then: $\frac{50 \text{ mm Hg}}{57 \text{ mm Hg}} = 0.87$; $X_{toluene} = (1\text{-}0.87)$ or 0.13

96. A 2.0 % aqueous solution of novocainium chloride(NC) is also 98.0 % in water. Assume that we begin with 100 g of solution. The molality of the solution is then:

$$\frac{2.0 \text{ g} \cdot \frac{1\text{mol NC}}{272.8 \text{ g NC}}}{0.0980 \text{ kg water}} = 0.075 \text{ m}$$

Using the "delta T" equation: $\Delta t = m \cdot K_{fp} \cdot i$, we can solve for i: $\frac{\Delta t}{m \cdot K_{fp}} = i$

$$i = \frac{-0.237 \text{ °C}}{0.075\text{m} \cdot -1.86 \text{ °C/m}} = 1.7$$

So approximately **2 moles of ions are present per mole of compound.**

98. a. We can calculate the freezing point of sea water if we calculate the molality of the solution. Let's imagine that we have 1,000,000 (or 10^6) g of sea water. The amounts of the ions are then equal to the concentration (in ppm). Calculating their concentrations we get:

Cl^- 1.95×10^4 g Cl^- $\bullet \dfrac{1 \text{mol } Cl^-}{35.45 \text{ g } Cl^-}$ = 550. mol Cl^-

Na^+ 1.08×10^4 g Cl^- $\bullet \dfrac{1 \text{mol } Na^+}{22.99 \text{ g } Na^+}$ = 470. mol Na^+

Mg^{+2} 1.29×10^3 g Mg^{+2} $\bullet \dfrac{1 \text{mol } Mg^{+2}}{24.31 \text{ g } Mg^{+2}}$ = 53.1 mol Mg^{+2}

SO_4^{-2} 9.05×10^2 g SO_4^{-2} $\bullet \dfrac{1 \text{mol } SO_4^{-2}}{96.06 \text{ g } SO_4^{-2}}$ = 9.42 mol SO_4^{-2}

Ca^{+2} 4.12×10^2 g Ca^{+2} $\bullet \dfrac{1 \text{mol } Ca^{+2}}{40.08 \text{ g } Ca^{+2}}$ = 10.3 mol Ca^{+2}

K^+ 3.80×10^2 g K^+ $\bullet \dfrac{1 \text{mol } K^+}{39.10 \text{ g } K^+}$ = 9.72 mol K^+

Br^- 67 g Br^- $\bullet \dfrac{1 \text{mol } Br^-}{79.90 \text{ g } Br^-}$ = 0.84 mol Br^-

For a total of: 1103 mol ions

The concentration **per gram is then:** $\dfrac{1103 \text{ mol ions}}{10^6 \text{ g } H_2O}$

The change in the freezing point of the sea water is then:

$$\Delta t = m \bullet K_{fp} = \frac{1103 \text{ mol ions}}{10^6 \text{ g } H_2O} \bullet \frac{1000 \text{ g } H_2O}{1 \text{ kg } H_2O} \bullet -1.86 \text{ °C/molal} = -2.05 \text{ °C}$$

So we expect this sea water to begin freezing at -2.05 °C.

b. The osmotic presure (in atmospheres) can be calculated if *we assume the density of sea water is 1.00 g/mL.*

$$\Pi = MRT = \frac{1.103 \text{ mol}}{1 \text{ L}} \bullet 0.082057 \frac{\text{L} \bullet \text{atm}}{\text{K} \bullet \text{mol}} \bullet 298 \text{ K} = 27.0 \text{ atm}$$

The pressure needed to purify sea water by reverse osmosis would then be a pressure **greater than 27.0 atm.**

100. A 2.00 % aqueous solution of sulfuric acid is also 98.00 % in water.

Assume that we begin with 100 g of solution.

a. The molality of the solution is then:

$$\frac{2.00 \text{ g} \bullet \dfrac{1 \text{mol } H_2SO_4}{98.06 \text{ g } H_2SO_4}}{0.09800 \text{ kg water}} = 0.208 \text{ m}$$

Using the "delta T" equation: $\Delta t = m \cdot K_{fp} \cdot i$, we can solve for i: $\frac{\Delta t}{m \cdot K_{fp}} = i$

$$i = \frac{-0.796\ °C}{0.208m \cdot -1.86\ °C/m} = 2.06$$

b. Given the van't Hoff factor of 2 (above), the best representation of a dilute solution of sulfuric acid in water has to be: $H^+ + HSO_4^-$.

Summary Problem

102. The vapor pressure data should permit us to calculate the molar mass of the boron compound.

$$P_{benzene} = X_{benzene} \cdot P°_{benzene}$$

94.16 mm Hg = $X_{benzene} \cdot$ 95.26 mm Hg , and rearranging:

$$X_{benzene} = \frac{94.16 \text{ mm Hg}}{95.26 \text{ mm Hg}} = 0.9885$$

Now we need to know the # of moles of the boron compound, using $X_{benzene}$ to find that:

$$10.0 \text{ g benzene} \cdot \frac{1 \text{ mol benzene}}{78.11 \text{ g benzene}} = 0.128 \text{ mol benzene}$$

$$0.9885 = \frac{0.128 \text{ mol benzene}}{0.128 \text{ mol benzene} + x \text{ mol BxF}_y}$$

$$0.9885(0.128 + x) = 0.128 \quad x = 0.00147 \text{ mol } B_xF_y$$

Knowing that this # of moles of compound has a mass of 0.146 g, we can calculate the molar mass:

$$\frac{0.146 \text{ g}}{0.00147 \text{ mol } B_xF_y} = 99.2 \text{ g/mol}$$

We can calculate the empirical formula, since we know that the compound is 22.1% boron and 77.9 % fluorine.

In 100 g of the compound there are: $22.1 \text{ g B} \cdot \frac{1 \text{ mol B}}{10.81 \text{ g B}} = 2.11 \text{ mol B}$ and

$$77.9 \text{ g F} \cdot \frac{1 \text{ mol F}}{19.00 \text{ g F}} = 4.10 \text{ mol F}$$

The empirical formula then is BF_2, which would have a formula weight of 48.8

Dividing the molar mass (found from the vapor pressure experiment) by the mass of the empirical formula, we get: $\frac{99.2}{48.8} = 2.0$

a. The molecular formula is then B_2F_4.

b. A Lewis structure for the molecule:

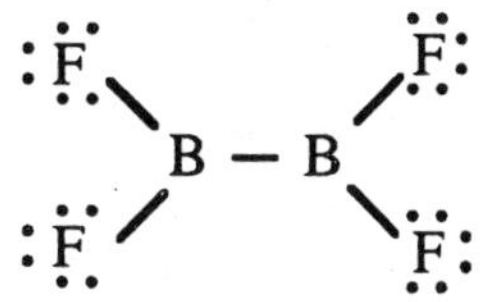

```
:F̤̈        F̤̈:
    \      /
     B — B
    /      \
:F̤̈        F̤̈:
```

We know that the molecule is nonpolar (does not have a dipole moment), hence the proposed structure. The F-B-F bond angles are 120°, as are the F-B-B bond angles. Hence the molecule is planar (flat). The hybridization of the boron atoms is sp^2.

Chapter 15
Principles of Reactivity: Chemical Kinetics

NUMERICAL QUESTIONS AND PROBLEMS

Reaction Rates

18. a. $2\ O_3\ (g) \rightarrow 3\ O_2\ (g)$

$$\text{Reaction Rate} = -\frac{1}{2} \cdot \frac{\Delta[O_3]}{\Delta t} = +\frac{1}{3} \cdot \frac{\Delta[O_2]}{\Delta t}$$

b. $2\ HOF\ (g) \rightarrow 2\ HF\ (g) + O_2\ (g)$

$$\text{Reaction Rate} = -\frac{1}{2} \cdot \frac{\Delta[HOF]}{\Delta t} = +\frac{1}{2} \cdot \frac{\Delta[HF]}{\Delta t} = +\frac{\Delta[O_2]}{\Delta t}$$

20. Plot the data for the hypothetical reaction $A \rightarrow 2\ B$

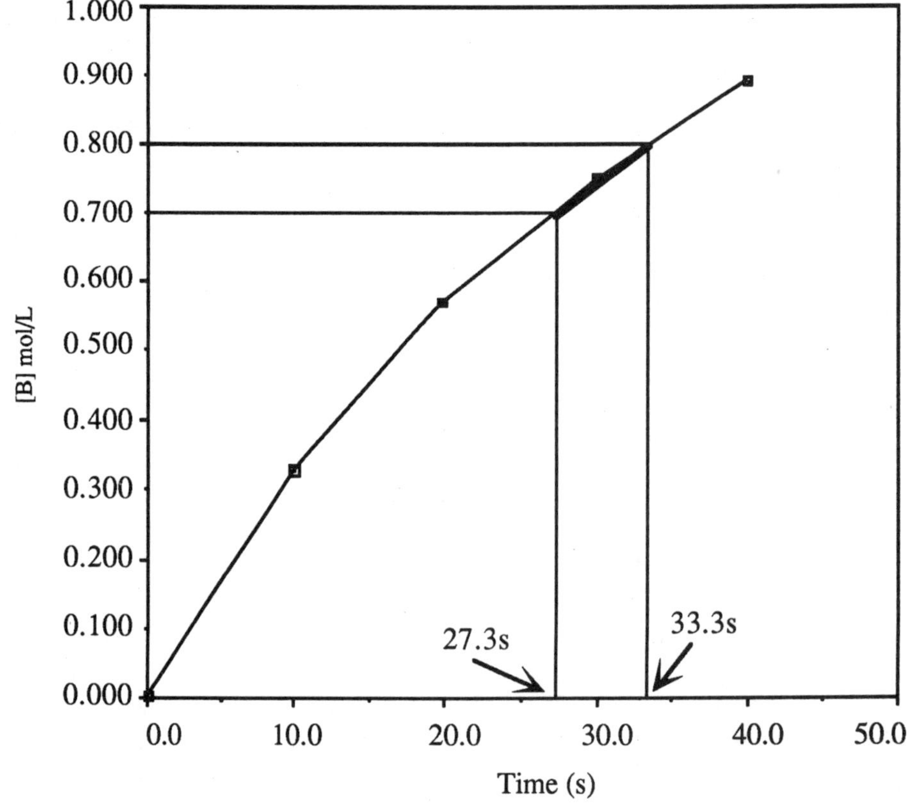

a. $\text{Rate} = \frac{\Delta[B]}{\Delta t} = \frac{(0.326 - 0.000)}{10.0 - 0.00} = +\frac{0.326}{10.0} = +0.0326\ \frac{\text{mol}}{\text{L}\bullet\text{s}}$

$$= \frac{(0.572 - 0.326)}{20.0 - 10.00} = +\frac{0.246}{10.0} = +0.0246 \frac{\text{mol}}{\text{L}\bullet\text{s}}$$

$$= \frac{(0.750 - 0.572)}{30.0 - 20.00} = +\frac{0.178}{10.0} = +0.0178 \frac{\text{mol}}{\text{L}\bullet\text{s}}$$

$$= \frac{(0.890 - 0.750)}{40.0 - 30.00} = +\frac{0.140}{10.0} = +0.0140 \frac{\text{mol}}{\text{L}\bullet\text{s}}$$

The rate of change decreases from one time interval to the next due to a continuing decrease in the amount of reacting material (A).

b. Since each A molecule forms 2 molecules of B, the concentration of A will decrease at a rate that is **half** of the rate at which B appears. The negative signs here indicate a **decrease in [A]--not a negative concentration of A!**

T	[B]	$[A] = 1/2([B]_0 - [B])$
10.0 s	0.326	-1/2(0.326) = - 0.163
20.0 s	0.572	-1/2(0.572) = - 0.286

$$\text{Rate at which A changes} = \frac{\Delta[A]}{\Delta t} = \frac{(-0.286 - -0.163)}{20.0 - 10.00}$$
$$= \frac{-0.123}{10.0} \text{ or } -0.0123 \frac{\text{mol}}{\text{L}\bullet\text{s}}$$

Note that the **negative sign** indicates a reduction in the concentration of A as the reaction proceeds. Compare this change with the change in [B] for the same interval above ($+0.0246 \frac{\text{mol}}{\text{L}\bullet\text{s}}$). The disappearance of A is half that of the appearance of B.

c. The instantaneous rate when [B] = 0.75 mol/L:

The instantaneous rate can be calculated by noting the tangent to the line at the point, [B] = 0.75 mol/L. Taking points equidistant ([B] = 0.70 mol/L and [B] = 0.80 mol/L) and determining the times associated with those concentrations, we can calculate the instantaneous rate.

$$\frac{\Delta[B]}{\Delta t} = \frac{(0.800\frac{\text{mol}}{\text{L}} - 0.700\ \frac{\text{mol}}{\text{L}})}{33.3 \text{ s} - 27.3 \text{ s}} = \frac{0.100 \text{ mol/L}}{6.0 \text{ s}} = 0.016 \frac{\text{mol}}{\text{L} \bullet \text{s}}$$

Using a greater time difference (Δt), one can increase the number of significant figures, and increase the number of significant figures reported in the answer.

Concentration and Rate Equations

22. a. The rate equation : Rate $= k[NO_2][O_3]$

b. Since k is constant, if $[O_3]$ is held constant, the rate would be tripled:
Let C represent the concentration of NO_2. Substituting into the rate equation:

$$Rate_1 = k[C][O_3]$$

$$Rate_2 = k[3C][O_3] = 3 \bullet k[C][O_3] \text{ or } 3 \bullet Rate_1$$

c. Halving the concentration of O_3—assuming $[NO_2]$ is constant, would halve the rate.

$$Rate_1 = k[NO_2][C]$$

$$Rate_2 = k[NO_2][1/2\ C] = 1/2[NO_2][C] \text{ or } 1/2 \bullet Rate_1$$

24. a. If we designate the three experiments (data sets in the table as i, ii, and iii respectively,

Experiment	[NO]	$[O_2]$	$-\frac{\Delta[NO]}{\Delta t}$ $(\frac{mol}{L \bullet s})$
i	0.010	0.010	2.5×10^{-5}
ii	0.020	0.010	1.0×10^{-4}
iii	0.010	0.020	5.0×10^{-5}

Note that experiment ii proceeds at a rate four times that of experiment i.

$$\frac{\text{experiment ii rate}}{\text{experiment i rate}} = \frac{1.0 \times 10^{-4} \frac{mol}{L \bullet s}}{2.5 \times 10^{-5} \frac{mol}{L \bullet s}} = 4$$

This rate change was the result of doubling the concentration of NO. The **order of dependence of NO must be second order**. Comparing experiments i and iii, we see that changing the concentration of O_2 by a factor of two, also affects the rate by a factor of two. **The order of dependence of O_2 must be first order.**

b. Using the results above we can write the rate equation: Rate $= k[NO]^2[O_2]^1$

c. To calculate the rate constant we have to have a rate. Note the data provided gives the rate of disappearance of NO. The relation of this concentration to the rate is:

$$Rate = -1/2 \bullet \frac{\Delta[NO]}{\Delta t}$$

Using experiment ii, the rate is $5.0 \times 10^{-5} \frac{mol}{L \bullet s}$

Substituting into the rate law

$$5.0 \times 10^{-5} \frac{mol}{L \bullet s} = k[0.020 \frac{mol}{L}]^2[0.010 \frac{mol}{L}]$$

$$12.5 \frac{L^2}{mol^2 \cdot s} = k$$

d. Rate when [NO] = 0.015 M and $[O_2]$ = 0.0050 M

$$\text{Rate} = k[NO]^2[O_2]$$
$$= 12.5 \frac{L^2}{mol^2 \cdot s} (0.015 \frac{mol}{L})^2 (0.0050 \frac{mol}{L})$$
$$= 1.4 \times 10^{-5} \frac{mol}{L \cdot s}$$

e. The relation between reaction rate and concentration changes:

$$\text{Rate} = -1/2 \cdot \frac{\Delta[NO]}{\Delta t} = -\frac{\Delta[O_2]}{\Delta t} = +1/2 \cdot \frac{\Delta[NO_2]}{\Delta t}$$

So when NO is reacting at $1.0 \times 10^{-4} \frac{mol}{L \cdot s}$ then O_2 will be reacting at $5.0 \times 10^{-5} \frac{mol}{L \cdot s}$ and NO_2 will be forming at $1.0 \times 10^{-4} \frac{mol}{L \cdot s}$

Concentration / Time Equations

26. Note that the reaction is first order. We can write:

$$\ln(\frac{[C_{12}H_{22}O_{11}]}{[C_{12}H_{22}O_{11}]_0}) = -kt.$$

Now substitute the concentrations of sucrose at t = 0 and t = 2.57 hr into the equation:

$\ln(\frac{[4.50 \text{ g/L}]}{[5.00 \text{ g/L}]_0}) = -k(2.57 \text{ hr})$ and solve for k to obtain $k = 0.0410 \text{ hr}^{-1}$

28. Since the reaction is first order,we can write:

$\ln(\frac{[SO_2Cl_2]}{[SO_2Cl_2]_0}) = -kt.$. Given the rate constant, $2.8 \times 10^{-3} \text{ min}^{-1}$, we can calculate the time required for the concentration to fall from 1.24×10^{-3} M to 0.31×10^{-3}M

$$\ln(\frac{[0.31 \times 10^{-3} M]}{[1.24 \times 10^{-3} M]_0}) = -(2.8 \times 10^{-3} \text{ min}^{-1})t.$$

$$\frac{\ln(0.25)}{(-2.8 \times 10^{-3} \text{ min}^{-1})} = t = 495 \text{ min or } 5.0 \times 10^2 \text{ min} \quad (2 \text{ sf})$$

30. The reaction is second order. So we use the integrated form of the rate law:

$$\frac{1}{[NH_4NCO]} - \frac{1}{[NH_4NCO]_0} = kt.$$

$$\frac{1}{[0.180 M]} - \frac{1}{[0.229 M]} = (0.0113 \frac{L}{mol \cdot min}) \cdot t$$

$$\frac{1.189}{(0.0113 \frac{L}{mol \cdot min})} = t = 105 \text{ min}$$

Half—Life

32. Given that the reaction is first order we can use the integrated form of the rate law:

$$\ln\left(\frac{[N_2O_5]}{[N_2O_5]_0}\right) = -kt.$$

a. Since the **definition of half-life** is "the time required for half of a substance to react", the fraction on the left side = 1/2, and ln(0.50)= - 0.693

Given the rate constant $5.0 \times 10^{-4}\ s^{-1}$ we can solve for t:

$- 0.693 = - (5.0 \times 10^{-4}\ s^{-1})t$ and $t = 1.4 \times 10^3$ seconds

b. Time required for the concentration to drop to 1/10 of the original value:

Substitute the ratio 1/10 for the concentration of N_2O_5:

$\ln(0.10) = - (5.0 \times 10^{-4}\ s^{-1})t$ and $t = 4.6 \times 10^3$ seconds

34. Since the decomposition is first order: $\ln \frac{[\text{azomethane}]}{[\text{azomethane}]_0} = - kt$

Converting 2.00 g of azomethane to **moles** :

$$2.00 \text{ g azomethane} \cdot \frac{1 \text{ mol azomethane}}{58.08 \text{ g azomethane}} = 0.0344 \text{ mol}$$

$$\ln \frac{[\text{azomethane}]}{[0.0344 \text{ mol}]_0} = - (40.8 \text{ min}^{-1})(0.0500 \text{ min})$$

$$\ln \frac{[\text{azomethane}]}{[0.0344 \text{ mol}]_0} = - 2.04$$

$$\frac{[\text{azomethane}]}{[0.0344 \text{ mol}]_0} = e^{-2.04} = 0.130$$

azomethane = 4.48×10^{-3} mol

or $4.48 \times 10^{-3} \text{ mol} \cdot \frac{58.08 \text{ g}}{1 \text{ mol}} = 0.260$ g azomethane remain

Since 1 mol N_2 is produced when 1 mol of azomethane decomposes, the amount of N_2 formed is: (0.0344 mol - 0.00448 mol) = 0.0300 mol N_2

36. Since this is a first-order process, $\ln \frac{[Cu^{2+}]}{[Cu^{2+}]_0} = -kt$ and $k = -\frac{0.693}{12.70 \text{ hr}}$; time = 64 hr

$$\text{so} \quad \ln \frac{[Cu^{2+}]}{[Cu^{2+}]_0} = -\frac{0.693}{12.70 \text{ hr}} \cdot 64 \text{ hr}$$

$$\ln \frac{[Cu^{2+}]}{[Cu^{2+}]_0} = -3.49 \text{ and } \frac{[Cu^{2+}]}{[Cu^{2+}]_0} = e^{-3.49} \text{ or } 0.030$$

or 3.0 % remains (2 sf)

Graphical Analysis of Rate Equations and k

38. a. Plot of ln[sucrose] and $\frac{1}{[\text{sucrose}]}$ versus time.

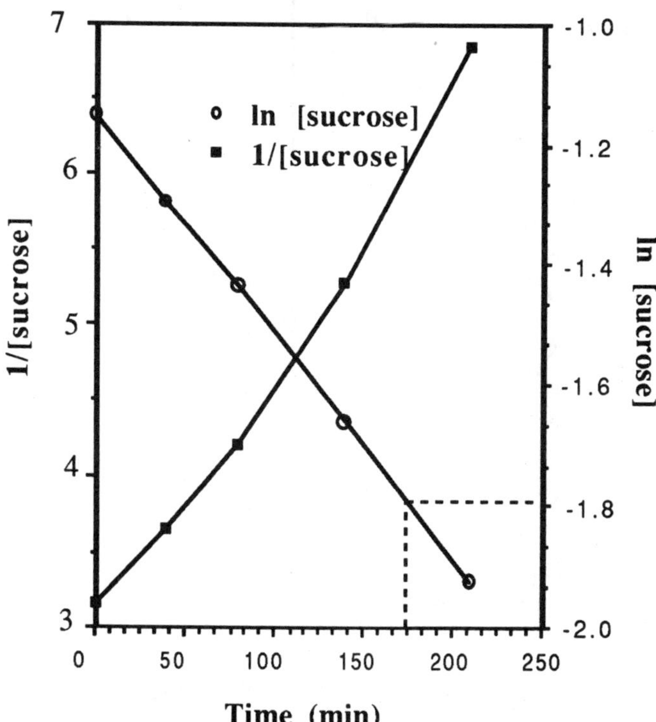

b. Since the reaction **is first order with respect to sucrose** (the plot of ln[sucrose] vs t is linear), the rate expression may be written: Rate = k [sucrose].

The rate constant can be calculated using two data points: Using the first two points yields:

$$\ln\left(\frac{[A]}{[A]_0}\right) = -kt \quad \text{and substituting:} \quad \ln\left(\frac{0.274}{0.316}\right) = -k(39 \text{ min})$$

$$\frac{\ln(0.867)}{39 \text{ min}} = -k \quad \text{and } 3.7 \times 10^{-3} \text{ min}^{-1} = k$$

c. Using the graph of ln[sucrose] vs time, an estimate at 175 minutes yields:

ln[sucrose] = -1.79 corresponding to [sucrose] = 0.167 M

40. For the decomposition of N_2O:

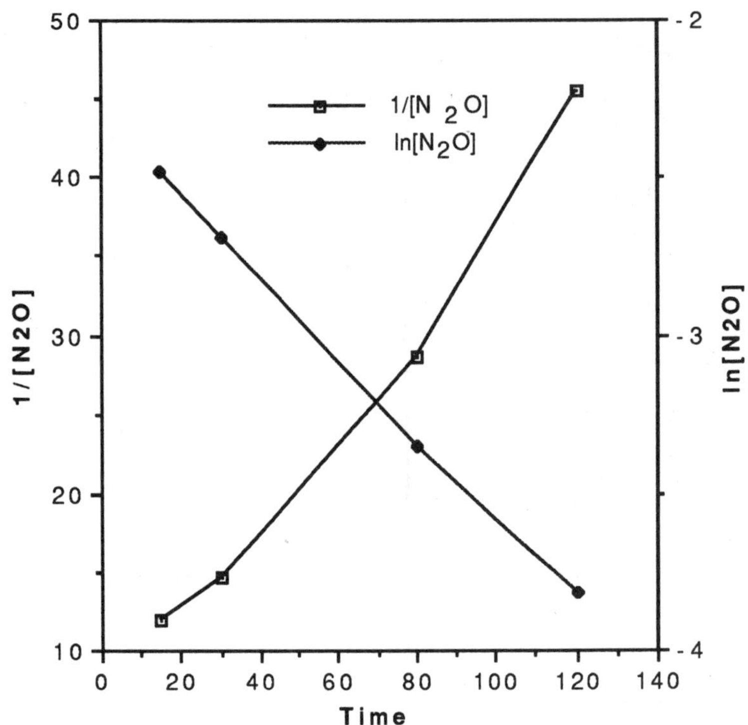

Since $\ln[N_2O]$ vs t gives a straight line, we know that the **reaction is first order**, and the line has a slope = - k. Taking the natural log (ln) of the concentrations at t = 120 min and t = 15.0 min gives ln(0.0220) = -3.8167; ln(0.0835) = -2.4829.

$$\text{slope} = -k = -\frac{(-3.8167) - (-2.4829)}{(120.0 - 15.0)\text{min}} = \frac{1.3338}{105.0 \text{ min}} = 0.0127 \text{ min}^{-1}$$

The rate equation is: Rate = k [N_2O] and

The rate of decomposition when [N_2O] = 0.035 mol/L :

$$\text{Rate} = (0.0127\ \text{min}^{-1})(0.035\ \frac{\text{mol}}{\text{L}}) = 4.4 \times 10^{-4} \frac{\text{mol}}{\text{L} \cdot \text{min}}$$

Kinetics and Energy

42. The E* for the reaction N_2O_5 (g) → 2 NO_2 (g) + 1/2 O_2 (g)

Given k at 25 °C = 3.46×10^{-5} s^{-1} and k at 55 °C = 1.5×10^{-3} s^{-1}

The rearrangement of the Arrhenius equation shown in your text as Equation 15.7 is helpful here.

$$\ln \frac{k_2}{k_1} = -\frac{E^*}{R}\left(\frac{1}{T_2} - \frac{1}{T_1}\right)$$

$$\ln \frac{1.5 \times 10^{-3}\ s^{-1}}{3.46 \times 10^{-5}\ s^{-1}} = -\frac{E^*}{8.31 \times 10^{-3}\ \text{kJ/mol} \cdot \text{K}}\left(\frac{1}{328} - \frac{1}{298}\right)$$

and solving for E* yields a value of **102 kJ/mol for E*.**

44. Using the Arrhenius equation: $\ln \frac{k_2}{k_1} = -\frac{E^*}{R}\left(\frac{1}{T_2} - \frac{1}{T_1}\right)$, T_1 = 800 K and T_2 = 850 K

Given E* = 260 kJ/mol and k_1 = 0.0315 s^{-1}, we can calculate k_1.

$$\ln \frac{k_2}{0.0315\ s^{-1}} = -\frac{260\ \text{kJ/mol}}{8.3145 \times 10^{-3}\ \text{kJ/mol} \cdot \text{K}}\left(\frac{1}{850\ \text{K}} - \frac{1}{800\text{K}}\right)$$

$\ln \frac{k_2}{0.0315\ s^{-1}} = 2.30$ and rearranging the "ln" side we get

$2.30 = \ln k_2 - \ln(0.0315\ s^{-1})$

$2.30 + \ln(0.0315\ s^{-1}) = \ln k_2 = 2.30 - 3.458 = -1.16$

$k_2 = e^{-1.16} = 0.3\ s^{-1}$ (800 K has 1 sf)

46. Energy progress diagram:

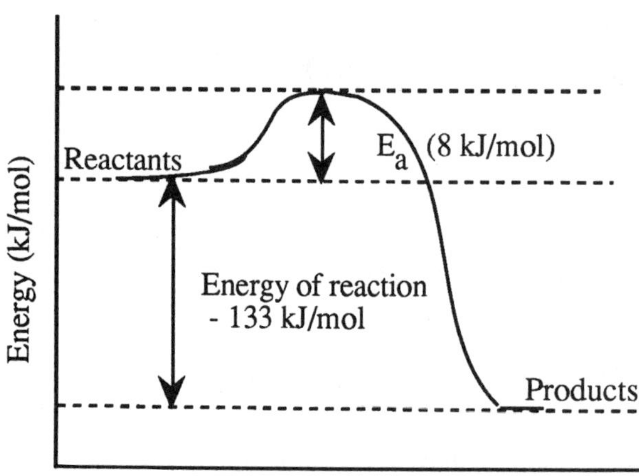

Mechanisms

48.

	Elementary Step	Rate law	
a.	$NO (g) + NO_3 (g) \rightarrow 2 NO_2 (g)$	Rate = $k[NO][NO_3]$	(bimolecular)
b.	$Cl (g) + H_2 (g) \rightarrow HCl (g) + H (g)$	Rate = $k[Cl][H_2]$	(bimolecular)
c.	$(CH_3)_3CBr (aq) \rightarrow (CH_3)_3C^+ (aq) + Br^- (aq)$	Rate = $k[(CH_3)_3CBr]$	(unimolecular)

50. a. The rate-determining step is **Step 2.** Recall that the slowest step in a mechanism is the "bottleneck" that will limit the rate of the reaction—hence rate-determining!

b. The rate equation for the rate-determining step must contain $CHCl_3$ and Cl. Since their coefficients are **1** in the elementary step, the rate equation must have the concentrations of these substances with exponents that match the coefficients in the equation. The rate low for the slow step would be: Rate = $k[CHCl_3][Cl]$.

52. a. Add the elementary steps:

$$\begin{array}{llll} \text{Slow} & NO_2 + \cancel{NO_2} & \rightarrow NO & + \cancel{NO_3} \\ \text{Fast} & \cancel{NO_3} + CO & \rightarrow \cancel{NO_2} & + CO_2 \\ \hline \text{Net} & NO_2 + CO & \rightarrow NO & + CO_2 \end{array}$$

Note that when the two steps are added, the desired overall equation results.

b. Molecularity: Both steps are bimolecular (two species reacting)

c. Rate law consistent with kinetic data: Rate= k $[NO_2]^2$

Since in the **slow** step we postulate the collision of two NO_2 molecules.

d. Intermediates are species that **do not appear** as reactants or products—for this equation, **NO_3 (g) is an intermediate**.

54. Rate law that is consistent with the mechanism:

The rate equation would be Rate = k[H][CO].

However the [H] is determined by an equilibrium from the previous step. The equilibrium for that can be written: $K = \frac{[H]^2}{[H_2]}$

and rearranging the equation for atomic H gives $K \cdot [H_2] = [H]^2$, and solving for [H] :

$K^{1/2} \cdot [H_2]^{1/2} = [H]$.

Substituting into the initial rate equation (for [H] gives :

Rate = $kK^{1/2} \cdot [H_2]^{1/2} \cdot [CO]$.

Since both k and $K^{1/2}$ are constants we can define a **new** k (call it k') and rewrite the rate equation as: Rate = $k'[H_2]^{1/2} \cdot [CO]$

Catalysis

56. a. True—The concentration of a catalyst may appear in the rate expression. The function may be to provide an alternative pathway or to assist in optimizing molecular orientations.

b. False—While a catalyst is involved in one or more steps of a process, the catalyst is regenerated in subsequent steps, being available for further reaction.

c. False—Homogeneous catalysts are always in the same phase as the reactants, but heterogeneous catalysts are in different phases. As an example, the Pt gauze used in the production of nitric acid is in the solid state, while the reactants are not.

58. One enzyme molecule hydrates 10^6 molecules of CO_2 / second.

so 1 L of 5 x 10^{-6}M enzyme would hydrate

$$5 \times 10^{-6} \frac{\text{mol enzyme}}{\text{L}} \cdot 1\text{ L} \cdot \frac{6.022 \times 10^{23} \text{ molecules enzyme}}{1 \text{ mol enzyme}} \cdot \frac{10^6 \text{ molecules } CO_2}{1 \text{ molecule enzyme}}$$

$$= 3 \times 10^{24} \text{ molecules } CO_2 \text{ in 1 second}$$

and in 1 hr,

$$\frac{3 \times 10^{24} \text{ molecules } CO_2}{1 \text{ s}} \cdot \frac{3600 \text{ s}}{1 \text{ hr}} \cdot \frac{7.308 \times 10^{-23} \text{ g } CO_2}{1 \text{ molecule } CO_2} \cdot \frac{1 \text{ kg } CO_2}{1 \times 10^3 \text{ g } CO_2}$$

$$= 792 \text{ kg } CO_2 \text{ or } 800 \text{ kg } CO_2 \quad (1 \text{ sf})$$

Note the mass of CO_2 per molecule (3rd factor) is determined by dividing the molar mass (44.01) by Avogadro's number.

General Questions

60. a. After you do part b, the rate expression may then be written as: Rate = k [CO][NO_2]

b. Using the kinetic data, note that the rate of the second experiment is two times that of the first experiment. Note that this corresponds to a doubling in [NO_2]—hence the reaction is **first order in [NO_2]**. If you compare the fourth and fifth experiments, note that the [CO] increases by a factor of 1.5 , and the rate increases by a factor of 10.2/6.8 = 1.5 .

The reaction is therefore **first order with respect to CO**, and second order overall.

c. Using the Rate expression from part a, substitute some of the data from the table.

$$\text{Rate} = k\,[CO][NO_2]$$

$$3.4 \times 10^{-8} \frac{\text{mol}}{\text{L} \cdot \text{hr}} = k \cdot 5.1 \times 10^{-4} \frac{\text{mol}}{\text{L}} \cdot 0.35 \times 10^{-4} \frac{\text{mol}}{\text{L}} \quad \text{and}$$

$$\frac{3.4 \times 10^{-8}\ \frac{\text{mol}}{\text{L} \bullet \text{hr}}}{1.785 \times 10^{-8}\ \frac{\text{mol}^2}{\text{L}^2}} = 1.9\ \frac{\text{L}}{\text{mol} \bullet \text{hr}}$$

62. a. What quantity of NO_x remains after 5.25 hr?

Average half-life is 3.9 hr. The rate constant $k = \frac{0.693}{3.9\ \text{hr}}$

$$\text{So } \ln \frac{[NO_x]}{[NO_x]0} = -\frac{0.693}{3.9\ \text{hr}} \bullet 5.25\ \text{hr} = -0.933$$

$$\text{and } \frac{[NO_x]}{[NO_x]0} = 0.393$$

So 39.3 % of the original remains or : 0.393 • 1.50 mg = 0.59 mg

b. Time to decrease 1.50 mg of NO_x to 2.50 x 10^{-6} mg:

$$\ln \frac{[2.50 \times 10^{-6}]}{[1.50]} = -\frac{0.693}{3.9\ \text{hr}} \bullet t \quad \text{and } t = 75\ \text{hr}$$

64. The net equation on summing three equations:

$$\cancel{Cl} + O_3 \longrightarrow \cancel{ClO} + O_2$$
$$\cancel{ClO} + \cancel{O} \longrightarrow \cancel{Cl} + O_2$$
$$O_3 \longrightarrow \cancel{O} + O_2$$
$$2\,O_3 \longrightarrow 3\,O_2$$

The overall **process consumes 2 molecules of ozone**, and since Cl atoms are not consumed the process can be repeated many, many times.

Since Cl is consumed (in the first step) and produced in a subsequent step, we denote **Cl as a catalyst. ClO is an intermediate.**

66. a. True—the rate-determining step in a mechanism **is the slowest step.**

b. True—the rate constant is a proportionality constant that relates the concentration and rate *at a given temperature* and varies with temperature.

c. False—As a reaction proceeds the concentration of reactants diminishes and the rate of the reaction slows as fewer collisions occur.

d. False—If the slow (single) step involves a termolecular process, then only one step is necessary. However termolecular collisions are **not highly likely**.

68. The data given are plotted as (ln k) vs (1/T)

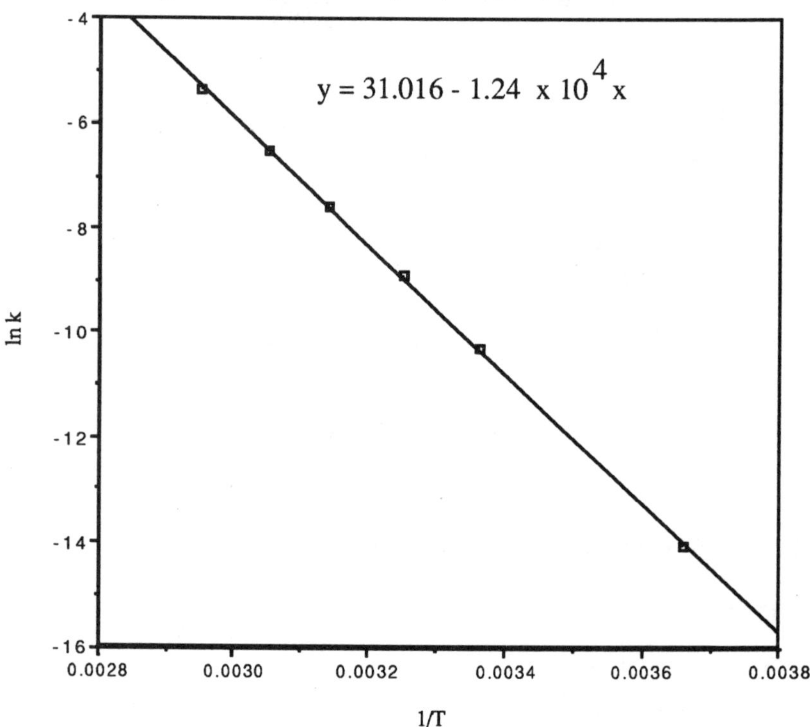

Such a plot has a slope (shown above as (-1.24×10^4)) = $-E_a/R$

and with $R = 8.31 \times 10^{-3}$ kJ/mol • K

$E_a = -(-1.24 \times 10^4 \text{ K})(8.31 \times 10^{-3} \text{ kJ/mol} \cdot \text{K}) = 103$ kJ/mol

70. The rate equation lets us know that the process is **first order in phosphine.**

Given that fact, we can calculate k:

$$k = \frac{0.693}{t_{1/2}} \text{ and since } t_{1/2} = 37.9 \text{ sec, then } k = \frac{0.693}{37.9 \text{ sec}} = 1.83 \times 10^{-2} \text{ sec}^{-1}.$$

a. If 3/4 of PH_3 decomposes, the fraction remaining is 1/4.

$$\ln\left[\frac{[PH_3]_t}{[PH_3]_0}\right] = \ln\left(\frac{1}{4}\right) = -(1.83 \times 10^{-2}\text{sec}^{-1})\ t$$

$$t = 7.58 \times 10^1 \text{ sec}$$

Alternately, for $[PH_3]$ to decrease to $\frac{1}{4}[PH_3]_o$ requires 2 half-lives:

$[(\frac{1}{2})^2 = \frac{1}{4}]$ so $t = 2 \cdot t_{1/2}$ or $2(37.9\text{ s}) = 75.8$ sec

b. Fraction of original sample of phosphine remaining after 1 min (60 minutes):

Using the equation above once more:

$\ln\frac{[PH_3]}{[PH_3]_0} = -(1.83 \times 10^{-2}\text{ sec}^{-1} \cdot 60\text{ sec}) = -1.098$ and solving for the ratio, $\frac{[PH_3]}{[PH_3]_0}$, $e^{-1.098}$ so $\frac{[PH_3]}{[PH_3]_0} = 0.333$—that is **1/3 of the original sample remains** after 1 minute.

72. a. Calculate ΔH for

1. $O_2(g) \longrightarrow O_2(\text{adsorbed})$: (-37) - (0) = -37 kJ
2. $O_2(\text{adsorbed}) \longrightarrow 2\,O(\text{adsorbed})$: (-251) - (-37) = -214 kJ
3. $O_2(g) \longrightarrow 2\,O(\text{adsorbed})$: (-251) - (0) = -251 kJ

or the sum of $\Delta H_1 + \Delta H_2$

b. What is E_a for $O_2(\text{adsorbed}) \longrightarrow 2\,O(\text{adsorbed})$?

The energy difference between O_2(adsorbed (-37 kJ) and the transition state (-17 kJ) is 20 kJ.

74. The rate equation would be Rate = $k[CHCl_3][Cl]$. However the [Cl] is determined by an equilibrium from the previous step. The equilibrium for that can be written:

$K = \frac{[Cl]^2}{[Cl_2]}$ and rearranging gives $K\cdot[Cl_2] = [Cl]^2$,

Solving for [Cl] we get: $K^{1/2}\cdot[Cl_2]^{1/2} = [Cl]$.

Substituting into the initial rate equation gives : Rate = $k[CHCl_3]K^{1/2}\cdot[Cl_2]^{1/2}$

and rearranging (to group the k's) gives: Rate = $K^{1/2}k[CHCl_3]\cdot[Cl_2]^{1/2}$.

Since both k and $K^{1/2}$ are constants we can define a **new** k (call it k') and rewrite the rate equation as: Rate = $k'[CHCl_3]\cdot[Cl_2]^{1/2}$

Conceptual Questions

76. Finely divided rhodium has **a larger surface area** than a block of the metal with the same mass. Since hydrogenation reactions depend on adsorption of H_2 on the catalyst—so the greater the surface area, the greater the locations for such adsorption to occur.

78. Assume that you begin with labeled oxygen in $CH_3O^{18}H$. If we represent the labeled oxygen with a • , you can see from the equation below that labeled oxygen (in the water) results **if the O originated from the methanol.** Experimental results indicate that the labeled oxygen from the alcohol *will be found on the ester.*

$$CH_3\overset{O}{\overset{\|}{C}}O\underline{H} + \underline{H\overset{\bullet}{O}}CH_3 \longrightarrow CH_3\overset{O}{\overset{\|}{C}}\overset{\bullet}{O}CH_3 + \underline{HO}\,\underline{H}$$

Challenging Questions

80. a. Volume of basement = 12 m x 7.0 m x 3.0 m = 252 m^3 or 252 x 10^3 L
[Recall that 1 L = 1 x 10^{-3} m^3]
Pressure of Rn = 1.0 x 10^{-6} mm Hg

We can calculate the moles of Rn with the Ideal Gas Law (assuming the basement is at 25°C or 298 K).

$$n = \frac{PV}{RT} = \frac{(1.0 \times 10^{-6}\ \text{torr})(252 \times 10^3\ \text{L})}{(62.4\ \frac{\text{L} \cdot \text{torr}}{\text{K} \cdot \text{mol}})(298\ \text{K})} = 1.4 \times 10^{-5}\ \text{mol Rn}$$

and multiplying by Avogadro's number — 8.2 x 10^{18} atoms Rn

So the number of atoms/L is: $\frac{8.2 \times 10^{18}}{252 \times 10^3\ \text{L}} = 3.2 \times 10^{13}\ \frac{\text{atoms}}{\text{L}}$

b. With $t_{1/2}$ of 3.82 days, the number of atoms/Liter after 31. days will be:

$$\ln\frac{[\text{Rn}]}{[\text{Rn}]_0} = -\frac{0.693}{3.82\ \text{days}} \cdot 31.\ \text{days} = -5.62$$

The fraction of radon after 31 days is: $\frac{[\text{Rn}]}{[\text{Rn}]_0} = 3.6 \times 10^{-3}$

So the number of $\frac{\text{atoms}}{\text{L}}$ will be $(3.6 \times 10^{-3})(3.3 \times 10^{13}\ \frac{\text{atoms}}{\text{L}}) = 1.2 \times 10^{11}\ \frac{\text{atoms}}{\text{L}}$

82. The reaction energy diagrams for an exothermic reaction might appear as follows:

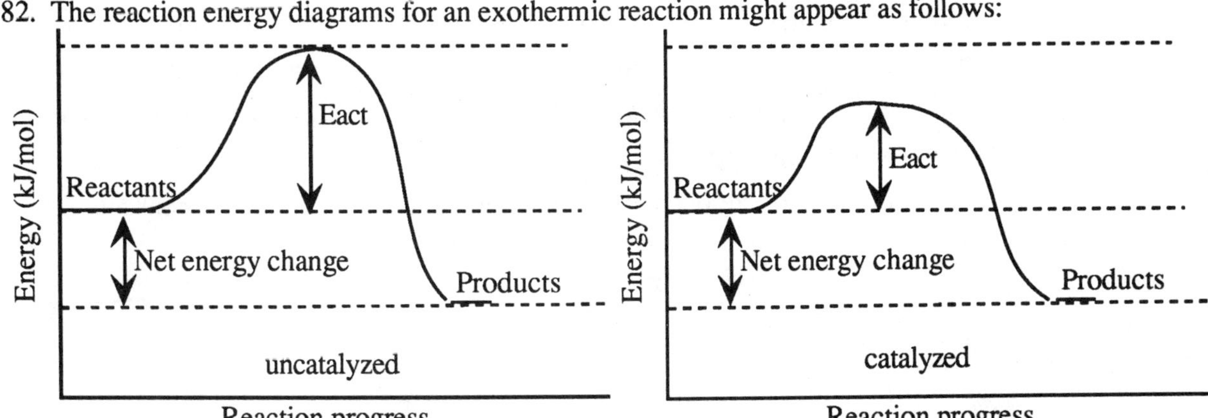

Note that the **Activation energy for the catalyzed reaction is less** than the Activation energy in the uncatalyzed reaction. Note also that the **net energy change for the reaction (catalyzed or uncatalyzed) is the same.**

84. Here's a chance to calculate the veracity of that old aphorism,"It takes longer to boil an egg on top of Pike's Peak". We can do so by recognizing that we're asking about the rates of the reaction (precipitating the egg protein albumin) at two different temperatures (90 and 100 °C). Obviously the slower the rate (at the lower T) the LONGER THE TIME needed. We can recognize the relationship between the time required (t) and the rate (and rate constant, k).

$$\frac{t_{90}}{t_{100}} = \frac{k_{100}}{k_{90}}$$

Remembering the relationship between the Arrhenius equation and k, we can write:

$k_{100} = Ae^{-\frac{Ea}{R \cdot 373K}}$ with a similar term for k_{90} (with T = 363K)

Substituting into the first equation for k_{100} and k_{90}:

$$\frac{t_{90}}{t_{100}} = \frac{Ae^{-\frac{Ea}{R \cdot 373K}}}{Ae^{-\frac{Ea}{R \cdot 363K}}} = e^{-\frac{Ea}{R}\left(\frac{1}{373K} - \frac{1}{363\ K}\right)}$$ Note that the factor A cancels.

$$\frac{t_{90}}{t_{100}} = e^{-\frac{Ea}{R}\left(\frac{1}{373K} - \frac{1}{363\ K}\right)} = e^{-\frac{52.0\ kJ/mol}{8.3145 \times 10\text{-}3\ kJ/K \bullet mol}\left(\frac{1}{373K} - \frac{1}{363\ K}\right)}$$

$\frac{t_{90}}{t_{100}} = e^{0.462} = 1.59$ and since $t_{100} = 3$ min then

$t_{90} = 1.59 \times 3$ min = 4.8 or about 5 minutes.

86. For the reaction HOF (g) → HF (g) + 1/2 O_2 (g) the half-life [for HOF (g)] at room temperature is 30 minutes. If the partial pressure of HOF is initially 100. mm Hg, after half of the HOF has decomposed, **its partial pressure will be 50.0 mm Hg**. The HF formed will contribute 50.0 mm Hg and the O_2 an additional 25.0 mm Hg for a **total pressure of 125.0 mm Hg**.

The first order kinetics for this decomposition allow us to calculate the rate constant, k.

$$\ln\left(\frac{[HOF]}{[HOF]_0}\right) = -kt$$

The [HOF] will be 1/2 of the original value, and substituting yields:

ln (0.5) = - k (30 min) and solving for k yields $k = 0.0231\ min^{-1}$

Substitution of the rate constant into the concentration/time equation permits us to calculate the fraction of HOF remaining (X) after 45 minutes.

$\ln X = -(0.0231 min^{-1})(45\ min)$ and $X = 0.354$

After 45 minutes the partial pressure of HOF is then 35.4 % of its original value or 35.4 mm Hg. The HF pressure will be 64.6 mm Hg (100.0 - 35.4) and the O_2 pressure approximately 32.3 mm Hg (1/2 O_2) for a total pressure (HOF + HF + O_2) of 132.3 mm Hg.

In summary:

Pressure of	30 minutes(exactly)	45 minutes
HOF	50.0 mm Hg	35.4 mm Hg
HF	50.0 mm Hg	64.6 mm Hg
O_2	25.0 mm Hg	32.3 mm Hg
Total	**125.0 mm Hg**	**132.3 mm Hg** (or 130 mm Hg)

Summary Problem

88. a. The molecularity of each elementary step: Slow step: unimolecular
Fast step: bimolecular

b. The doubling of the rate with the doubling of the concentration of $Ni(CO)_4$ indicates a first-order dependence on $Ni(CO)_4$ while the lack of affect of [L] indicates a zero-order dependence on L. This is consistent with the SLOW step in the proposed mechanism shown. The rate equation is then: Rate = k $[Ni(CO)_4]$.

c. First order kinetics obey the equation: $\ln\frac{[Ni(CO)_4]}{[Ni(CO)_4]_0} = -kt$ with $k = 9.3 \times 10^{-3}\ s^{-1}$ and $t = 3.0 \times 10^2 s$ (Remember that units of time must cancel, so convert 5.0 minutes to seconds.)

$$\ln\frac{[Ni(CO)_4]}{[Ni(CO)_4]_0} = -(9.3 \times 10^{-3}\ s^{-1})(3.0 \times 10^2 s) = -2.79$$ and taking the inverse natural logarithm of both sides we get:

$$\frac{[Ni(CO)_4]}{[Ni(CO)_4]_0} = 0.0614.$$

Since the initial concentration of $Ni(CO)_4$ is 0.025M we can solve for the concentration after 5.0 minutes: $[Ni(CO)_4] = 0.0614 \cdot 0.025 = 1.5 \times 10^{-3}$ M

d. To determine the maximum amount of $Ni(CO)_4$ possible, determine the limiting reactant:

moles of CO = $n = \frac{P \cdot V}{R \cdot T} = \frac{(1.50 atm)(0.750\ L)}{(0.082057 \frac{L \cdot atm}{K \cdot mol})(295\ K)} = 0.0465$ mol CO

moles of Ni = $0.125\ g\ Ni \cdot \frac{1\ mol\ Ni}{58.69\ g\ Ni} = 0.00213$ mol Ni

The balanced equation is $Ni + 4\ CO \rightarrow Ni(CO)_4$. This requires **4 mol of CO** for each **1 mol of Ni**. If we divide the # of moles of CO by 4 (or multiply the # of moles of Ni x 4) we find that Ni is the limiting reactant. Now we can calculate the maximum amount of $Ni(CO)_4$.

$$0.00213\ mol\ Ni \cdot \frac{1\ mol\ Ni(CO)_4}{1\ mol\ Ni} \cdot \frac{170.74\ g\ Ni(CO)_4}{1\ mol\ Ni(CO)_4} = 0.364\ g\ Ni(CO)_4$$

Pressure of CO in the 750-mL flask at 29 °C:

Excess CO = 0.0465 mol CO - (4)(0.00213 mol Ni) = 0.0380 mol CO

The pressure is then:

$$P = \frac{n \cdot R \cdot T}{V} = \frac{(0.0380\ mol\ CO)(0.082057 \frac{L \cdot atm}{K \cdot mol})(302K)}{0.750L} = 1.25\ atm$$

e. Enthalpy change for $Ni(CO)_4\ (g) \rightarrow Ni\ (s) + 4\ CO\ (g)$

$\Delta H_{reaction} = [4\ mol \cdot \Delta H^\circ_f\ CO(g) + 1 mol \cdot \Delta H^\circ_f\ Ni(s)] - [1\ mol \cdot \Delta H^\circ_f\ Ni(CO)_4(g)]$

$\Delta H_{reaction} = [(4\ mol \cdot -110.525\ kJ/mol) + (1 mol \cdot 0)] - [(1\ mol \cdot -602.91\ kJ/mol)] = 160.81\ kJ$

Chapter 16
Principles of Reactivity : Chemical Equilibria

NUMERICAL AND OTHER QUESTIONS

Writing and Manipulating Equilibrium Constant Expressions

8. Equilibrium constant expressions:

a. $K_c = \frac{[H_2O]^2[O_2]}{[H_2O_2]^2}$ b. $K_c = \frac{[CO_2]}{[CO][O_2]^{1/2}}$

c. $K_c = \frac{[CO]^2}{[CO_2]}$ d. $K_c = \frac{[CO_2]}{[CO]}$

10. Comparing the two equilibria:

1 $SO_2 (g) + \frac{1}{2} O_2 (g) \rightleftharpoons SO_3 (g)$ K_1
2 $2 SO_3 (g) \rightleftharpoons 2 SO_2 (g) + O_2 (g)$ K_2

the expression that relates K_1 and K_2 is (e) : $K_2 = \frac{1}{K_1^2}$

Reversing equation 1 gives: $SO_3 (g) \rightleftharpoons SO_2 (g) + \frac{1}{2} O_2 (g)$ with the eq. constant $\frac{1}{K_1}$

Multiplying the modified equation 1x 2 gives: $2 SO_3 (g) \rightleftharpoons 2 SO_2 (g) + O_2 (g)$ and the modified eq. constant $(\frac{1}{K_1})^2$

12. Calculate K_c for the reaction:

$SnO_2 (s) + 2 CO (g) \rightleftharpoons Sn (s) + 2 CO_2 (g)$ given:

1 $SnO_2 (s) + 2 H_2 (g) \rightleftharpoons Sn (s) + 2 H_2O (g)$ $K = 8.12$
2 $H_2 (g) + CO_2 (g) \rightleftharpoons H_2O (g) + CO (g)$ $K = 0.771$

Take equation 2 and reverse it, then multiply by 2 to give:

$2 H_2O (g) + 2 CO (g) \rightleftharpoons 2 H_2 (g) + 2 CO_2 (g)$ $K = (\frac{1}{0.771})^2 = 1.68$

add equation 1: $SnO_2 (s) + 2 H_2 (g) \rightleftharpoons Sn (s) + 2 H_2O (g)$ $K = 8.12$

$SnO_2 (s) + 2 CO (g) \rightleftharpoons Sn (s) + 2 CO_2 (g)$ $K_{net} = 8.12 \cdot 1.68 = 13.7$

14. For the equation $H_2\ (g) + Cl_2\ (g) \rightleftharpoons 2\ HCl\ (g)$ $K = 4.8 \times 10^{10}$

K for: $\frac{1}{2}\ H_2\ (g) + \frac{1}{2}\ Cl_2\ (g) \rightleftharpoons HCl\ (g)$ $K = (4.8 \times 10^{10})^{1/2} = 2.2 \times 10^5$

K for: $HCl\ (g) \rightleftharpoons \frac{1}{2}\ H_2\ (g) + \frac{1}{2}\ Cl_2\ (g)$ $K = \dfrac{1}{2.2 \times 10^5} = 4.6 \times 10^{-6}$

The Reaction Quotient

16. The equilibrium expression for the reaction : $I_2\ (g) \rightleftharpoons 2\ I\ (g)$ and $K = 5.6 \times 10^{-12}$

Substituting the molar concentrations into the equilibrium expression:

$$Q_c = \frac{[I]^2}{[I_2]} = \frac{(2.0 \times 10^{-8})^2}{(2.0 \times 10^{-2})} = 2.0 \times 10^{-14}$$

Since $Q_c < K$, the system **is not at equilibrium** and will move to **the right to make more product (and reach equilibrium).**

18. $K = \dfrac{[N_2O_4]}{[NO_2]^2} = 170.$

Note that the quantities given are moles **not** molar concentrations. First calculate the concentrations, and then substitute the molar concentrations into the equilibrium expression:

$$Q_c = \frac{[N_2O_4]}{[NO_2]^2} = \frac{[1.5 \times 10^{-4} \text{moles}]}{[2.0 \times 10^{-4} \text{moles}]^2} = 3.8 \times 10^3$$

Since $Q_c > K$, the **system is not at equilibrium** and the **concentration of NO_2 will increase** as the system moves "to the left" to reach equilibrium.

Calculating Equilibrium Constants

20. For the equilibrium: $PCl_5\ (g) \rightleftharpoons PCl_3\ (g) + Cl_2\ (g)$, calculate K_c

The equilibrium concentrations are: $[PCl_5] = 4.2 \times 10^{-5}$ M

$[PCl_3] = 1.3 \times 10^{-2}$ M

$[Cl_2] = 3.9 \times 10^{-3}$ M

The equilibrium expression is: $\dfrac{[PCl_3][Cl_2]}{[PCl_5]} = \dfrac{[1.3 \times 10^{-2}][3.9 \times 10^{-3}]}{[4.2 \times 10^{-5}]} = 1.2$

22. The equilibrium expression is : $K_c = \frac{[CO]^2}{[CO_2]}$

The quantities given are **moles**, so first calculate the **concentrations at equilibrium.**

$$[CO] = \frac{0.10 \text{ moles}}{2.0 \text{ L}} = 0.050 \text{ M} \qquad [CO_2] = \frac{0.20 \text{ moles}}{2.0 \text{ L}} = 0.10 \text{ M}$$

a. $K_c = \frac{[CO]^2}{[CO_2]} = \frac{[0.050]^2}{[0.10]} = 2.5 \times 10^{-2}$

b. The only change here is in the amount of carbon. Since C does not appear in the equilibrium expression, K_c would be the same as in a.

24. K_c for the equation: $H_2O\ (g) \rightleftharpoons H_2\ (g) + \frac{1}{2}\ O_2\ (g)$

If $[H_2O]_i$ = 2.0 M and 10. % dissociates to H_2 and O_2.

	H_2O	H_2	O_2
Initial	2.0 M	0	0
Change	- 0.20	+ 0.20	+ 0.10
Equilibrium	1.8	+ 0.20	+ 0.10

$$\text{and } K_c = \frac{[H_2][O_2]^{1/2}}{[H_2O]} = \frac{(0.20)(0.10)^{1/2}}{(1.8)} = 0.035$$

26. Calculate K_c for : $2\ SO_3 \rightleftharpoons 2\ SO_2 + O_2$

Note that the reaction occurs in an 8.00 L flask. The molar concentrations are found by dividing the # of moles by 8.00 L.

	SO_3	SO_2	O_2
Initial	0.375 M	0	0
Change	- 0.145	+ 0.145	+ 0.0725
Equilibrium	+0.230	+ 0.145	+ 0.0725

$$\text{and } K_c = \frac{[SO_2]^2[O_2]}{[SO_3]} = \frac{(+0.145)^2(0.0725)}{(+0.230)^2} = 0.029 \qquad \text{(2 sf)}$$

28. For the system: $NH_4I(s) \rightleftharpoons NH_3\ (g) + HI\ (g)$, what is K_c ?

Given that the total pressure is attributable **only** to NH_3 + HI, **and** that the stoichiometry of the equation tells us that equal amounts of the two substances are formed, we can easily determine the pressure of **each** of the gases. Since we desire K_p in atmospheres, we must first convert 705 mm Hg to units of atmospheres:

$$705 \text{ mm Hg} \cdot \frac{1 \text{ atm}}{760 \text{ mm Hg}} = 0.928 \text{ atm}$$

$P_{total} = P(NH_3) + P(HI) = 0.928$ atm so the pressure of each gas is $\frac{0.928}{2}$ or 0.464 atm.

$K_p = P(NH_3) \cdot P(HI) = (0.464 \text{ atm})^2 = 0.215$

Using Equilibrium Constants in Calculations

30. For the system butane $\rightleftharpoons$ isobutane $K_c = 2.5$

Equilibrium concentrations may be found:

	butane	isobutane
Initial	0.034	0
Change	- x	+ x
Equilibrium	0.034 - x	x

$$K_c = \frac{x}{0.034 - x} = 2.5 \text{ and } x = 2.4 \times 10^{-2}$$

and [isobutane] = 2.4×10^{-2} M, [butane] = 0.034 - x = 1.0×10^{-2} M

32. Given at equilibrium : $\frac{[CO][Br_2]}{[COBr_2]} = 0.190$

First, calculate concentrations: $[COBr_2] = \frac{0.015 \text{ mol}}{2.5 \text{ L}} = 6.0 \times 10^{-3}$

Substituting the$[COBr_2]$ into the equilibrium expression we get: $\frac{[CO][Br_2]}{(6.0 \times 10^{-3})} = 0.190$

Note that the stoichiometry of the equation tells us that for **each CO, we obtain 1 Br_2**.

We can rewrite the expression to read:

$$\frac{[CO]^2}{6.0 \times 10^{-3}} \text{ or } \frac{[Br_2]^2}{6.0 \times 10^{-3}} = 0.190$$

If we then take the square root of both sides of the equation, we obtain:

[CO] or $[Br_2]$ = $(6.0 \times 10^{-3} \cdot 0.190)^{1/2} = 0.034$ M

34. For the equilibrium the equilibrium constant is $\frac{[I]^2}{[I_2]} = 3.76 \times 10^{-3}$

The initial concentration of I_2 is $\frac{0.105 \text{ mol}}{12.3 \text{ L}}$ or 8.54×10^{-3} M.

The equation indicates that 2 mol of I form for each mol of I_2 that reacts.

If some amount, say x M, I_2 reacts then the amount of I that forms is 2x, and the amount of I_2 remaining at equilibrium is then $(8.54 \times 10^{-3} - x)$

The equilibrium concentrations can then be substituted into the equilibrium expression.

$$\frac{(2x)^2}{8.54 \times 10^{-3} - x} = 3.76 \times 10^{-3}$$

$$\text{or } 4x^2 = 3.76 \times 10^{-3}(8.54 \times 10^{-3} - x)$$

$$4x^2 + 3.76 \times 10^{-3}x - 3.21 \times 10^{-5} = 0.$$

Now we need to solve this using the quadratic equation:

Using the positive solution to that equation we find that $x = 2.40 \times 10^{-3}$

So at equilibrium:

$$[I_2] = 8.54 \times 10^{-3} - 2.40 \times 10^{-3} = 6.14 \times 10^{-3} \text{ M}$$

$$[I] = 2(2.40 \times 10^{-3}) = 4.79 \times 10^{-3} \text{ M} \quad (\text{rounding gives } 4.80 \times 10^{-3})$$

36. Given the equilibrium constant for the system, and the pressure of N_2O_4 at equilibrium, we can easily calculate the pressure of NO_2 at equilibrium.

$$K_p = \frac{P^2(NO_2)}{P(N_2O_4)} = \frac{P^2(NO_2)}{0.85} = 0.15$$

The pressure of NO_2 is then 0.36 atm. According the Dalton's Law the total pressure of the gas mixture at equilibrium will be the **sum of the pressure of NO_2 + N_2O_4.**

$P_{total} = 0.85$ atm + 0.36 atm = 1.21 atm

Disturbing a Chemical Equilibrium : Le Chatelier's Principle

38. The equilibrium may be represented :

$$N_2O_3\,(g) + \text{heat} \rightleftharpoons NO_2\,(g) + NO\,(g)$$

Since the process is **endothermic** ($\Delta H = +$), heat is absorbed in the "left to right" reaction. The effect of:

a. Adding more N_2O_3 (g): An increase in the Pressure of N_2O_3 (adding more N_2O_3) will shift the equilibrium to the **right**, producing more NO_2 and NO.

b. Adding more NO_2 (g): An increase in the Pressure of NO_2 (adding more NO_2) will shift the equilibrium to the **left**, producing more N_2O_3.

c. Increasing the volume of the reaction flask: If the volume of the flask is increased, the pressure will drop. (Remember that P for gases is inversely related to volume.) A drop in pressure will favor that side of the equilibrium with the "larger total number of moles of gas"—so this equilibrium will shift to the **right**, producing more NO_2 and NO.

d. Lowering the temperature: You should note that a change in T (up or down) will result in a **change in the equilibrium constant**. (None of the three changes mentioned above change K!) However, the same principle applies. The removal of heat (a decrease in T) favors the exothermic process (shifts the equilibrium to the **left**) producing more N_2O_3.

40. K_c for butane $\rightleftharpoons$ iso-butane is 2.5.

a. Equilibrium concentration if 0.50 mol/L of iso-butane is added:

	butane	iso-butane
Original concentration	1.0	2.5
Change immediately after addition	1.0	2.5 + 0.50
Change (going to equilibrium)	+ x	- x
Equilibrium concentration	1.0 + x	3.0 - x

$$K_c = \frac{[\text{iso-butane}]}{[\text{butane}]} = \frac{3.0 - x}{1.0 + x} = 2.5$$

$$3.0 - x = 2.5\,(1.0 + x)$$

and $0.14 = x$

The equilibrium concentrations are:

[butane] = 1.0 + x = 1.1 M and [iso-butane] = 3.0 - x = 2.9 M

b. Equilibrium concentrations if 0.50 mol /L of butane is added:

	butane	iso-butane
Original concentration	1.0	2.5
Change immediately after addition	1.0 + 0.50	2.5
Change (going to equilibrium)	- x	+ x
Equilibrium concentration	1.5 - x	2.5 + x

$$K_c = \frac{[\text{iso-butane}]}{[\text{butane}]} = \frac{2.5 + x}{1.5 - x} = 2.5$$

$$2.5 + x = 2.5\,(1.5 - x) \qquad \text{and } x = 0.36$$

The equilibrium concentrations are:

[butane] = 1.5 - x = 1.1 M and [iso-butane] = 2.5 + x = 2.9 M

General Questions

42. $K_c = \frac{[NO]^2}{[N_2][O_2]} = 1.7 \times 10^{-3}$ at 2300 K.

a. What is K_p for the reaction?

$K_p = K_c \cdot (RT)^{\Delta n}$ (Equation 16.2)

Since $\Delta n = 0$

$N_2 + O_2 \rightleftharpoons 2\,NO$

2 mol gas total $\rightleftharpoons$ 2 mol gas total

$K_p = 1.7 \times 10^{-3}$

b. K_c for the reaction when all coefficients are halved?

For the original equation, $K_c = \frac{[NO]^2}{[N_2][O_2]}$

and with coefficients halved, $K'_c = \frac{[NO]}{[N_2]^{1/2}\,[O_2]^{1/2}}$

so $K_c = (K'_c)^2$ or $K'_c = (K_c)^{1/2} = (1.7 \times 10^{-3})^{1/2}$ or 4.1×10^{-2}

c. K_c for the reversed reaction is the reciprocal of the original K_c

$$\frac{1}{1.7 \times 10^{-3}} = 5.9 \times 10^2$$

44. For the equilibrium: $Br_2\,(g) \rightleftharpoons 2\,Br\,(g)$, we must first calculate the concentration of Br_2

$$[Br_2] = \frac{0.086 \text{ mol}}{1.26 \text{ L}} = 0.068 \text{ M}$$

Then we can complete an equilibrium table:

	Br_2	Br
Initial	0.068 M	0
change	- (0.037)(0.068) M	+2 (0.037)(0.068) M
Equilibrium	0.066 M	+0.0051 M

Substituting into the K_c expression: $K_c = \frac{[Br]^2}{[Br_2]} = \frac{(0.0051)^2}{(0.066)} = 3.9 \times 10^{-4}$

46. First calculate the concentration of NOCl (2.00 mol/1.00L).

The stoichiometry indicates for each mole of NOCl that reacts, one mol of NO and 1/2 mol of Cl_2 are formed.

	NOCl	NO	Cl_2
Initial	2.00 M	0	0
change	-0.66 M	+0.66 M	+0.33 M
Equilibrium	1.34 M	+0.66 M	+0.33 M

$$\text{So } K_c = \frac{[0.66]^2[0.33]}{[1.34]^2} = 8.0 \times 10^{-2}$$

48. $P_{total} = P_{NO_2} + P_{N_2O_4} = 1.5 \text{ atm}$; $P_{N_2O_4} = 1.5 - P_{NO_2}$

$$K_p = \frac{P_{N_2O_4}}{P^2_{NO_2}} = 6.75 = \frac{(1.5 - P_{NO_2})}{P^2_{NO_2}}$$

or $1.5 - P_{NO_2} = 6.75\ P^2_{NO_2}$ and rearranging: $6.75\ P^2_{NO_2} + P_{NO_2} - 1.5 = 0$

and solving for P_{NO_2} (with the quadratic equation) yields:

$P_{NO_2} = 0.40$ atm and $P_{N_2O_4} = 1.1$ atm

50. $K_c = \dfrac{[\text{dimer}]}{[\text{monomer}]^2} = 3.2 \times 10^4$

	monomer	dimer
Initial pressure	5.4×10^{-4}	0
Equilibrium	x	½ $(5.4 \times 10^{-4} - x)$

$$K_c = \frac{½\,(5.4 \times 10^{-4} - x)}{x^2} = 3.2 \times 10^4$$ and multiplying by x^2:

$2.7 \times 10^{-4} - 0.5\,x = 3.2 \times 10^4\,x^2$

rearranging $3.2 \times 10^4\,x^2 + 0.5\,x - 2.7 \times 10^{-4} = 0$

and solving via the quadratic equation yields: $x = 8.4 \times 10^{-5}$

a. The % of acetic acid converted to the dimer is

$$\frac{5.4 \times 10^{-4} - 8.4 \times 10^{-5}}{5.4 \times 10^{-4}} \times 100 = 84\ \%$$

b. As the temperature increases, the equilibrium would **shift to the left**, producing more monomeric acetic acid. This change reflects the move to **favor the endothermic process** as T increases.

52. $K_c = \frac{[SO_3]^2}{[SO_2]^2[O_2]} = 279$; $K_p = \frac{P^2_{SO_3}}{P^2_{SO_2} \cdot P_{O_2}} = ?$

Remember that $K_p = K_c \cdot (RT)^{\Delta n}$ [Equation 16.2]

and $\Delta n = 2 - 3$ or -1

$$K_p = 279 \cdot \frac{1}{(0.821 \frac{L \cdot atm}{K \cdot mol})(1000K)} = 3.40 \text{ (or 3 to 1 sf)}$$

54. For zinc carbonate dissolving in water, the equilibrium constant expression is:

$K_c = [Zn^{2+}][CO_3^{2-}] = 1.5 \times 10^{-11}$ [Solids are omitted]

At equilibrium $[Zn^{2+}] = [CO_3^{2-}]$ so $[Zn^{2+}]^2 = [CO_3^{2-}]^2 = 1.5 \times 10^{-11}$

and $[Zn^{2+}] = [CO_3^{2-}] = 3.9 \times 10^{-6}$ M

Conceptual Questions

56. As the liquid water evaporates, the process H_2O (l) $\rightarrow$ H_2O (g) proceeds. As more gaseous water is formed, the process H_2O (g) $\rightarrow$ H_2O (l) (also known as condensation)—initially slow owing to a scarcity of gaseous water—begins to increase (since more gaseous water is present as the evaporation of water continues). The l$\rightarrow$ g transition (evaporation) slows as the g $\rightarrow$ l transition (condensation) increases. When these two opposing processes are proceeding **at the same rate**, equilibrium is achieved. Once equilibrium has been achieved, the two opposing processes continue to occur—at the same rate. The net effect is that "nothing appears to be happening".

58. Consider the equilibrium : O_3 (g) + NO (g) $\rightleftharpoons$ O_2 (g) + NO_2 (g) for which $K_c = 6.0 \times 10^{34}$.

a. If a system is at equilibrium, the ratios of "reactants" and "products" , when substituted into the equilibrium expression will have a value equal to K_c.

Substituting we get: $\frac{[O_2][NO_2]}{[O_3][NO]} = \frac{[8.2 \times 10^{-3}][2.5 \times 10^{-4}]}{[1.0 \times 10^{-6}][1.0 \times 10^{-5}]} = 2.1 \times 10^5$

Q_c is considerably smaller than K, so the reaction will proceed **to the right** (forming more O_2 and NO_2)

b. Determine if the "left to right" process is endo- or exothermic:

$$\Delta H^\circ_{reaction} = [\Delta H^\circ_f O_2 + \Delta H^\circ_f NO_2] - [\Delta H^\circ_f O_3 + \Delta H^\circ_f NO]$$
$$= [0 \text{ kJ/mol (1mol)} + 33.18 \text{ kJ/mol(1mol)}] - [142.7 \text{ kJ/mol(1mol)} + 90.25 \text{ kJ/mol(1mol)}]$$
$$= -199.8 \text{ kJ}$$

The "left to right" process is exothermic. On a warm day (when T is increased), the equilibrium would shift **to the left**, making more O_3 (g) + NO (g).

60. For the equilibrium : $(NH_3)[B(CH_3)_3] \rightleftharpoons B(CH_3)_3 + NH_3$ $K_c = 0.15$
Substituting $(CH_3)_3P$ for NH_3 gives an equilibrium with $K_c = 4.2 \times 10^{-3}$
while substituting $(CH_3)_3N$ gives an equilibrium with $K_c = 1.5 \times 10^{-2}$.

a. To determine which of the three systems would provide the largest concentration of $B(CH_3)_3$, one needs to ask "What does K tell me?" One answer to that question is "the extent of reaction—that is the larger the value of K, the more products are formed. Hence **the system with the largest K_c would give the largest concentration of $B(CH_3)_3$ at equilibrium.**

b. Given

	$(NH_3)[B(CH_3)_3]$ $\rightleftharpoons$	$B(CH_3)_3$ +	NH_3	$K_c = 0.15$
Initial:	$\frac{0.010\text{mol}}{0.100\text{L}} = 0.10$ M	0 M	0 M	
Change	- x	+ x	+ x	
Equilibrium	0.10 - x	+x	+ x	

Substituting into the equilibrium expression we have: $K_c = \frac{(x)(x)}{(0.10 - x)} = 0.15$

and simplifying gives : $x^2 = 0.15(0.10 - x)$ and $x^2 + 0.15x - 0.015 = 0$

Solving via the quadratic formula gives x = 0.069 M (0.0686 to 3 sf)

So the **concentrations at equilibrium** are:

$[B(CH_3)_3] = [NH_3] = 0.069$ M and $[(NH_3)[B(CH_3)_3]] = 0.031$ M

The percent dissociation of $(NH_3)[B(CH_3)_3]$ is:

$$\frac{\text{amount changed}}{\text{original amount}} = \frac{0.068 \text{ M}}{0.10 \text{ M}} \times 100 = 68\ \% \text{ dissociated}$$

62. For the reaction of hemoglobin (Hb) with CO we can write:

$$K = \frac{[HbCO][O_2]}{[HbO_2][CO]} = 200$$

If $\frac{[HbCO]}{[HbO_2]} = 1$ then $\frac{[O_2]}{[CO]} = 200$

If $[O_2] = 0.2$ atm then $\frac{0.2 \text{ atm}}{[CO]} = 200$ and solving for [CO] we obtain

$\frac{0.2 \text{ atm}}{200} = [CO] = 1 \times 10^{-3}$ atm.

A partial pressure of [CO] $= 1 \times 10^{-3}$ atm would likely be fatal.

Summary Question

64. For the system of $2 \text{ NO (g)} + Br_2 \text{ (g)} \rightarrow 2 \text{ NOBr (g)}$:

a. The mass of NOBr that could be prepared when 3.50 g NO and 9.67 g Br_2 are mixed?

$$3.50 \text{ g NO} \bullet \frac{1 \text{ mol NO}}{30.01 \text{ g NO}} = 0.117 \text{ mol NO}$$
$$9.67 \text{ g } Br_2 \bullet \frac{1 \text{ mol } Br_2}{159.8 \text{ g } Br_2} = 0.0605 \text{ mol } Br_2$$

The stoichiometry dictates that we have 2 mol of NO for each 1 mol of Br_2, we find that **NO is the limiting reagent**—since the # of mol of NO is **less than** 2 x the # of mol of Br_2:

$$0.117 \text{ mol NO} \bullet \frac{2 \text{ mol NOBr}}{2 \text{ mol NO}} \bullet \frac{109.9 \text{ g NOBr}}{1 \text{ mol NOBr}} = 12.8 \text{ g NOBr}$$

b. The electron dot structure for nitrosyl bromide:

:Br—N=O

c. The electron-pair geometry of NOBr: trigonal planar (three groups around the N)

The molecular geometry of NOBr: bent (two atoms and 1 lone pair attached to the N)

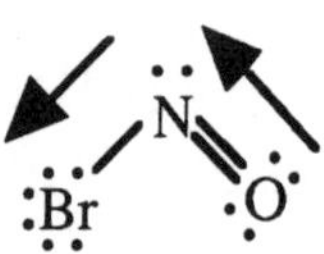

In the diagram, the arrows point toward the more electronegative atom of the two bonds. Note the net dipole moment .(The molecule would be polar.)

d. Calculate K_p for the equilibrium: $2\ NOBr\ (g) \rightleftharpoons 2\ NO\ (g) + Br_2\ (g)$

If the total pressure of the gases at equilibrium = 190 mm Hg and the compound (NOBr) is 34% dissociated.

What Dalton's Law tells us is that 190 mm Hg = P (NOBr) + P (NO) + P (Br_2)

Let's assume that we began with 1.00 mol NOBr (though the initial amount isn't critical). If 34% of the compound dissociates, at equilibrium we have:

	2 NOBr (g) ⇌	2 NO (g) +	Br_2 (g)
Initial:	1.00	0	0
change:	- 0.34	+ 0.34	+ 0.17
Equilibrium:	0.66	0.34	0.17

With these data, we can calculate the **mole fraction** of each of the gases present:

0.66 mol NOBr + 0.34 mol NO + 0.17 mol Br_2 = 1.17 mol (total)

Then $X_{NOBr} = \dfrac{0.66\text{ mol}}{1.17\text{ mol}}$; $X_{NO} = \dfrac{0.34\text{ mol}}{1.17\text{ mol}}$ and $X_{Br_2} = \dfrac{0.17\text{ mol}}{1.17\text{ mol}}$

We know that the pressure a gas exerts is proportional to the mole fraction of that gas present, so we can calculate the pressure that each gas exerts!

$$P_{NOBr} = \frac{0.66\text{ mol}}{1.17\text{ mol}} \cdot 190\text{ mm Hg} = 107\text{ mm Hg} = 0.14\text{ atm}$$

$$P_{NO} = \frac{0.34\text{ mol}}{1.17\text{ mol}} \cdot 190\text{ mm Hg} = 55.2\text{ mm Hg} = 0.073\text{ atm}$$

$$P_{Br_2} = \frac{0.17\text{ mol}}{1.17\text{ mol}} \cdot 190\text{ mm Hg} = 27.6\text{ mm Hg} = 0.036\text{ atm}$$

Now we can substitute these pressures into the K_p expression for the equilibrium:

$$2\ NOBr\ (g) \rightleftharpoons 2\ NO\ (g) + Br_2\ (g)$$

$$Kp = \frac{P^2_{NO} \cdot P_{Br_2}}{P^2_{NOBr}} = \frac{(0.073)^2 \cdot (0.036)}{(0.14)^2} = 9.8 \times 10^{-3} \qquad \text{(2 sf)}$$

Chapter 17
Principles of Reactivity: The Chemistry of Acids and Bases

NUMERICAL AND OTHER QUESTIONS

The Brønsted Concept of Acids and Bases

12.

Conjugate Base of:	Formula	Name
a. HCN	CN^-	cyanide ion
b. HSO_4^-	SO_4^{2-}	sulfate ion
c. HF	F^-	fluoride ion
d. HNO_2	NO_2^-	nitrite ion
e. HCO_3^-	CO_3^{2-}	carbonate ion

14. Products of acid-base reactions:

a. $HNO_3 + H_2O \rightarrow H_3O^+ + NO_3^-$

acid, base, conjugate acid, conjugate base

b. $HSO_4^- + H_2O \rightarrow H_3O^+ + SO_4^{2-}$

acid, base, conjugate acid, conjugate base

c. $H_3O^+ + F^- \rightarrow HF + H_2O$

acid, base, conjugate acid, conjugate base

16. The equation for potassium carbonate dissolving in water:

$$K_2CO_3\ (aq) \rightarrow 2\ K^+\ (aq) + CO_3^{2-}\ (aq)$$

Soluble salts--like K_2CO_3--dissociate in water. The carbonate ion formed in this process is a base, and reacts with the acid, water.

$$CO_3^{2-}\ (aq) + H_2O\ (\ell) \rightarrow HCO_3^-\ (aq) + OH^-\ (aq)$$

The production of the hydroxide ion, a strong base, in this second step is responsible for the basic nature of solutions of this carbonate salt.

18. HPO_4^{2-} (aq) + H_2O (ℓ) $\rightleftharpoons$ PO_4^{3-} (aq) + H_3O^+(aq) (Acid)
HPO_4^{2-} (aq) + H_2O (ℓ) $\rightleftharpoons$ $H_2PO_4^-$ (aq) + OH^-(aq) (Base)

20. a. HCO_2H (aq) + H_2O (ℓ) $\rightleftharpoons$ HCO_2^- (aq) + H_3O^+ (aq)

acid	base	conjugate of HCO_2H	conjugate of H_2O

b. H_2S (aq) + NH_3 (aq) $\rightleftharpoons$ NH_4^+ (aq) + HS^- (aq)

acid	base	conjugate of NH_3	conjugate of H_2S

c. HSO_4^- (aq) + OH^- (aq) $\rightleftharpoons$ SO_4^{2-} (aq) + H_2O (ℓ)

acid	base	conjugate of HSO_4^-	conjugate of OH^-

22. a. strongest acid: HF largest K_a
weakest acid: HPO_4^{2-} smallest K_a
b. conjugate base of HF: F^-
c. acid with weakest conjugate base: HF conjugate of strongest acid → weakest base.
d. acid with strongest conjugate base : HPO_4^{2-} conjugate of weakest acid → strongest base

24. a. strongest base: NH_3 (largest K_b) weakest base: C_5H_5N (smallest K_b)
b. conjugate acid of C_5H_5N: $C_5H_5NH^+$
c. base with strongest conjugate acid: C_5H_5N conjugate of weakest base → strongest acid
base with weakest conjugate acid: NH_3 conjugate of strongest base → weakest acid

26. The substance which has the smallest value for K_a will have the strongest conjugate base. One can prove this quantitatively with the relationship: $K_a \cdot K_b = K_w$. An examination of Table 17.4 shows that—of these three substances—HClO has the smallest K_a, and ClO^- will therefore be the strongest conjugate base.

Writing Acid-Base Reactions

28. Since both NH_4Cl and NaH_2PO_4 are soluble in water, we should view this in two steps:

    [1] dissociation into component ions: $NH_4Cl \rightarrow NH_4^+ (aq) + Cl^- (aq)$

    $$NaH_2PO_4 \rightarrow Na^+ (aq) + H_2PO_4^- (aq)$$

    [2] transfer of proton:

    $$NH_4^+ (aq) + H_2PO_4^- (aq) \rightarrow NH_3 (aq) + H_3PO_4 (aq)$$

    One might suggest that the proton would be transferred from the ammonium ion to the dihydrogen phosphate ion, as shown above. However, examination of Table 17.4 in your text shows H_3PO_4 to be more acidic than NH_4^+, making this process unlikely.

30. Predict whether the equilibrium lies predominantly to the left or to the right:

    a. $NH_4^+ (aq) + Br^- (aq) \rightleftharpoons NH_3 (aq) + HBr (aq)$

    HBr is a stronger acid than NH_4^+; equilibrium lies to **left**.

    b. $HPO_4^{2-} (aq) + CH_3CO_2^- (aq) \rightleftharpoons PO_4^{3-} (aq) + CH_3CO_2H (aq)$

    CH_3CO_2H is a stronger acid than HPO_4^{2-} ; equilibrium lies to **left**,

    c. $NH_2^- (aq) + H_2O (\ell) \rightleftharpoons NH_3 (aq) + OH^- (aq)$

    NH_2^- is a stronger base than OH^- ; equilibrium lies to **right**.

    d. $Fe(H_2O)_6^{3+} (aq) + HCO_3^- (aq) \rightleftharpoons Fe(H_2O)_5(OH)^{2+} (aq) + H_2CO_3 (aq)$

    $Fe(H_2O)_6^{3+}$ is a stronger acid than H_2CO_3; equilibrium lies to **right**.

pH Calculations

32. Since pH = 3.40, $[H_3O^+] = 10^{-pH}$ or $10^{3.40}$ or 4.0×10^{-4} M

    Since the $[H_3O^+]$ is greater than 1×10^{-7} M ($pH<7$), the solution **is acidic**.

34. pH of a solution of 0.0013 M HNO_3:

    Since HNO_3 is considered a strong acid, a solution of 0.0013 M HNO_3 has

    $[H_3O^+] = 0.0013$ or 1.3×10^{-3} M.

    $$pH = -\log[H_3O^+] = -\log[1.3 \times 10^{-3}] = 2.89$$

    The hydroxide ion concentration is readily determined since $[H_3O^+] \cdot [OH^-] = 1.0 \times 10^{-14}$

    $$[OH^-] = \frac{1.0 \times 10^{-14}}{1.3 \times 10^{-3}} = 7.7 \times 10^{-12} \text{ M}$$

36. The pH of a solution of $Ca(OH)_2$ can be calculated by remembering that this base dissolves in water to provide **two OH^- ions for each formula unit.**

So $[Ca(OH)_2] = 0.0015$ M $\Rightarrow [OH^-] = 0.0030$ M

$pOH = -\log[OH^-] = -\log[0.0030] = 2.52$ and

$pH = 14.00 - pOH = 11.48$

38. Interconversions: (Emboldened data is given in problem statement).

	pH	$[H_3O^+]$ M	$[OH^-]$ M	Solution character
a.	**1.00**	1.0×10^{-1}	1.0×10^{-13}	acidic
b.	**10.50**	3.2×10^{-11}	3.2×10^{-4}	basic
c.	4.89	**1.3×10^{-5}**	7.7×10^{-10}	acidic
d.	10.36	4.3×10^{-11}	**2.3×10^{-4}**	basic

Using pH to Calculate Ionization Constants

40. a. With a pH = 3.80 the solution has a $[H_3O^+] = 10^{-3.80}$ or 1.6×10^{-4} M

b. Writing the equation for the unknown acid, HA, in water we obtain:

$HA (aq) + H_2O (\ell) \rightleftharpoons H_3O^+ (aq) + A^- (aq)$

$[H_3O^+] = 1.6 \times 10^{-4}$ M implying that $[A^-]$ is also 1.6×10^{-4} M. Therefore the equilibrium concentration of acid, HA, is $(2.5 \times 10^{-3} - 1.6 \times 10^{-4})$ or $\approx 2.3 \times 10^{-3}$ M.

$$K_a = \frac{[H_3O^+][A]}{[HA]} = \frac{(1.6 \times 10^{-4})^2}{2.3 \times 10^{-3}} = 1.1 \times 10^{-5}$$

We would classify this acid as a **moderately weak acid.**

42. a. If we write the formula for the acid as HA, then [HA] = 0.015 M initially.

Since pH = 2.67, $[H_3O^+] = 10^{-2.67}$ or 2.1×10^{-3} M.

b. Substituting into the equilibrium expression for a weak monoprotic acid we obtain:

$$K_a = \frac{[H_3O^+][A^-]}{[HA]} = \frac{(2.1 \times 10^{-3})^2}{(0.015 - 0.0021)} = 3.6 \times 10^{-4}$$

44. With a pH = 9.11, the solution has a pOH of 4.89 and $[OH^-] = 10^{-4.89} = 1.3 \times 10^{-5}$ M.

The equation for the base in water can be written:

$H_2NOH (aq) + H_2O (\ell) \rightleftharpoons H_3NOH^+ (aq) + OH^- (aq)$

At equilibrium, $[H_2NOH]$ = $[H_2NOH] - [OH^-]$ = $(0.025 - 1.3 \times 10^{-5})$ or approximately 0.025 M.

$$K_b = \frac{[H_3NOH^+][OH^-]}{[H_2NOH]} = \frac{(1.3 \times 10^{-5})^2}{0.025} = 6.6 \times 10^{-9}$$

Using Ionization Constants to Calculate pH

46. The equilibrium concentrations of the species may be found as follows:

If we assume that some small amount (x) of the acid HA dissociates, the equilibrium concentration of the acid HA will be $[HA]_i$ - x. The amount of H_3O^+ will be equal to the amount of molecular acid which dissociates, x. This will also be equal to the concentration of the anion, A^-, present at equilibrium. Substitution into the K_a expression yields:

$$K_a = \frac{[A^-][H_3O^+]}{[HA]} = \frac{(x)^2}{(0.040 - x)} = 4.0 \times 10^{-9}$$

Since the K_a for the acid is small, assume that the denominator may be simplified to yield:

$$\frac{x^2}{0.040} = 4.0 \times 10^{-9}$$

and x = 1.3×10^{-5} (2 sf) so $[H_3O^+] = [A^-] = 1.3 \times 10^{-5}$ M and [HA] = 0.040 M.

pH = $-\log(1.3 \times 10^{-5}) = 4.89$

48. Using the same logic as in question 46 above, we can write:

	$HCN\ (aq) + H_2O\ (\ell)$	$\rightleftharpoons H_3O^+\ (aq)$	$+\ CN^-\ (aq)$
Initial concentration:	0.025	0	0
Change (going to eq.)	-x	+x	+x
Eq. concentrations	0.025 - x	x	x

Substituting these values into the K_a expression for HCN:

$$K_a = \frac{[CN^-][H_3O^+]}{[HCN]} = \frac{(x)^2}{(0.025 - x)} = 4.0 \times 10^{-10}$$

Assuming that the denominator may be approximated as 0.025 M, we obtain:

$$\frac{x^2}{0.025} = 4.0 \times 10^{-10}$$

and x = 3.2×10^{-6}. The equilibrium concentrations of $[H_3O^+] = [CN^-] = 3.2 \times 10^{-6}$ M.

The equilibrium concentration of HCN = $(0.025 - 3.2 \times 10^{-6})$ or 0.025 M.

Since x represents $[H_3O^+]$ the pH = $-\log(3.2 \times 10^{-6})$ or 5.50.

50. The concentration of $KH_8C_5O_4$:

$$1.28 \text{ g } KC_8H_5O_4 \bullet \frac{1 \text{ mol } KC_8H_5O_4}{204.2 \text{ g } KC_8H_5O_4} \bullet \frac{1}{0.125 \text{ L}} = 0.0500 \text{ M}$$

For the equilibrium of the phthalate ion in water we can write:

$$C_8H_5O_4^- (aq) + H_2O (\ell) \rightleftharpoons H_3O^+ (aq) + C_8H_4O_4^- (aq)$$

	$C_8H_5O_4^-$	H_3O^+	$C_8H_4O_4^-$
Initial concentration:	0.0500	0	0
Change (going to eq.)	-x	+x	+x
Eq. concentrations	0.0500 - x	x	x

Substituting into the K_a expression:

$$K_a = \frac{[C_8H_4O_4^-][H_3O^+]}{[C_8H_5O_4^-]} = \frac{(x)^2}{(0.0500 - x)} = 2.0 \times 10^{-7}$$

Assuming that the denominator may be approximated as 0.0500 M, we obtain:

$$\frac{x^2}{0.0500} = 2.0 \times 10^{-7} \quad \text{and } x = 1.0 \times 10^{-4}$$

Since x represents the$[H_3O^+]$, then pH = $-\log(1.0 \times 10^{-4})$ or 4.00

52. a. 4-chlorobenzoic acid has a larger K_a than benzoic acid, so it is the stronger acid.

b. A 0.010 M solution of benzoic acid would produce a lower concentration of hydronium ions than would a similar solution of 4-chlorobenzoic acid, and would be therefore less acidic—have a higher pH.

54. The equilibrium of ammonia in water can be written:

$$NH_3 + H_2O \rightleftharpoons NH_4^+ + OH^- \qquad K_b = 1.8 \times 10^{-5}$$

	NH_3	NH_4^+	OH^-
Initial concentration:	0.15	0	0
Change (going to eq.)	-x	+x	+x
Eq. concentrations	0.15 - x	x	x

Substituting into the K_b expression:

$$K_b = \frac{[NH_4^+][OH^-]}{[NH_3]} = \frac{(x)^2}{(0.15 - x)} = 1.8 \times 10^{-5}$$

With K_b small, the extent to which ammonia reacts with water is slight. We can approximate the denominator (0.15 - x) as 0.15 M, and solve the equation:

$$\frac{(x)^2}{0.15} = 1.8 \times 10^{-5} \text{ and } x = 1.6 \times 10^{-3}$$

So $[NH_4^+] = [OH^-] = 1.6 \times 10^{-3}$ M and $[NH_3] = (0.15 - 1.6 \times 10^{-3}) \approx 0.15$ M

Then: pOH = $-\log(1.6 \times 10^{-3}) = 2.78$ and the pH = (14.00-2.78) = 11.22

56. For $CH_3NH_2 + H_2O \rightleftharpoons CH_3NH_3^+ + OH^-$ $K_b = 4.2 \times 10^{-4}$

Using the approach of problem 54, the equilibrium expression can be written:

$$K_b = \frac{x^2}{0.25 - x} = 4.2 \times 10^{-4} \approx \frac{x^2}{0.25}$$

Solving the expression **with approximations** gives $x = 1.0 \times 10^{-2}$

$[OH^-] = 1.02 \times 10^{-2}$ M (1.0×10^{-2} M rounded to 2 sf)

pOH would then be 1.99 and pH 12.01.

58. pH of 1.0×10^{-3} M HF; $K_a = 7.2 \times 10^{-4}$

Using the same method as in question 50, we can write the expression:

$$K_a = \frac{x^2}{(1.0\times10^{-3} - x)} = 7.2 \times 10^{-4}$$

The concentration of the HF and K_a preclude the use of our usual approximations ($1.0 \times 10^{-3} - x \approx 1.0 \times 10^{-3}$)

So we multiply both sides of the equation by the denominator to get:

$$x^2 = 7.2 \times 10^{-4}(1.0 \times 10^{-3} - x)$$

$$x^2 = 7.2 \times 10^{-7} - 7.2 \times 10^{-4}\,x$$

Using the quadratic equation we solve for x:

and $x = 5.6 \times 10^{-4}$ M $= [F^-] = [H_3O^+]$ and pH = 3.25

Acid-Base Properties of Salts

60. Most of the salts shown are sodium salts. Since Na^+ does not hydrolyze, we can estimate the acidity (or basicity) of such solutions by looking at the extent of reaction of the anions with water (hydrolysis).

The Al^{3+} ion is acidic, as is the $H_2PO_4^-$ ion. From Table 17.4 we see that the K_a for the hydrated aluminum ion is greater than that for the $H_2PO_4^-$ ion, making the **Al^{3+} solution more acidic —lower pH** than that of $H_2PO_4^-$. All the other salts will produce basic solutions and since the S^{2-} ion has the largest K_b, we will anticipate that the **Na_2S solution will be most basic—i.e. have the highest pH.**

62. Hydrolysis of the NH_4^+ produces H_3O^+ according to the equilibrium:

$$NH_4^+ (aq) + H_2O (\ell) \rightleftharpoons H_3O^+ (aq) + NH_3 (aq)$$

With the ammonium ion acting as an acid, to donate a proton, we can write the K_a expression:

$$K_a = \frac{[NH_3][H_3O^+]}{[NH_4^+]} = \frac{K_w}{K_b} = \frac{1.0 \times 10^{-14}}{1.8 \times 10^{-5}} = 5.6 \times 10^{-10}$$

The concentrations of both terms in the numerator are equal, and the concentration of ammonium ion is 0.20 M. (Note the approximation for the **equilibrium** concentration of NH_4^+ to be equal to the **initial** concentration.)

Substituting and rearranging we get

$$[H_3O^+] = \sqrt{0.20 \cdot 5.6 \times 10^{-10}} = 1.1 \times 10^{-5}\ M$$

and the pH = 4.98.

64. The hydrolysis of CN^- produces OH^- according to the equilibrium:

$$CN^- (aq) + H_2O (\ell) \rightleftharpoons HCN (aq) + OH^- (aq)$$

Calculating the concentrations of Na^+ and CN^- :

$$[Na^+]_i = [CN^-]_i = \frac{10.8 \text{ g NaCN}}{0.500 \text{ L}} \cdot \frac{1 \text{ mol NaCN}}{49.01 \text{ g NaCN}} = 4.41 \times 10^{-1}\ M$$

$$\text{Then } K_b = \frac{K_w}{K_a} = \frac{1.0 \times 10^{-14}}{4.0 \times 10^{-10}} = 2.5 \times 10^{-5} = \frac{[HCN][OH^-]}{[CN^-]}$$

Substituting the $[CN^-]$ concentration into the K_b expression and noting that:

$[OH^-]_e = [HCN]_e$ we may write

$$[OH^-]_e = [(2.5 \times 10^{-5})(4.41 \times 10^{-1})]^{1/2} = 3.3 \times 10^{-3}\ M$$

$$[H_3O^+] = \frac{1.0 \times 10^{-14}}{3.3 \times 10^{-3}} = 3.0 \times 10^{-12}\ M$$

Conjugate Acid-Base Pairs

66. a. The K_a for the conjugate acid of a base can be calculated if we remember the relationship:

$$K_a \cdot K_b = K_w$$

$$K_a\text{ (anilinium)} = \frac{K_w}{K_b\text{(aniline)}} = \frac{1.0 \times 10^{-14}}{4.0 \times 10^{-10}} = 2.5 \times 10^{-5}$$

b. The pH of 0.080 M anilinium hydrochloride

$$K_a = \frac{[C_6H_5NH_2][H^+]}{[C_6H_5NH_3^+]} = 2.5 \times 10^{-5}$$

Using the normal approach:

$$\frac{x^2}{0.080 - x} = 2.5 \times 10^{-5}$$

Solving for x yields $x = 1.4 \times 10^{-3}$

The approximation yields 1.41×10^{-3} while the quadratic equation gives 1.40×10^{-3}

So $[H^+] = 1.4 \times 10^{-3}$ M and a pH of 2.85.

68. The sodium salt would exist in water as sodium and saccharide ions. The weakly acidic nature of saccharin would indicate some reaction of the saccharide ion with water.

$$C_7H_4NO_3S^- + H_2O \rightleftharpoons HC_7H_4NO_3S + OH^-$$

$$\text{with a } K_b = \frac{1.0 \times 10^{-14}}{2.1 \times 10^{-12}} = 4.8 \times 10^{-3}$$

$$K_b = \frac{[HC_7H_4NO_3S][OH^-]}{[C_7H_4NO_3S^-]} = \frac{(x)(x)}{0.10 - x} = 4.8 \times 10^{-3}$$

The approximation ($0.10 - x \approx 0.10$) gives a value for x of 2.2×10^{-2}.

Using the quadratic equation, we get $x = 1.96 \times 10^{-2}$ (2.0×10^{-2} to 2 sf) as $[OH^-]$ and pOH of 1.71 and a pH of 12.29.

Polyprotic Acids and Bases

70. a. pH of 0.45 M H_2SO_3:

The equilibria for the diprotic acid are:

$$K_{a1} = \frac{[HSO_3^-][H_3O^+]}{[H_2SO_3]} = 1.2 \times 10^{-2} \quad \text{and } K_{a2} = \frac{[SO_3^{2-}][H_3O^+]}{[HSO_3^-]} = 6.2 \times 10^{-8}$$

For the first step of dissociation:

	H_2SO_3	HSO_3^-	H_3O^+
Initial concentration	0.45		
Change	-x	+x	+x
Equilibrium concentration	0.45 - x	+x	+x

$$K_{a1} = \frac{(x)(x)}{(0.45\text{-}x)} = 1.2 \times 10^{-2}$$

We must solve this expression with the quadratic equation since ($0.45 < 100 \cdot K_{a1}$). The equilibrium concentrations for HSO_3^- and H_3O^+ ions are found to be 0.0677 M. The further dissociation is indicated by K_{a2}. Using the equilibrium concentrations obtained in the first step, we substitute into the K_{a2} expression.

	HSO_3^-	SO_3^{2-}	H_3O^+
Initial concentration	0.0677	0	0.0677
Change	-x	+x	+x
Equilibrium concentration	0.0677 - x	+x	0.0677 + x

$$K_{a2} = \frac{[SO_3^{2-}][H_3O^+]}{[HSO_3^-]} = \frac{(+x)(0.0677 + x)}{(0.0677 - x)} = 6.2 \times 10^{-8}$$

We note that x will be small in comparison to 0.0677, and we simplify the expression:

$$K_{a2} = \frac{(+x)(0.0677)}{(0.0677)} = 6.2 \times 10^{-8}$$

In summary the **concentrations of HSO_3^- and H_3O^+ ions have been virtually unaffected** by the second dissociation, so $[H_3O^+] = 0.0677$ M and pH = 1.17

b. The equilibrium concentration of SO_3^{2-} :

From the K_{a2} expression above: $[SO_3^{2-}] = 6.2 \times 10^{-8}$ M

72. a. Concentrations of OH^-, $N_2H_5^+$, and $N_2H_6^{2+}$ in 0.010 M N_2H_4:

The K_{b1} equilibrium allows us to calculate $N_2H_5^+$ and OH^- , formed by the reaction of N_2H_4 with H_2O.

$$K_{b1} = \frac{[N_2H_5^+][OH^-]}{[N_2H_4]} = 8.5 \times 10^{-7}$$

	N_2H_4	$N_2H_5^+$	OH^-
Initial concentration	0.010		
Change	-x	+x	+x
Equilibrium concentration	0.010 - x	+x	+x

Substituting into the K_{b1} expression : $\frac{(x)(x)}{0.010 - x} = 8.5 \times 10^{-7}$

We can simplify the denominator ($0.010 > 100 \cdot K_{b1}$).

$$\frac{(x)(x)}{0.010} = 8.5 \times 10^{-7} \text{ and } x = 9.2 \times 10^{-5} \text{ M} = [N_2H_5^+] = [OH^-]$$

The second equilibrium (K_{b2}) indicates further reaction of the $N_2H_5^+$ ion with water. The step should consume some $N_2H_5^+$ and produce more OH^-. The magnitude of K_{b2} indicates that the equilibrium "lies to the left" and we anticipate that not much $N_2H_6^{2+}$ (or additional OH^-) will be formed by this interaction.

	$N_2H_5^+$	$N_2H_6^{2+}$	OH^-
Initial concentration	9.2×10^{-5}		9.2×10^{-5}
Change	-x	+x	+x
Equilibrium concentration	9.2×10^{-5} - x	x	9.2×10^{-5} + x

$$K_{b2} = \frac{[N_2H_6^{2+}][OH^-]}{[N_2H_5^+]} = 8.9 \times 10^{-16} = \frac{x \cdot (9.2 \times 10^{-5} + x)}{(9.2 \times 10^{-5} - x)}$$

Simplifying yields $\frac{x \cdot (9.2 \times 10^{-5})}{(9.2 \times 10^{-5})} = 8.9 \times 10^{-16}$; $x = 8.9 \times 10^{-16}$ M = $[N_2H_6^{2+}]$

In summary we see that the second stage produces a negligible amount of OH^- and consumes very little $N_2H_5^+$ ion. The equilibrium concentrations are:

$[N_2H_5^+]$: 9.2×10^{-5} M; $[N_2H_6^{2+}]$: 8.9×10^{-16} M; $[OH^-]$: 9.2×10^{-5} M

b. The pH of the 0.010 M solution: $[OH^-] = 9.2 \times 10^{-5}$ M so pOH = 4.04
and pH = 14.0 - 4.04 = 9.96

Lewis Acids and Bases

74.

a. Mn^{2+}	electron deficient	Lewis acid
b. $:NH_2(CH_3)$	electron rich	Lewis base
c. H_2NOH	electron rich	Lewis base
d. SO_2	donates an electron pair	Lewis base
e. $Zn(OH)_2$	accepts electron pairs	Lewis acid

76. **BH_3 is a Lewis acid** in that it seeks to complete the electron octet on B by forming a new bond with the lone pair on the nitrogen atom of NH_3.

78. ICl_3 accepts an electron pair from the chloride ion in forming the ICl_4^- ion. This behavior is that of a **Lewis acid**. The Lewis dot structure for ICl_3 is:

$$:\ddot{Cl}-\ddot{\underset{|}{I}}-\ddot{Cl}: \qquad :\underset{..}{Cl}:$$

ICl_3 will have a T-shape and ICl_4^- will be a square planar complex.
(6 electron pairs around the I atom, four of which are bonding pairs to Cl atoms.)

I in ICl_3 will be $\mathbf{sp^3d}$ hybridized and in ICl_4^-, $\mathbf{sp^3d^2}$.

General Questions

80. $H^- (s) + H_2O (\ell) \rightarrow OH^- (aq) + H_2 (g)$

The hydride ion will act as a Lewis base (donate an electron pair) and water will act as a Lewis acid(accept an electron pair). The resulting solution of sodium hydroxide would be basic.

82. For the equilibrium: $HC_9H_7O_4 (aq) + H_2O (\ell) \rightleftharpoons C_9H_7O_4^- (aq) + H_3O^+ (aq)$,

we can write the K_a expression:

$$K_a = \frac{[C_9H_7O_4^-][H_3O^+]}{[HC_9H_7O_4]} = 3.27 \times 10^{-4}$$

The initial concentration of aspirin is:

$$2 \text{ tablets} \cdot \frac{0.325 \text{ g}}{1 \text{ tablet}} \cdot \frac{1 \text{ mol } HC_9H_7O_4}{180.2 \text{ g } HC_9H_7O_4} \cdot \frac{1}{0.225 \text{ L}} = 1.60 \times 10^{-2} \text{ M } HC_9H_7O_4$$

Substituting into the K_a expression:

$$K_a = \frac{[H_3O^+]^2}{1.60 \times 10^{-2} - x} = 3.27 \times 10^{-4} = \frac{x^2}{1.60 \times 10^{-2} - x}$$

Since $100 \cdot K_a \approx$ [aspirin], the quadratic equation will provide a "good" value.
Using the quadratic equation, $[H_3O^+] = 2.13 \times 10^{-3}$ M and pH = 2.672.

84. The reaction of H_2S and $NaCH_3CO_2$ involves the reaction of H_2S with $CH_3CO_2^-$ ion:

$$H_2S(aq) + CH_3CO_2^- (aq) \rightleftharpoons HS^- (aq) + CH_3CO_2H(aq)$$

$H_2S(aq)$	$CH_3CO_2^-(aq)$	$HS^-(aq)$	$CH_3CO_2H(aq)$
acid	base	conjugate base	conjugate acid

From the Table 17.4, we see that CH_3CO_2H is a stronger acid than H_2S and HS^- is a stronger base than $CH_3CO_2^-$. Hence the **reaction does not occur** to any appreciable extent.

86. $0.50 \text{ g Ca(OH)}_2 \cdot \frac{1 \text{ mol}}{74.1 \text{ g}} = 0.0067$ mol and when dissolved in water to make 1.0 L, the concentration of $Ca(OH)_2$ is 0.0067 M. However, $[OH]^- = 2 \cdot 0.0067 \text{ M} = 0.014 \text{ M}$ (When $Ca(OH)_2$ dissolves, there are 2 OH^- per formula)
pOH = 1.87 and so pH = 12.13.

88. If the pH = 3.44 then $[H_3O^+] = 10^{-pH} = 10^{-3.44} = 3.6 \times 10^{-4}$ M

$$\text{Acid(aq)} + H_2O(\ell) \rightleftharpoons \text{Conjugate Base (aq)} + H_3O^+(aq)$$

	[Acid]	[Conjugate Base]	$[H_3O^+]$
Initial	0.010	0	0
Change	-0.00036	+0.00036	+0.00036
Equilibrium	0.010-0.00036	0.00036	0.00036

$$K_a = \frac{[\text{Conjugate Base}][H_3O^+]}{[\text{Acid}]} = \frac{(0.00036)^2}{0.010 - 0.00036} = 1.4 \times 10^{-5}$$

90. a. The K_b for butylamine is:

$$K_b = \frac{K_w}{K_a} = \frac{1.0 \times 10^{-14}}{2.3 \times 10^{-11}} = 4.3 \times 10^{-4}$$

b. With a K_a of 2.3×10^{-11}, the butylammonium ion would go immediately above the phosphate ion (PO_4^{3-}) and below $Ni(H_2O)_5OH^+$. The **PO_4^{3-} ion is a stronger base** than butylamine (as are all the other bases below it in the list.)

c. pH of a 0.015 M butylammonium ion solution:

$$C_4H_9NH_3^+ (aq) + H_2O (\ell) \rightleftharpoons C_4H_9NH_2 (aq) + H_3O^+ (aq) \text{ and}$$

$$K_a = \frac{[C_4H_9NH_2][H_3O^+]}{[C_4H_9NH_3^+]} = \frac{[x]^2}{[0.015 - x]} = 2.3 \times 10^{-11}$$

making the approximation that the denominator $(0.015 - x) \cong 0.015$
(Recall that this is reasonable when the K is much much smaller than the solution concentration). $x^2 = (0.015)(2.3 \times 10^{-11})$ and $x = 5.87 \times 10^{-7}$ M
$pH = -\log(5.87 \times 10^{-7}) = 6.23$

92. pH of aqueous solutions of

	reaction	pH
a. $NaHSO_4$	hydrolysis of HSO_4^- produces H_3O^+	< 7
b. NH_4Br	hydrolysis of HSO_4^- produces H_3O^+	< 7
c. $KClO_4$	no hydrolysis occurs	= 7
d. Na_2CO_3	hydrolysis of CO_3^{2-} produces OH^-	> 7
e. $(NH_4)_2S$	hydrolysis of S^{2-} produces OH^-	> 7
f. $NaNO_3$	no hydrolysis occurs	= 7
g. Na_2HPO_4	hydrolysis of HPO_4^{2-} produces OH^-	> 7
h. LiBr	no hydrolysis occurs	= 7
i. $FeCl_3$	hydrolysis of Fe^{3+} produces H_3O^+	< 7

94. $$NH_2OH(aq) + H_2O(\ell) \rightleftharpoons NH_3OH^+(aq) + OH^-(aq)$$

	$[NH_2OH]$	$[NH_3OH^+]$	$[OH^-]$
Initial:	0.051	0	0
Change:	- x	+ x	+ x
Equilibrium:	0.051 - x	x	x

$$K_b = \frac{[NH_3OH^+][OH^-]}{[NH_2OH]} = \frac{x^2}{0.051 - x} = 6.6 \times 10^{-9}$$

Making the very reasonable assumption that x is very small relative to 0.051 (since K_b is small), we find $x = 1.8 \times 10^{-5}$ M = $[NH_3OH^+] = [OH^-]$.
Therefore, the hydroxide ion concentration gives pOH = 4.74 and pH = 9.26.

96. Of the solutions listed:

a. Acidic

0.1 M CH_3CO_2H	a weak acid
0.1 M NH_4Cl	a salt of a strong acid and weak base

b. Basic

0.1 M NH_3	a weak base
0.1 M Na_2CO_3	a salt of a strong base and a weak acid
0.1 M $NaCH_3CO_2$	a salt of a strong base and a weak acid

NaCl	neutral—NaCl is the salt of a strong acid and a strong base.
$NH_4CH_3CO_2$	neutral—both NH_4^+ and $CH_3CO_2^-$ ions hydrolyze—but to the same extent, hence the solution remains neutral.

c. The most acidic:

CH_3CO_2H	K_a for acetic acid is greater than the corresponding K_a of the ammonium ion.

98. The solutions may be arranged in order of increasing pH by examining K_a's and K_b's in Table 17.4. From the most acidic (lowest pH) to the least acidic (highest pH):

$$HCl < CH_3CO_2H < NaCl < NH_3 < NaCN < NaOH$$

{Recall that the larger the K_a, the greater the dissociation of the acid into (anion + H^+). The larger the K_a, the greater [H^+], and the lower the pH!}

100. For oxalic acid $K_{a1} = 5.9 \times 10^{-2}$ and $K_{a2} = 6.4 \times 10^{-5}$

Representing oxalic as H_2A. The first step can be written as:

$$H_2A + H_2O \rightleftharpoons HA^- + H_3O^+ \qquad K_{a1} = 5.9 \times 10^{-2}$$

We can write the second step:

$$HA^- + H_2O \rightleftharpoons A^{2-} + H_3O^+ \qquad K_{a2} = 6.4 \times 10^{-5}$$

If we add the two equations we get:

$$H_2A + 2\,H_2O \rightleftharpoons A^{2-} + 2\,H_3O^+ \qquad K_{net} = (5.9 \times 10^{-2})(6.4 \times 10^{-5})$$

or $H_2C_2O_4\,(aq) + 2\,H_2O\,(\ell) \rightleftharpoons C_2O_4^{2-}\,(aq) + 2\,H_3O^+$ $\qquad = 3.8 \times 10^{-6}$

102. The reaction: $I_2 + I^- \rightleftharpoons I_3^-$ can be pictured as below:

I_2 = Lewis acid, I^- = Lewis base

$$:\ddot{I}-\ddot{I}: + [:\ddot{I}:]^- \rightleftharpoons [:\ddot{I}-\ddot{I}-\ddot{I}:]^-$$

In forming the triiodide ion, the iodide ion donates a pair of electrons to one of the iodine atoms in the I_2 molecule.

Conceptual Questions

104. a. As hydrogen atoms are successively replaced by the very electronegative chlorine atoms, an increase in acidity is seen. This is quite understandable if you remember that an increasing "pull" on the electrons in the "carboxylate" end of the molecule--brought about by an **increasing number of chlorine atoms**—will weaken the O-H bond and **increase the acidity of the specie.**

b. The acid with the largest K_a (Cl_3CCO_2H) would be the strongest acid and would have the lowest pH. The acid with the smallest K_a (CH_3CO_2H) would have the highest pH.

106. a. The reaction of perchloric acid with sulfuric acid:

$$HClO_4 + H_2SO_4 \rightleftharpoons ClO_4^- + H_3SO_4^+$$

b. Lewis dot structure for sulfuric acid:

```
       ..
      :O:                              ..        +
   ..  |  ..                          :O:
 H:O - S - O:                    ..    |   ..
   ..  |  ..                   H:O  -  S - O:H
      :O:                        ..    |   ..
       ..                             :O:
       H                               ..
                                       H
```

and the protonated form would look like: (bracketed structure above, with overall + charge)

Sulfuric acid has two oxygen atoms capable of "donating" an electron pair, e.g. to a proton, H^+, thereby acting as a base.

Challenging Questions

108. a. As the K_a values increase, the concentration of Hydronium ion that the acid would produce increases—reducing the pH of the solution. So the **nitropyridinium hydrochloride (x = NO_2) solution would have the lowest pH** and the **methylpyridinium hydrochloride (x = CH_3) solution would provide the highest pH.**

b. Strongest and weakest Brønsted base:

Remembering the relationship that $K_a \cdot K_b = K_w$, it follows that the conjugate acid with the largest K_a (nitropyridinium hydrochloride) will have the conjugate base (**nitropyridine**) with the smallest K_b (**weakest base**) and similarly **methyl pyridine** would be the **strongest base.**

110. What is K_a for the conjugate acid of caffeine?

a. If $pK_a = 10.4$ then $K_a = 10^{-10.4}$ or 4×10^{-11} (1 sf)

b. Remembering the relationship that $K_a \cdot K_b = K_w$, then

$$K_b = \frac{K_w}{K_a} = \frac{1.0 \times 10^{-14}}{3.98 \times 10^{-11}} = 2.5 \times 10^{-4} \qquad (3 \times 10^{-4} \text{ to 1 sf})$$

Summary Questions

112. a. BF_3 is electron-deficient and is the **Lewis acid**. Dimethyl ether, $(CH_3)_2O$, has electron pairs (on the oxygen atom) and is the **Lewis base.**

b. What is the F-B-F angle in BF_3? in $(CH_3)_2O\text{-}BF_3$?

Boron trifluoride has three groups distributed around the B atom, and the resulting **trigonal planar** geometry gives a 120° bond angle. In the $(CH_3)_2O\text{-}BF_3$ complex, there are four groups distributed around the B atom, so the resulting F-B-F bond angle will be 109°.

c. What is the hybridization of O in the complex ? the hybridization of B?

The O will have four "groups" around it: 3 atoms-C,C,B- and a lone pair of electrons, so it will use **sp^3 hybridization**. As mentioned in (b) above, the four groups around the boron will also result in **sp^3 hybridization**.

d. Calculate the # of moles of complex:

$$1.00 \text{ g complex} \cdot \frac{1 \text{ mol complex}}{113.9 \text{ g complex}} = 0.00878 \text{ mol complex}$$

Since the complex is a gas, we can calculate the initial pressure of the complex using the Ideal Gas Law:

$$P = \frac{n \cdot R \cdot T}{V} = \frac{0.00878 \text{ mol} \cdot 0.082057 \frac{\text{L} \cdot \text{atm}}{\text{K} \cdot \text{mol}} \cdot 298 \text{ K}}{0.565 \text{ L}} = 0.380 \text{ atm}$$

For the equilibrium we can write:

	$(CH_3)_2O\text{-}BF_3$ (g) $\rightleftharpoons$	BF_3 (g) +	$(CH_3)_2O$ (g)	$K_p = 0.17$
Initial	0.380	0	0	
Change	- x	+ x	+ x	
Equilibrium	0.380 - x	x	x	

$$K_p = \frac{P_{BF_3} \cdot P_{(CH_3)_2O}}{P_{(CH_3)_2O\text{-}BF_3}} = \frac{x^2}{0.380 - x} = 0.17$$

Given the magnitude of K_p and the concentration of complex (0.380 atm), our usual simplifying assumptions won't suffice here. Solve the equation, using the quadratic equation.

$x^2 = 0.17(0.380 - x)$ and $x^2 + 0.17x - 6.46 \times 10^{-2} = 0$

$x = 0.18$ atm (2 sf)

So the pressure of: $BF_3 = 0.18$ atm

$(CH_3)_2O = 0.18$ atm

$(CH_3)_2O\text{-}BF_3 = (0.380 - 0.18) = 0.20$ atm

and the **total pressure** = $P_{BF_3} + P_{(CH_3)_2O} + P_{(CH_3)_2O\text{-}BF_3}$

= 0.18 + 0.18 + 0.20

= 0.56 atm

Chapter 18
Principles of Reactivity : Reactions Between Acids and Bases

Acid and Base Reactions

8. The value of the equilibrium constant for:

$$C_6H_5CO_2H + OH^- \rightarrow H_2O + C_6H_5CO_2^-$$

Note that this equilibrium can be represented as the sum of two reactions

$C_6H_5CO_2H + H_2O \rightarrow C_6H_5CO_2^- + H_3O^+$	K_a
$H_3O^+ + OH^- \rightleftharpoons 2\ H_2O$	$\frac{1}{K_w}$
$C_6H_5CO_2H + OH^- \rightarrow H_2O + C_6H_5CO_2^-$	K_{net}

and $K_{net} = K_a \cdot \frac{1}{K_w} = 6.3 \times 10^{-5} \cdot \frac{1}{1.0 \times 10^{-14}} = 6.3 \times 10^{9}$

This equilibrium **lies** predominantly **to the right**. We take advantage of this when we react weak acids with strong bases.

10. The net reaction is:

$CH_3CO_2H\ (aq) + NaOH\ (aq) \rightarrow CH_3CO_2^-\ (aq) + Na^+\ (aq) + H_2O\ (\ell)$

The addition of 22.0 mL of 0.15 M NaOH (3.3 mmol NaOH) to 22.0 mL of 0.15 M CH_3CO_2H (3.3 mmol CH_3CO_2H) produces water and the soluble salt, sodium acetate (3.3 mmol $CH_3CO_2^-\ Na^+$). The acetate ion is the anion of a weak acid and reacts with water according to the equation:

$$CH_3CO_2^-\ (aq) + H_2O\ (\ell) \rightarrow CH_3CO_2H\ (aq) + OH^-\ (aq)$$

The equilibrium constant expression may be written:

$$K_b = \frac{[CH_3CO_2H][OH^-]}{[CH_3CO_2^-]} = 5.6 \times 10^{-10}$$

The concentration of acetate ion is: $\frac{3.3 \text{ mmol}}{(22.0 + 22.0) \text{ ml}} = 0.075$ M

Equilibrium concentrations:

	$CH_3CO_2^-$	CH_3CO_2H	OH^-
Initial concentration	0.075		
Change (going to equilibrium)	-x	+x	+x
Equilibrium concentration	0.075 - x	+x	+x

$$K_b = \frac{[CH_3CO_2H][OH^-]}{[CH_3CO_2^-]} = \frac{x^2}{0.075 - x} = 5.6 \times 10^{-10}$$

Simplifying ($100 \cdot K_b << 0.075$) we get $\frac{x^2}{0.075} = 5.6 \times 10^{-10}$

$$x = 6.5 \times 10^{-6} = [OH^-]$$

The hydrogen ion concentration is related to the hydroxyl ion concentration by the equation:

$$K_w = [H_3O^+][OH^-] = 1.0 \times 10^{-14}$$

$$[H_3O^+] = \frac{1.0 \times 10^{-14}}{[OH^-]} = \frac{1.0 \times 10^{-14}}{6.5 \times 10^{-6}} = 1.5 \times 10^{-9}\text{ M}$$

pH = 8.81

The pH is greater than 7, as we expect for a salt of a strong base and weak acid.

12. Equal numbers of moles of acid and base are added in each case, leaving only the salt of the acid and base. The reaction (if any) of that salt with water (hydrolysis) will affect the pH.

pH of solution	Reacting Species	Reaction controlling pH
a. >7	CH_3CO_2H/KOH	Hydrolysis of $CH_3CO_2^-$
b. < 7	HCl/NH_3	Hydrolysis of NH_4^+
c. = 7	$HNO_3/NaOH$	No hydrolysis

14. We can calculate the amount of phenol present by converting mass to moles:

$$0.515\text{ g } C_6H_5OH \cdot \frac{1\text{ mol } C_6H_5OH}{94.11\text{ g } C_6H_5OH} = 5.47 \times 10^{-3}\text{ mol phenol}$$

At the equivalence point 5.47×10^{-3} mol of NaOH will have been added. Phenol is a monoprotic acid, that is one mol of phenol reacts with one mol of sodium hydroxide. The volume of 0.123 NaOH needed to provide this amount of base is:

$$\text{moles} = M \times V$$

$$5.47 \times 10^{-3}\text{ mol NaOH} = \frac{0.123\text{ mol NaOH}}{\text{L}} \times V$$

or 44.5 mL of the NaOH solution.

The total volume would be (125 + 44.5) or 170. mL solution.

Sodium phenoxide is a soluble salt, hence the initial concentration of both sodium and phenoxide ions will be equal to:

$$\frac{5.47 \times 10^{-3}\ \text{mol}}{0.170\ \text{L}} = 3.23 \times 10^{-2}\ \text{M}$$

The phenoxide ion however is the conjugate ion of a weak acid and undergoes hydrolysis.

$$C_6H_5O^- (aq) + H_2O (\ell) \rightleftharpoons C_6H_5OH (aq) + OH^- (aq)$$

	$C_6H_5O^-$	C_6H_5OH	OH^-
Initial	3.23×10^{-2} M		
Change	-x	+x	+x
Equilibrium	3.23×10^{-2} M - x	x	x

$$K_b = \frac{[C_6H_5OH][OH^-]}{[C_6H_5O^-]} = \frac{1.0 \times 10^{-14}}{1.3 \times 10^{-10}} = \frac{(x)(x)}{3.23 \times 10^{-2} - x} = 7.7 \times 10^{-5}$$

Since $7.7 \times 10^{-5} \cdot 100 < 3.23 \times 10^{-2}$ we simplify: $\frac{(x)(x)}{3.23 \times 10^{-2}} = 7.7 \times 10^{-5}$

$$x = 1.5 \times 10^{-3}$$

At the equivalence point: **$[OH^-] = 1.5 \times 10^{-3}$ M**

and $[H_3O^+] = \frac{1.0 \times 10^{-14}}{1.5 \times 10^{-3}} =$ **6.5×10^{-12}**

and **pH = 11.19**

The **Na^+** is a "spectator ion" and its **concentration remains unchanged at 3.23×10^{-2} M. The concentration of phenoxide is** reduced (albeit slightly) so that at equilibrium its concentration is also **3.23×10^{-2} M.**

16. At the equivalence point the moles of acid = moles of base.

$$(0.03678\ \text{L})\ (0.0105\ \text{M HCl}) = 3.86 \times 10^{-4}\ \text{mol HCl}$$

If this amount of base were contained in 25.0 mL of solution, **the concentration of NH_3 in the original solution was 0.0154 M.**

At the equivalence point NH_4Cl will hydrolyze according to the equation:

$$NH_4^+ (aq) + H_2O (\ell) \rightleftharpoons NH_3 (aq) + H_3O^+ (aq)$$

$$K_a = \frac{[NH_3][H_3O^+]}{[NH_4^+]} = \frac{1.0 \times 10^{-14}}{1.8 \times 10^{-5}} = 5.6 \times 10^{-10}$$

The salt (3.86×10^{-4} mol) is contained in (25.0 + 36.78) 61.78 mL. Its concentration will be $\frac{3.69 \times 10^{-4}\text{ mol}}{0.06178\text{ L}}$ or 6.25×10^{-3} M.

Substituting into the K_a expression :

$$\frac{[H_3O^+]^2}{6.25 \times 10^{-3}} = 5.6 \times 10^{-10}$$ and $\mathbf{[H_3O^+] = 1.9 \times 10^{-6}\ M}$

and **pH = 5.73.**

Since $[H_3O^+][OH^-] = 1.0 \times 10^{-14}$ then $[OH^-] = \frac{1.0 \times 10^{-14}}{1.9 \times 10^{-6}} = \mathbf{5.3 \times 10^{-9}\ M}$

and the $[NH_4^+]$ = $\mathbf{6.25 \times 10^{-3}\ M}$

The Common Ion Effect and Buffer Solutions

18. To determine how pH is expected to change, examine the equilibria in each case:

a. $NH_3\ (aq) + H_2O\ (\ell) \rightleftharpoons NH_4^+\ (aq) + OH^-\ (aq)$

As the added NH_4Cl dissolves, ammonium ions are liberated—increasing the ammonium ion concentration and shifting the position of equilibrium to the left—reducing OH^-, and **decreasing the pH.**

b. $CH_3CO_2H + H_2O \rightleftharpoons CH_3CO_2^- + H_3O^+$

As sodium acetate dissolves, the additional acetate ion will shift the position of equilibrium to the left—reducing H_3O^+, and **increasing the pH.**

c. $NaOH \rightarrow Na^+ + OH^-$

NaOH is a strong base, and as such is totally dissociated. Since the added NaCl does not hydrolyze to any appreciable extent—**no change in pH occurs.**

20. As in study question 18, examine the equilibria involved:

a. Adding HCl will consume NH_3, **lowering the pH.**

b. Adding NaOH will consume acetic acid, **raising the pH.**

22. The pH of the buffer solution is:

$$K_b = \frac{[NH_4^+][OH^-]}{[NH_3]} = \frac{(0.20)[OH^-]}{(0.20)} = 1.8 \times 10^{-5}$$

and solving for hydroxyl ion yields: $[OH^-] = 1.8 \times 10^{-5}$ M

pOH = 4.74 pH = 9.26

24. The original pH of the 0.12 M NH_3 solution will be:

$$\frac{[NH_4^+][OH^-]}{[NH_3]} = 1.8 \times 10^{-5} \quad \text{and} \quad \frac{x^2}{0.12 - x} = 1.8 \times 10^{-5}$$

Assuming $(0.12 - x \approx 0.12)$, $x = 1.47 \times 10^{-3}$

$[OH^-] = 1.47 \times 10^{-3}$ M and pOH = 2.83 with pH = 11.17

Adding 2.2 g of NH_4Cl (0.041 mol) to 250 mL will produce an immediate increase of 0.16 M NH_4^+ (0.041 mol/0.250 L).

Substituting into the equilibrium expression as we did earlier, we get:

$$\frac{(x + 0.16)(x)}{0.12 - x} = 1.8 \times 10^{-5}$$

Assuming $(x + 0.16 \approx 0.16)$ and $(0.12 - x \approx 0.12)$

$$\frac{0.16(x)}{0.12} = 1.8 \times 10^{-5} \quad x = 1.35 \times 10^{-5} = [OH^-]$$

Note the hundred-fold decrease in $[OH^-]$ over the initial ammonia solution, as predicted by LeChatelier's principle.

So pOH = 4.88 and pH = 9.12 **(lower than original)**

If one uses K_a of 5.6×10^{-10}, the pH will be 9.11.

26. Mass of sodium acetate needed to change 1.00 L solution of 0.10 M CH_3CO_2H to pH = 4.5:

The equilibrium affected is that of acetic acid in water:

$$CH_3CO_2H + H_2O \rightleftharpoons CH_3CO_2^- + H_3O^+ \qquad K_a = 1.8 \times 10^{-5}$$

The equilibrium expression is:

$$K_a = \frac{[CH_3CO_2^-][H_3O^+]}{[CH_3CO_2H]} = 1.8 \times 10^{-5}$$

We know the concentration of acetic acid (0.10M), and we know the desired $[H_3O^+]$:

pH = 4.5 so $[H_3O^+] = 10^{-4.5} = 3.2 \times 10^{-5}$

Substituting these values into the equilibrium expression gives:

$$\frac{[CH_3CO_2^-][H_3O^+]}{[CH_3CO_2H]} = 1.8 \times 10^{-5} = \frac{[CH_3CO_2^-][3.2 \times 10^{-5}]}{[0.10]}.$$

We can solve for $[CH_3CO_2^-] = 0.057$ M

What mass of $NaCH_3CO_2$ would give this concentration of $CH_3CO_2^-$?

$$\frac{0.057 \text{ mol}}{1 \text{ L}} \bullet \frac{1\text{L}}{1} \bullet \frac{82.0 \text{ g } NaCH_3CO_2}{1 \text{ mol } NaCH_3CO_2} = 4.7 \text{ g } NaCH_3CO_2 \text{ (or 5 g to 1 sf)}$$

28. pH of buffer with 12.2 g $C_6H_5CO_2H$ and 7.20 g $C_6H_5CO_2Na$ in 250. mL of solution:

1. Calculate concentrations for the acid and its salt:

$$12.2 \text{ g } C_6H_5CO_2H \bullet \frac{1 \text{ mol } C_6H_5CO_2H}{122.1 \text{ g } C_6H_5CO_2H} \bullet \frac{1}{250 \text{ L}} = 0.400 \text{ M } C_6H_5CO_2H$$

$$7.20 \text{ g } C_6H_5CO_2Na \bullet \frac{1 \text{ mol } C_6H_5CO_2Na}{144.1 \text{ g} C_6H_5CO_2Na} \bullet \frac{1}{250 \text{ L}} = 0.200 \text{ M } C_6H_5CO_2Na$$

2. Substitute equilibrium concentrations into the equilibrium constant expression:

$$C_6H_5CO_2H \text{ (aq)} + H_2O(\ell) \rightleftharpoons C_6H_5CO_2^- \text{ (aq)} + H_3O^+ \text{ (aq)}$$

	$C_6H_5CO_2H$	$C_6H_5CO_2^-$	H_3O^+
Initial concentration	0.400	0.200	
Change (going to equilibrium)	-x	+x	+x
Equilibrium concentration	0.400 - x	0.200 + x	+x

$$K_a = \frac{[C_6H_5CO_2^-][H_3O^+]}{[C_6H_5CO_2H]} = \frac{(0.200 + x)(x)}{0.400 - x} = 6.3 \times 10^{-5}$$

Simplifying the equation: $\frac{(0.200)(x)}{(0.400)} = 6.3 \times 10^{-5}$

$$x = 1.26 \times 10^{-4} \text{ M} = [H_3O^+] \text{ and pH} = 3.90$$

Diluting this solution to 0.500 L **will not change the pH**. Notice that the concentrations of both acid and salt are changed, leaving the ratio of salt : acid and the pH unchanged.

30. The best combination to provide a buffer solution of pH 9 is b, the NH_3/NH_4^+ system. Note that K_a (for NH_4^+) is approximately 10^{-10}. Buffer systems are good when the desired pH is ± 1 unit from pK_a (10 in this case).

The HCl and NaCl don't form a buffer. The acetic acid/sodium acetate system would form an acidic buffer ($pK_a \approx 5$) in the pH range 4 - 6.

32. a. Initial pH

Need to know the concentrations of the conjugate pairs:

The equilibrium expression shows the *ratio of the conjugate pairs*, we can calculate **moles** of the conjugate pairs, and know that the ratio of the # of moles of the species will have the same value as the ratio of their concentrations!

$CH_3CO_2H = 0.250 \text{ L} \cdot 0.150 \text{ M} = 0.0375 \text{ mol}$

$NaCH_3CO_2 = 4.95 \text{ g} \cdot \frac{1 \text{ mol}}{82.07 \text{ g}} = 0.0603 \text{ mol}$

Substituting into the K_a expression:

$$\frac{[CH_3CO_2^-][H_3O^+]}{[CH_3CO_2H]} = 1.8 \times 10^{-5} = \frac{[0.0603][H_3O^+]}{[0.0375]}$$

and solving for $[H_3O^+] = 1.1 \times 10^{-5}$ M ; pH = 4.95

b. pH after 82 mg NaOH is added to 100. mL of the buffer. The amount of the conjugate pairs in 100/250 of the buffer is (100/250)(0.0375 mol) = 0.0150 mol CH_3CO_2H

and (100/250)(0.0603 mol) = 0.0241 mol $CH_3CO_2^-$

$82 \text{ mg NaOH} \cdot \frac{1 \text{ mmol NaOH}}{40.0 \text{ mg NaOH}} = 2.05 \text{ mmol NaOH or } 0.00205 \text{ mol NaOH}$

or 2.1 mmol NaOH (to 2 significant figures)

This base would **consume** an equivalent amount of $\mathbf{CH_3CO_2H}$ and **produce** an equivalent amount of $\mathbf{CH_3CO_2^-}$. After that process: (0.0150- 0.0021) or 0.0129 mol CH_3CO_2H and (0.0241 + 0.0021) or 0.0262 mol $CH_3CO_2^-$ are present. Substituting into the K_a expression as in part a:

$$\frac{(0.0262)[H_3O^+]}{(0.0129)} = 1.8 \times 10^{-5}$$

and $[H_3O^+] = 8.9 \times 10^{-6}$ M and pH = 5.05

34. a. The pH of the buffer solution is:

$$K_b = \frac{[NH_4^+][OH^-]}{[NH_3]} = \frac{(0.250)[OH^-]}{(0.500)} = 1.8 \times 10^{-5}$$

Note : Here the data presented are given as moles (in the case of ammonium chloride) and molar concentration (in the case of ammonia). In solution #32 we substituted the # moles of the conjugate pairs into the K expression. Here we must first *decide* whether to substitute # *moles* or *molar concentrations* into the K_b expression. **Either would work**. What is critical to remember is that we have to have both species expression in one

or the other form—not a mix of the two. Here I chose to convert moles of NH_4Cl into molar concentrations, and substitute.

Solving for hydroxyl ion in the K_b expression above yields: $[OH^-] = 3.6 \times 10^{-5}$ M

pOH = 4.44 pH = 9.56

If one uses K_a of 5.6×10^{-10}, the pH will be 9.55.

b. pH after addition of 0.0100 mol HCl:

The basic component of the buffer (NH_3) will react with the HCl, producing more ammonium ion. The composition of the solution is then:

	NH_3	NH_4Cl
Moles present (before HCl is added)	0.250	0.125
Change (reaction)	- 0.0100	+ 0.0100
Following reaction	0.240	+ 0.135

The amounts of NH_3 and NH_4Cl following the reaction with HCl are only slightly different from those amounts prior to reaction. Converting these numbers into molar concentrations (volume is 500. mL) and substituting the concentrations into the K_b expression yields:

$$K_b = \frac{[NH_4^+][OH^-]}{[NH_3]} = \frac{(0.270)[OH^-]}{(0.480)} = 1.8 \times 10^{-5}$$

$[OH^-] = 3.2 \times 10^{-5}$ M; pOH = 4.50, and the new pH = 9.50.

Using the Henderson-Hasselbalch Equation

36. The pK_a for acetic acid is = $-\log(K_a)$ or 4.74

$$pH = pK_a + \log\frac{[\text{conjugate base}]}{[\text{acid}]}$$
$$= 4.74 + \log\frac{0.075}{0.050} = 4.74 + 0.18 \text{ or } 4.92$$

38. a. The pK_a for formic acid is = $-\log(K_a)$ = $-\log(1.8 \times 10^{-4})$ or 3.74

b. $pH = pK_a + \log \frac{[\text{conjugate base}]}{[\text{acid}]}$

$= 3.74 + \log \frac{0.035}{0.050} = 3.74 - 0.15 \text{ or } 3.59$

c. Ratio of conjugate pairs to increase pH by 0.5 (to pH = 4.09)

Substituting into the Henderson-Hasselbalch equation

$4.09 = 3.74 + \log \frac{[\text{conjugate base}]}{[\text{acid}]}$

$+0.35 = \log \frac{[\text{conjugate base}]}{[\text{acid}]}$

$-0.35 = \log \frac{[\text{acid}]}{[\text{conjugate base}]}$

$0.4 = \frac{[\text{acid}]}{[\text{conjugate base}]}$

Titration Curves and Indicators

40. The titration of 0.10 M NaOH with 0.10 M HCl (a strong base vs a strong acid)

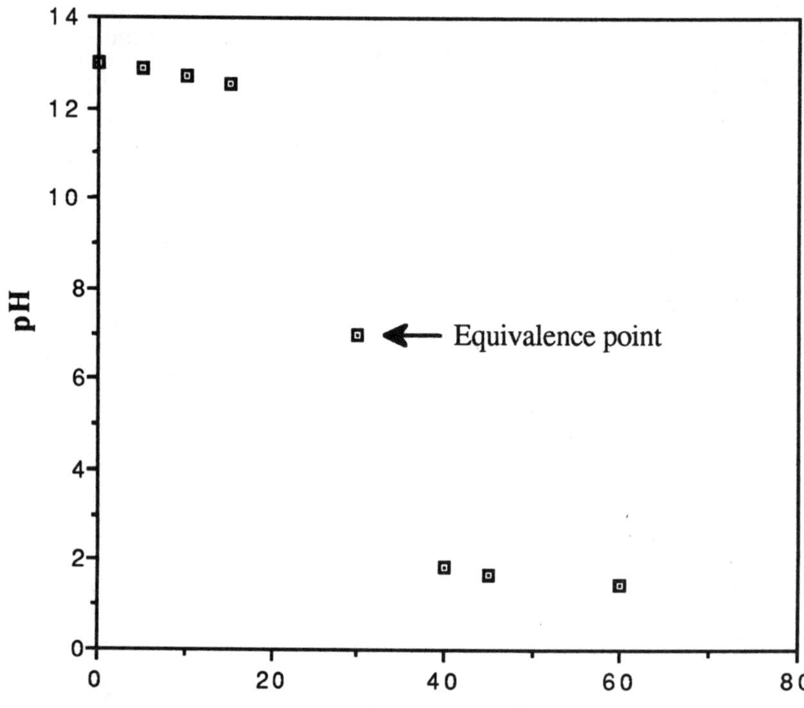

The initial pH of a 0.10 M NaOH would be pOH = - log[0.1]
pOH = 1 and pH = 13.

When 15.0 mL of 0.10 M HCl have been added, one-half of the NaOH initially present will be consumed, leaving 0.5 (0.030 L • 0.10 mol/L) or 1.50×10^{-3} mol NaOH in 45.0 mL—therefore a concentration of 0.0333 M NaOH. pOH = 1.48 and pH = 12.52.

At the equivalence point (30.0 mL of the 0.10 M acid are added) there is only NaCl present. Since this salt does not hydrolyze, the pH at that point is exactly 7.0. The total volume present at this point is 60 mL.

Once a total of 60.0 mL of acid are added, there is an excess of 3.0×10^{-3} mol of HCl. Contained in a total volume of 90.0 mL of solution, the [HCl] = 0.0333 and the pH =1.5.

42. a. pH of 25.0 mL of 0.11 M NH_3:

For the weak base, NH_3, the equilibrium in water is represented as:

$$NH_3\ (aq) + H_2O\ (\ell) \rightleftharpoons NH_4^+\ (aq) + OH^-\ (aq)$$

The slight dissociation of NH_3 would form equimolar amounts of NH_4^+ and OH^- ions.

$$K_b = \frac{[NH_4^+][OH^-]}{[NH_3]} = \frac{(x)(x)}{0.11 - x} = 1.8 \times 10^{-5}$$

Simplifying, we get : $\frac{x^2}{0.11} = 1.8 \times 10^{-5}$ $x = 1.4 \times 10^{-3} = [OH^-]$

pOH = 2.85 and pH = 11.15

Addition of HCl will consume NH_3 and produce NH_4^+ (the conjugate) according to the net equation: $NH_3\ (aq) + H^+\ (aq) \rightleftharpoons NH_4^+\ (aq)$

The strong acid will drive this equilibrium to the right so we will assume this reaction to be complete. Let us first calculate the moles of NH_3 initially present:

$$(0.0250\ L)\ (0.11\ \frac{mol\ NH_3}{L}) = 0.0028\ mol\ NH_3$$

Reaction with the HCl will produce the conjugate acid, NH_4^+. The task is two-fold. First calculate the amounts of the conjugate pair present. Second substitute the concentrations into the K_b expression. [One time-saving hint: **The ratio of concentrations** and the **ratio of the amounts (moles)** will have the same numerical value. One can substitute the amounts of the conjugate pair into the K_b expression.]

$$K_b = \frac{[NH_4^+][OH^-]}{[NH_3]} = 1.8 \times 10^{-5}$$

Present initially: 2.75 millimol NH_3

mL of 0.10 M HCl added	millimol HCl added	millimol NH_3 after reaction	millimol NH_4^+ after reaction	$[OH^-]$ after reaction	pH
5.00	0.50	2.25	0.50	8.1×10^{-5}	9.91
11.0	1.1	1.65	1.1	2.7×10^{-5}	9.43
12.5	1.25	1.50	1.25	2.2×10^{-5}	9.33
15.0	1.5	1.25	1.5	1.5×10^{-5}	9.18
20.0	2.0	0.75	2.0	6.8×10^{-6}	8.83
22.0	2.2	0.55	2.2	4.5×10^{-6}	8.65

b. When 28.0 mL of the 0.10 M HCl has been added (total solution volume = 53.0 mL), the reaction is at the equivalence point. All the NH_3 will be consumed, leaving the salt, NH_4Cl. The NH_4Cl (2.75 millimol) has a concentration of 5.2×10^{-2} M. This <u>salt,</u> being formed from a <u>weak base</u> and <u>strong acid,</u> undergoes <u>hydrolysis.</u>

$$NH_4^+(aq) + H_2O\ (\ell) \rightleftharpoons NH_3\ (aq) + H_3O^+\ (aq)$$

	NH_4^+	NH_3	H_3O^+
Initial concentration	5.2×10^{-2}		
Change (going to equilibrium)	-x	+x	+x
Equilibrium concentration	5.2×10^{-2} - x	+x	+x

$$K_a = \frac{[NH_3][H_3O^+]}{[NH_4^+]} = 5.6 \times 10^{-10} = \frac{x^2}{5.2 \times 10^{-2} - x}$$

$$\approx \frac{x^2}{5.2 \times 10^{-2}} = 5.6 \times 10^{-10} \text{ and } x = 5.4 \times 10^{-6} = [H_3O^+]$$

and pH = 5.27 (equivalence point)

c. The midpoint of the titration occurs when 13.75 (13.8) mL of the acid have been added. At that point the amount of base and salt present are equal. An examination of the K_b expression will show that under these conditions the $[OH^-] = K_b$. Hence a pOH of 4.74 and a pH = 9.26.

d. From the table of indicators in your text we see that one indicator to use is Methyl Red. This indicator would be yellow prior to the equivalence point and red past that point. Bromcresol green would also be suitable, being blue prior to the equivalence point and yellow-green after.

e. All points except the one for pH after the addition of 30.0 mL have been listed above. Addition of acid in excess of 27.50 mL will result in a solution which is essentially a strong acid.

After the addition of 30.0 mL, substances present are:

millimol HCl added:	3.00
millimol NH_3 present:	2.75
excess HCl present :	0.25 millimol

This HCl is present in a total volume of 55.0 mL of solution, hence the calculation for a strong acid proceeds as follows:

$$[H_3O^+] = \frac{0.25 \text{ millimol HCl}}{55.0 \text{ milliliters}} = 4.5 \times 10^{-3} \text{ M} \quad \text{and pH} = 2.34.$$

The graph for these data is:

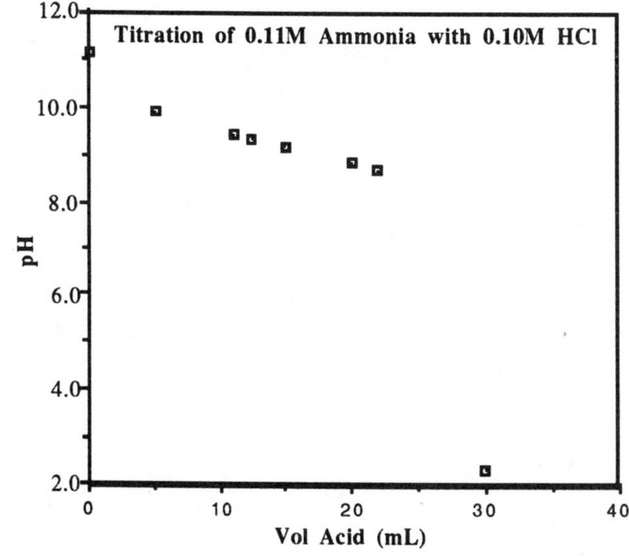

A Summary

mL acid	pH
0.00	11.15
5.00	9.91
11.0	9.43
12.5	9.33
15.0	9.18
20.0	8.83
22.0	8.65
30.0	2.34

44. a. pH of 25.0 mL of 0.010 M $HOCH_2CH_2NH_2$:

For the weak base, NH_3, the equilibrium in water is represented as:

$HOCH_2CH_2NH_2\ (aq) + H_2O\ (\ell) \rightleftharpoons HOCH_2CH_2NH_3^+\ (aq) + OH^-\ (aq)$

Note : The K_b for ethanolamine is given as 3.2×10^{-5}.

The slight dissociation of $HOCH_2CH_2NH_2$ would form equimolar amounts of $HOCH_2CH_2NH_3^+$ and OH^- ions.

$$K_b = \frac{[HOCH_2CH_2NH_3^+][OH^-]}{[HOCH_2CH_2NH_2]} = \frac{(x)(x)}{0.010 - x} = 3.2 \times 10^{-5}$$

Simplifying, we get : $\frac{x^2}{0.010} = 3.1 \times 10^{-5}$ and $x = [OH^-] = 5.7 \times 10^{-4}$ M

pOH = 3.25 and pH = 10.75

b. pH at the equivalence point:

At the equivalence point, there are equal # of moles of acid and base. The number of moles of ethanolamine = (0.025 L)(0.010 M) = 2.5×10^{-4} mol ethanolamine .

That amount of HCl would be:

$$2.5 \times 10^{-4} \text{ mol HCl} \bullet \frac{1 \text{ L}}{9.5 \times 10^{-3} \text{ mol}} = 0.02632 \text{ L (or 26.32 mL)}$$

This amount of acid when added to the volume of base would be : 26.32 + 25.0 = 51.3 mL. Since we have added equal amounts of acid and base, the reaction between the two will result in the existence of **only the salt.**

and the concentration of salt would be: $\frac{2.5 \times 10^{-4} \text{ mol}}{5.13 \times 10^{-2} \text{ L}} = 4.9 \times 10^{-3}$ M

$HOCH_2CH_2NH_3^+\ (aq) + H_2O\ (\ell) \rightleftharpoons HOCH_2CH_2NH_2\ (aq) + H_3O^+\ (aq)$

The equilibrium constant (K_a) would be $\frac{K_w}{K_b} = \frac{1.0 \times 10^{-14}}{3.2 \times 10^{-5}} = 3.1 \times 10^{-10}$

The equilibrium expression would be:

$$\frac{[HOCH_2CH_2NH_2][H_3O^+]}{[HOCH_2CH_2NH_3^+]} = 3.1 \times 10^{-10}$$

Given that the salt would hydrolyze to form equal amount of ethanolamine and hydronium ion (as shown by the equation above), if we represent the concentrations of those species as x, then we can write (using our usual approximation):

$$\frac{x^2}{4.9 \times 10^{-3}} = 3.1 \times 10^{-10}$$ and solving for x = 1.23×10^{-6}

Since x represents $[H_3O^+]$, then pH = - log(1.23×10^{-6}) = 5.91

c. pH at the midpoint of the titration:

The volume of acid added at the midpoint would be 1/2 the volume of acid calculated in (b) above, or 13.16 mL. This volume isn't **that important** since if you recall that at this point in the titration, the amounts of base and conjugate acid will be equal. Substituting that fact into the equilibrium expression we obtain:

$$\frac{[HOCH_2CH_2NH_3^+][OH^-]}{[HOCH_2CH_2NH_2]} = \frac{(x)[OH^-]}{(x)} = 3.2 \times 10^{-5}$$ so we can see that the

$[OH^-] = 3.2 \times 10^{-5}$ M and pOH = 4.49 and pH = (14.00-4.49) = 9.51

d. Suitable indicator for endpoint: Bromcresol purple changes color about pH = 5.9, being purple above this pH and yellow at pH values less than 5.9

e. pH after adding volumes of 0.0095 M HCl (*See Question 42 for additional explanation*)

Present initially: 0.25 millimol base

mL of 0.0095M HCl added	millimol HCl added	millimol base after reaction	millimol Hbase+ after reaction	$[OH^-]$ after reaction	pOH	pH
5.00	0.0475	0.2025	0.0475	1.4×10^{-4}	3.89	10.13
10.0	0.0950	0.155	0.0950	5.2×10^{-5}	4.28	9.72
20.0	0.1900	0.060	0.1900	1.0×10^{-5}	5.00	9.00
30.0†	0.2850	**-0.035**	**0.035**	*******	****	3.20

Substituting into the K_b expression we have: $$\frac{[Hbase^+][OH^-]}{[base]} = 3.2 \times 10^{-5}$$

Values in the table are calculated as follows:

millimol base after reaction = (0.25 millimol - millimol HCl added)

millimol HBase+ after reaction = # millimol of acid acid

$[OH^-]$ calculated by substitution of millimol base and millimol Hbase+ into K_b

Note: Ratios of # of moles are equal to the *ratios* of the concentrations of the conjugate pairs!!

† Note that the addition of 30.0 mL of the acid results in adding **more moles of HCl than moles of ethanolamine initially present.** Hence, once all the ethanolamine has been consumed, the HCl is left *in excess* and the concentration of this acid may be calculated:

$$\frac{3.5 \times 10^{-5} \text{ mol HCl}}{0.055 \text{ L}} = 6.4 \times 10^{-4} = [H_3O^+] \text{ and the pH} = 3.20$$

A plot summarizing these calculations follows:

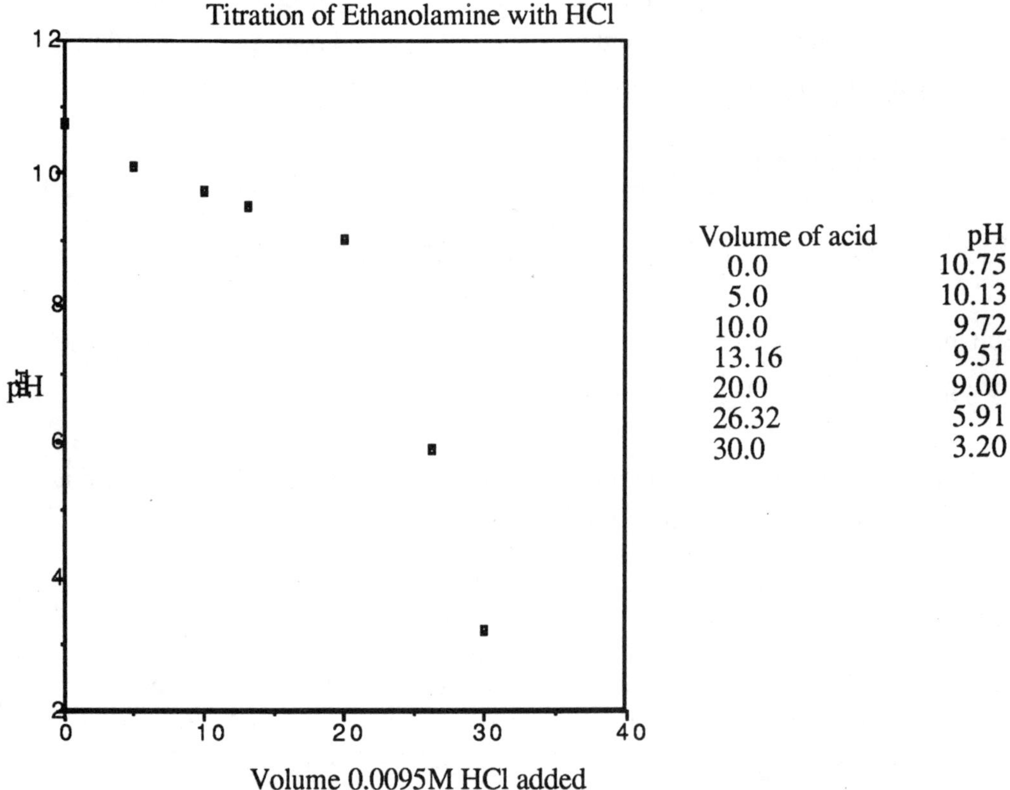

Volume of acid	pH
0.0	10.75
5.0	10.13
10.0	9.72
13.16	9.51
20.0	9.00
26.32	5.91
30.0	3.20

46. Suitable indicators for titrations:

a. HCl with pyridine: A solution of pyridinium chloride would have a pH of approximately 5.1 at the equivalence point.

This can be seen by writing the equation for the pyridinium ion reacting in water: [Representing the pyridinium ion as PH^+]

$$PH^+ (aq) + H_2O (\ell) \rightleftharpoons P (aq) + H_3O^+ (aq) \qquad K_a = \frac{1 \times 10^{-14}}{1.5 \times 10^{-5}}$$

Let's assume that we begin with 0.1 M PH^+.

The equilibrium expression would be:

$$\frac{[P] \cdot [H_3O^+]}{[PH^+]} = 6.7 \times 10^{-10},$$

If we express $[P] = [H_3O^+] = x$ and $[PH^+] = 0.1 - x$ then we can write:

$\frac{x \cdot x}{0.1 - x} = 6.7 \times 10^{-10}$ and with the usual approximations we can solve for $x = 8.2 \times 10^{-6}$ M and $pH = -\log(8.2 \times 10^{-6}) = 5.1$. **Methyl red** would be a good indicator for this equivalence point.

b. NaOH with formic acid: The salt formed at the equivalence point is sodium formate. Hydrolysis of the formate ion would give rise to a basic solution (pH ≈ 8.5). **Phenolphthalein** would be a suitable indicator.

c. Hydrazine and HCl: The salt, hydrazinium hydrochloride, hydrolyzes. The hydrazinium ion ($N_2H_6^{2+}$) would produce an acidic solution (pH ≈ 2). You can do this calculation as we did in part (a) using as $K_a = \frac{1 \times 10^{-14}}{8.9 \times 10^{-16}}$. Since K_a is large, then we can calculate the $[H^+]$. Assume we begin with 0.1 M hydrazinium chloride. Since each mole of the chloride furnishes a mole of H^+ we get 0.1 M $[H^+]$, or a pH = 1. Suitable indicators would be **cresol red or thymol blue**.

General Questions

48. a. $CH_3CH_2CO_2H(aq) + OH^-(aq) \rightleftharpoons CH_3CH_2CO_2^-(aq) + H_2O(\ell)$

b. Since the equilibrium constant is large for this reaction , the reaction proceeds by having all the base consumed, consuming 0.40 mol of acid (leaving 0.60 mol of acid) and producing 0.40 mol of the conjugate base (propanoate ion).

c. The pH of the solution:

$$\frac{[CH_3CH_2CO_2^-][H_3O^+]}{[CH_3CH_2CO_2H]} = 1.3 \times 10^{-5}$$

$$[H_3O^+] = 1.3 \times 10^{-5} \cdot \frac{0.60}{0.40}$$

$$[H_3O^+] = 2.0 \times 10^{-5} \text{ and pH} = 4.71$$

d. This mass of NaOH corresponds to 0.01 mol of base. If we add this amount to the solution, we'll have 0.59 mol of acid and 0.41 mol of salt present. Substituting these data into the expression above we get

$$[H_3O^+] = 1.3 \times 10^{-5} \frac{(0.59)}{(0.41)} = 1.9 \times 10^{-5} \text{ M and a pH of 4.7.}$$

The pH would remain essentially the same.

50. For the equilibrium : $NH_3\,(aq) + H_2O\,(\ell) \rightleftharpoons NH_4^+\,(aq) + OH^-\,(aq)$

The NH_3 would have the equilibrium:

$$K_b = \frac{[NH_4^+][OH^-]}{[NH_3]} = \frac{(x)(x)}{(0.20 - x)} = 1.8 \times 10^{-5}$$

The 0.040 M NaOH would change $[OH^-]$—from the reaction of ammonia with water—to (x + 0.040).

$$\frac{(x)(x + 0.040)}{(0.20 - x)} = 1.8 \times 10^{-5}$$

Assuming that x is negligible with respect to 0.20 and 0.040, we can simplify the expression to:

$$\frac{(x)(0.040)}{(0.20)} = 1.8 \times 10^{-5} \quad \text{and} \quad x = 9.0 \times 10^{-5}$$

So $x = [NH_4^+] = 9.0 \times 10^{-5}$ M; $[NH_3] = 0.20$ M and $[Na^+] = [OH^-] = 0.040$ M

52. pH of 0.160 M CH_3CO_2H:

The equilibrium may be written: $CH_3CO_2H + H_2O \rightleftharpoons CH_3CO_2^- + H_3O^+$

$$K_a = \frac{[CH_3CO_2^-][H_3O^+]}{[CH_3CO_2H]} = 1.8 \times 10^{-5}$$

$$= \frac{[H_3O^+]^2}{(0.160)} = 1.8 \times 10^{-5} \quad \text{and } [H_3O^+] = 1.70 \times 10^{-3} \text{ and pH} = 2.77$$

pH after 56.8 g of $NaCH_3CO_2$ is added:

The concentration of sodium acetate is :

$$\frac{56.8 \text{ g } NaCH_3CO_2}{1.50 \text{ L}} \bullet \frac{1 \text{ mol } NaCH_3CO_2}{82.03 \text{ g } NaCH_3CO_2} = 0.462 \text{ M } NaCH_3CO_2$$

Assuming that there is no volume change when the sodium acetate is added, the concentrations can be substituted into the K_a expression:

$$K_a = \frac{(x + 0.462)(H_3O^+)}{0.160 - x} = 1.8 \times 10^{-5}$$

Simplifying: $K_a = \frac{0.462[H_3O^+]}{0.160} = 1.8 \times 10^{-5}$ and $[H_3O^+] = 6.2 \times 10^{-6}$ M; pH = 5.20

54. At the equivalence point, the number of moles of acid and base are equal. 25.0 mL of 0.120 M HCO_2H contain 0.00300 moles. The volume of 0.105 M NaOH which contains

that amount of HCO_2H is $\frac{(0.025)(0.120)}{0.105} = 0.0286$ L.

The total volume of solution is (0.025 L + 0.0286) 0.0536 L. The concentration of the salt formed when the NaOH reacts with HCO_2H is then $\frac{0.00300 \text{ mol}}{0.0536 \text{ L}}$ or 0.0560 M.

Substituting into the K_b expression for the formate ion gives:

$$K_b = \frac{[HCO_2H][OH^-]}{[HCO_2^-]} = \frac{[OH^-]^2}{0.0560} = \frac{1.0 \times 10^{-14}}{1.8 \times 10^{-4}}$$

$[OH^-] = 1.7 \times 10^{-6}$ M and pH = 8.25

56. Arranged in order of increasing pH:

$HCl < CH_3CO_2H < CH_3CO_2H/NaCH_3CO_2 < NaCl < NH_3/NH_4Cl < NH_3$

Since all solutions are equimolar, we need only concentrate upon the degree of dissociation of the compounds, or their reaction with water.

- HCl is a strong acid, so would generate the greatest $[H_3O^+]$ and the lowest pH.
- CH_3CO_2H is a weak acid, and would generate a lower $[H_3O^+]$ than HCl and a higher pH.
- The $CH_3CO_2H/NaCH_3CO_2$ buffer (since it contains the basic acetate ion) would have a slightly higher pH than acetic acid.
- NaCl is a salt (of a strong acid and strong base) that will not react with water and therefore gives a pH = 7.
- NH_3 is a weak base, and would have the highest pH of all the substances listed.
- The buffer (since it contains the acidic ammonium ion) would have a slightly lower pH than ammonia.

58. Given that a 0.30 M solution of the acid has a pH = 2.25, we know that $[H_3O^+] = 10^{-2.25}$ or 5.6×10^{-3} M. If we write the equilibrium of the weak acid in water :

$$HA + H_2O \rightleftharpoons A^- + H_3O^+$$

We can write the equilibrium expression:

$$K_a = \frac{[H_3O^+][A^-]}{[HA]}$$

The stoichiometry indicates that for each H_3O^+ we get one A^-.

So $[H_3O^+] = [A^-] = 5.6 \times 10^{-3}$ M

Substituting these values into the K_a expression yields:

$$K_a = \frac{[H_3O^+][A^-]}{[HA]} = \frac{[5.6 \times 10^{-3}][5.6 \times 10^{-3}]}{[0.30 - 5.6 \times 10^{-3}]} = 1.07 \times 10^{-4}$$

Now that we know K_a, AND the ratio of the acid and its conjugate base ($[HA] = [A^-]$), we can substitute these values into the K_a expression and solve for $[H_3O^+]$. Let's rewrite the K_a expression:

$$\frac{[HA] \cdot K_a}{[A^-]} = [H_3O^+] \text{ so } K_a = [H_3O^+] = 1.07 \times 10^{-4} \text{ M and}$$

$\text{pH} = -\log (1.07 \times 10^{-4}) = 3.97$

60. The ratio of the two buffer components can be calculated from the equilibrium expression:

$$K_a = \frac{[H_3O^+][HPO_4^{2-}]}{[H_2PO_4^-]} = 6.2 \times 10^{-8}$$

With pH = 7.40, $[H_3O^+] = 4.0 \times 10^{-8}$ M. Substituting this value into the K_a expression and rearranging gives:

$$\frac{[HPO_4^{2-}]}{[H_2PO_4^-]} = \frac{6.2 \times 10^{-8}}{4.0 \times 10^{-8}} = 1.6$$

The requested ratio, however, is the reciprocal of $\frac{[HPO_4^{2-}]}{[H_2PO_4^-]}$.

Taking the reciprocal gives a ratio of 0.65.

62. a. What is the pH of the $[NH_3OH]Cl$ solution before the titration begins?

$$NH_3OH^+(aq) + H_2O(\ell) \rightleftharpoons NH_2OH(aq) + H_3O^+(aq)$$

$$K_a = \frac{K_w}{K_b} = \frac{1.0 \times 10^{-14}}{6.6 \times 10^{-9}} = 1.5 \times 10^{-6}$$

$$K_a = \frac{[NH_2OH][H_3O^+]}{[NH_3OH^+]} = \frac{[H_3O^+]^2}{0.155 - [H_3O^+]} = 1.5 \times 10^{-6}$$

$[H_3O^+] = 4.8 \times 10^{-4}$ M $\quad$ pH = 3.31

b. What is the pH at the equivalence point?

$0.025 \text{ L} \cdot 0.155 \text{ mol/L} = 0.0039 \text{ mol } NH_3OH^+$

0.0039 mol NH_3OH^+ requires 0.0039 mol NaOH and produces 0.0039 mol NH_2OH.

The volume of NaOH that contains 0.0039 mol NaOH is:

$$0.0039 \text{ mol NaOH} \cdot \frac{1.00 \text{ L}}{0.108 \text{ mol}} = 0.036 \text{ L}$$

Total volume at the equivalence point = 61 mL (25 + 36)

$$[NH_2OH] = \frac{0.0039 \text{ mol}}{0.061 \text{ L}} = 0.064 \text{ M}$$

$$K_b = \frac{[NH_3OH^+][OH^-]}{[NH_2OH]} = \frac{[OH^-]^2}{0.064 - [OH^-]} = 6.6 \times 10^{-9}$$

$[OH^-] = 2.1 \times 10^{-5}$ M pOH = 4.69 and pH = 9.31

c. What is the pH at the mid-point of the titration?

At the midpoint of the titration, $[NH_2OH] = [NH_3OH^+]$.

Therefore $[H_3O^+] = K_a$ for NH_3OH^+.

$[H_3O^+] = Ka = 1.5 \times 10^{-6}$ M and so pH = 5.82

Conceptual Questions

64 a. Examine the equilibrium of acetic acid in water

$$CH_3CO_2H + H_2O \rightleftharpoons CH_3CO_2^- + H_3O^+$$

LeChatelier's principle tells us that if we remove H_3O^+ (raise the pH) the equilibrium will shift to the right—with the effect of producing more acetate ion (and consuming acetic acid).

b. Predominant species at pH 4 is acetic acid (~ 85 %)

At pH 6, acetate ion predominates (~ 95 %)

c. Remembering the Henderson-Hasselbalch equation

$$pH = pK_a + \log\frac{[\text{conjugate base}]}{[\text{acid}]}$$

At the point in which the concentrations of acid and conjugate base are equal, the log term vanishes. So the pH = pK_a. The pK_a for acetic acid is 4.75, hence the pH of the solution is anticipated to be 4.75.

66. a. HB is the stronger acid.

At the equivalence point, the conjugate base of the weak acid is present.

$$B + H_2O \rightleftharpoons HB^+ + OH^-$$

The stronger the acid, the farther this equilibrium will lie to the left, reducing OH^- —providing a lower pH.

b. Since HA is the weaker of the two acids, **A^- is the stronger base.**

Challenging Questions

68. Prove that *at the mid-point of the titration of a weak base with a strong acid* $pH = 14.00 + \log K_b$.

For the weak base, let's call it B:

$$B + H_2O \rightleftharpoons HB^+ + OH^- \text{ and } K_b = \frac{[HB^+][OH^-]}{[B]}$$

At the mid-point of the titration, 1/2 of the base (B) will have been consumed, producing an equal amount of the protonated base (HB^+) so $[B] = [HB^+]$

We can then simplify the K_b expression:

$K_b = [OH^-]$ taking the logarithm of both sides:

$\log K_b = \log [OH^-]$.

Multiply by -1 to obtain: $-\log K_b = -\log [OH^-]$.

The **definition of pOH** $= -\log [OH^-]$, so we can state :

$-\log K_b = pOH$

At 25 °C, $pH + pOH = 14.00$ so $pH = 14.00 - pOH$.

Substitute for pOH to obtain: $pH = 14.00 - pOH$

$pH = 14.00 - (-\log K_b)$ or

$$\mathbf{pH = 14.00 + \log K_b}$$

70. At the second equivalence point in the titration of oxalic acid, there is only $C_2O_4^{2-}$ ion (and water). This equilibrium can be written:

$$C_2O_4^{2-} + H_2O \rightleftharpoons HC_2O_4^- + OH^-$$

This K_b can be calculated by $\frac{K_w}{K_{a2}}$ or $\frac{1.0 \times 10^{-14}}{6.4 \times 10^{-5}} = 1.6 \times 10^{-10}$

The K_b expression is :

$$\frac{[HC_2O_4^-][OH^-]}{[C_2O_4^{2-}]} = 1.6 \times 10^{-10}$$

Assume we begin with 100. mL of 0.100 M oxalic acid. At the second equivalence point, we will have added ~200 mL of 0.100 M NaOH (Since each mole of oxalic acid would require **two** moles of NaOH) reducing the [$C_2O_4^{2-}$] to approximately 1/3 the original concentration (or ~0.0333 M). Substituting into the K_b expression

$$\frac{x^2}{0.0333 - x} = 1.6 \times 10^{-10} \text{ and}$$

$x = 2.31 \times 10^{-6} = [OH^-]$ and pOH $= +5.64$

So pH $= (14.00 - 5.64) = 8.36$ or approximately 8.4

72. Number of moles of HCl needed to decrease the pH of a buffer solution by one pH unit:

The buffer consists of 0.150 M NH_3 and 10.0 g NH_4Cl

The amount of ammonium chloride is : $10.0 \text{ g } NH_4Cl \cdot \frac{1 \text{ mol } NH_4Cl}{53.49\text{g } NH_4Cl} = 0.187$ mol and

since this is in 1.0 L, the concentration of ammonium ion is $\frac{0.187 \text{ mol}}{1.0 \text{ L}}$ or 0.187 M

The pH of this buffer is then:

$$K_b = \frac{[NH_4^+][OH^-]}{[NH_3]} = 1.8 \times 10^{-5} = \frac{(0.187)(OH^-)}{(0.150)} = \text{so } [OH^-] = 1.4 \times 10^{-5} \text{ M and}$$

pOH = 4.85 or pH = 9.15

If we wish to **decrease the pH** by 1 unit, we also want to **increase the pOH** by 1 unit. This is also equivalent to $[OH^-] = 1.4 \times 10^{-6}$ M.

Remember that $1.8 \times 10^{-5} = \frac{(0.187)(OH^-)}{(0.150)}$

Rewriting the K_b expression we get : $\frac{1.8 \times 10^{-5}}{[OH^-]} = \frac{0.187}{0.150}$ and substituting the

$[OH^-] = 1.4 \times 10^{-6}$ we get $\frac{1.8 \times 10^{-5}}{1.4 \times 10^{-6}} = \frac{0.187 + x}{0.150 - x} = 12.85$

Note that the HCl added would **reduce the concentration of NH_3 and increase the concentration of NH_4^+.** (Hence the 0.150 - x and 0.187 + x) in the equation above. Solving for x we get x = **0.125 mol HCl.**

74. For the amino acid, glycine, a plot of the distribution of the three forms of glycine.

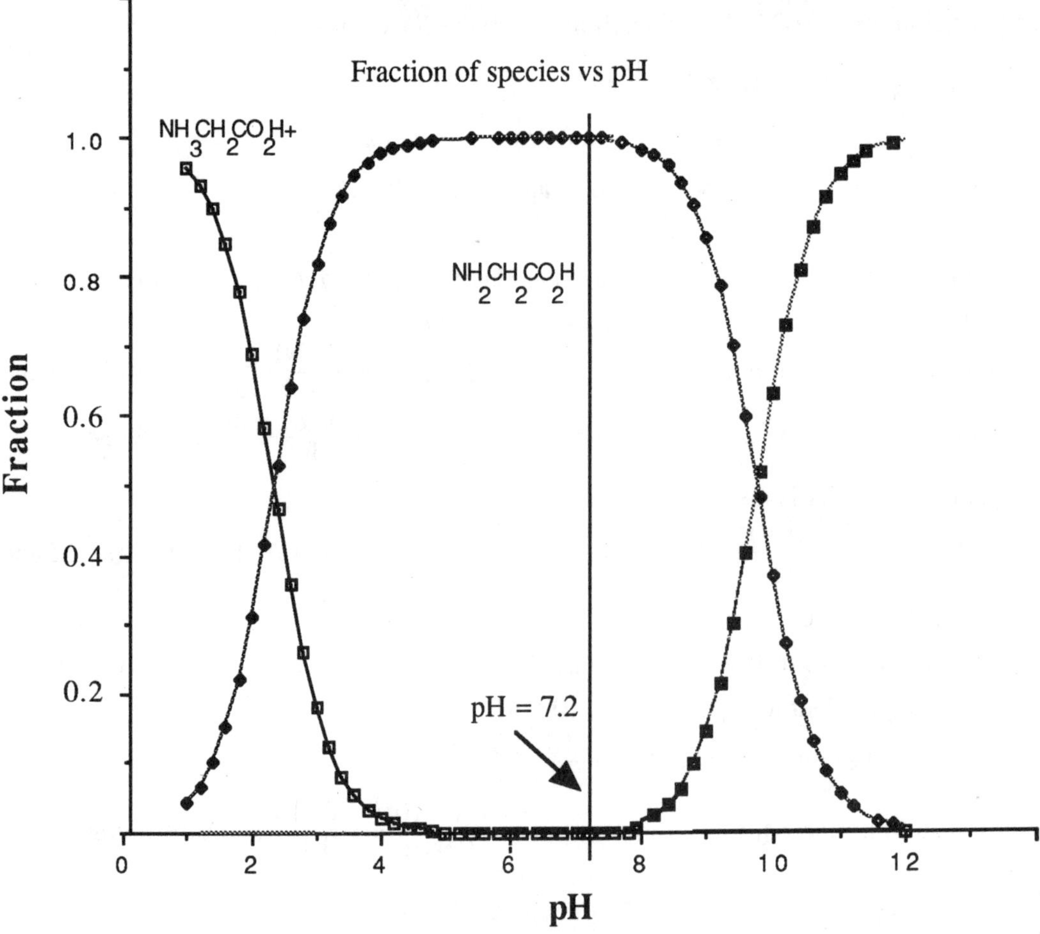

For purposes of our discussion, let's use a shortcut notation:

The cationic form ($NH_3CH_2CO_2H^+$) = HA

The neutral zwitterion form ($NH_2CH_2CO_2H$) = A

The anionic form ($NH_2CH_2CO_2^-$) = A^-

The forms of glycine can be calculated by examination of the two Ka expressions:

$$K_{a1} = \frac{[A][H^+]}{[HA]} = 4.5 \times 10^{-3} \quad \text{and } K_{a2} = \frac{[A^-][H^+]}{[A]} = 1.7 \times 10^{-10}$$

Assume that the concentration of the initial specie (HA for step 1 or A for step 2) is unity 1. Consider step 1:

$K_{a1} = \frac{[x][H^+]}{[1-x]} = 4.5 \times 10^{-3}$ Substituting 0.1 (for pH = 1) for $[H^+]$, one can solve for x to find x= 0.043. So the fraction of the acid present as the cation (HA) = 0.957 and as the neutral zwitterion (A) as 0.043. Substituting various values of $[H^+]$ one obtains the alpha plot. Note that when the fraction is 0.50, the pH corresponds to the pK_{a1}.

Using the process described above, the distribution between the neutral zwitterion (A) and the anion (A^-) is found in the similar fashion. Once again, note that when the fraction is 0.50 (about pH 9.8) the pH = pK_{a2}. The vertical line at pH = 7.2 indicates that the form of the amino acid at this point is the **neutral zwitterion.**

Summary Question

75. a. Approximate bond angles for:

(a) C- C-C in the ring : 120° for sp^2 hybridized C
(b) O-C=O: 120° for sp^2 hybridized C
(c) C-O-H: 109° for sp^3 hybridized C
(d) C-C-H: 120° for sp^2 hybridized C

b. C atoms in the ring are sp^2 hybridized as is the C atom in $-CO_2H$ group.

c. pH of the solution containing 1.00 g of salicylic acid in 460 mL of water
The concentration of the salicylic acid is:

$$\frac{1.00 \text{ g salicylic acid}}{0.46 \text{ L}} \bullet \frac{1 \text{ mol salicylic acid}}{138.1 \text{ g salicylic acid}} = 1.6 \times 10^{-2} \text{ mol}$$

$$K_a = \frac{[A^-][H_3O^+]}{[HA]} = \frac{[x][x]}{1.6 \times 10^{-2} - x} = 1.1 \times 10^{-3}$$

Since $100 \bullet K_a \approx [HA]$, the quadratic equation must be used to find an exact solution.

$$x^2 = 1.73 \times 10^{-5} - 1.1 \times 10^{-3}(x)$$
$$x^2 + 1.1 \times 10^{-3}(x) - 1.73 \times 10^{-5} = 0$$
$$x = [H_3O^+] = 3.65 \times 10^{-3} \text{ M and the pH} = 2.44$$

d. The percentage of the acid in the form of salicylate ion :

From the equilibrium expression:

$$\frac{[H_3O^+][\text{salicylate}]}{[\text{acid}]} = 1.1 \times 10^{-3}$$

At a pH = 2.0, $[H_3O^+] = 1 \times 10^{-2}$ M

Substituting into the equilibrium expression:

$$\frac{[1 \times 10^{-2}][\text{salicylate}]}{[\text{acid}]} = 1.1 \times 10^{-3} \text{ or}$$

$$\frac{[\text{salicylate}]}{[\text{acid}]} = \frac{1.1 \times 10^{-3}}{1 \times 10^{-2}} = 0.11$$

or 11 % of the acid as the salicylate ion.

e. The pH at the midpoint of the titration :

At the midpoint of the titration, the concentration of the acid and its conjugate ion are equal. Therefore $[H_3O^+] = K_a = 1.1 \times 10^{-3}$ M and pH = 2.96

The pH at the equivalence point can be calculated after the concentration of the sodium salicylate is calculated.

$$0.0250 \text{ mL} \bullet 0.014 \text{ M salicylic acid} = 3.5 \times 10^{-4} \text{ moles}$$

The volume of 0.010 M NaOH needed to neutralize this amount is 35.0 mL.

The concentration of the salt is: $\frac{3.5 \times 10^{-4}}{(0.0250 + 0.0350)} = 5.8 \times 10^{-3}$ M

Substituting into the K_b expression for the salicylate ion:

$$K_b = \frac{[\text{salicylic acid}][OH^-]}{[\text{salicylate}]} = \frac{[OH^-]^2}{5.83 \times 10^{-3}} = \frac{1.0 \times 10^{-14}}{1.1 \times 10^{-3}} \begin{array}{l} \leftarrow K_w \\ \leftarrow K_a \end{array}$$

and $[OH^-] = 2.3 \times 10^{-7}$ M; pOH = 6.64 and pH = 7.36

Chapter 19
Principles of Reactivity: Precipitation Reactions

Solubility Guidelines

8. Two insoluble salts of

a. Cl^-	AgCl and $PbCl_2$
b. Zn^{2+}	ZnS and $ZnCO_3$
c. Fe^{2+}	FeS and FeC_2O_4

10. Using the table of solubility guidelines, predict water solubility for the following:

a. $(NH_4)_2CO_3$	Ammonium salts are **soluble.**
b. $ZnSO_4$	Sulfates are generally **soluble.**
c. NiS	Sulfides are generally **insoluble.**
d. $BaSO_4$	Sr^{2+}, Ba^{2+}, and Pb^{2+} form **insoluble** sulfates.

12. Solution producing precipitate

a. $Ag^+ (aq) + Br^- (aq) \rightarrow AgBr (s)$

b. $Pb^{2+} (aq) + 2\ Cl^- (aq) \rightarrow PbCl_2 (s)$

c. No precipitate is expected.

Writing Solubility Product Constant Expressions

14.

Salt dissolving	K_{sp} expression	K_{sp} values
a. $AgCN(s) \rightleftharpoons Ag^+(aq) + CN^-(aq)$	$K_{sp} = [Ag^+][CN^-]$	1.2×10^{-16}
b. $NiCO_3(s) \rightleftharpoons Ni^{2+}(aq) + CO_3^{2-}(aq)$	$K_{sp} = [Ni^{2+}][CO_3^{2-}]$	6.6×10^{-9}
c. $AuBr_3(s) \rightleftharpoons Au^{3+}(aq) + 3\ Br^-(aq)$	$K_{sp} = [Au^{3+}][Br^-]^3$	4.0×10^{-36}

Calculating K_{sp}

16. $K_{sp} = [Tl^+][Br^-] = (1.8 \times 10^{-3})(1.8 \times 10^{-3}) = 3.2 \times 10^{-6}$

18. For the process in which silver acetate dissolves, we can write:

$AgCH_3CO_2\ (s) \rightleftharpoons Ag^+\ (aq) + CH_3CO_2^-\ (aq)$

The concentration of the solid that dissolved is: $1.0\ g \cdot \frac{1\ mol}{166.91\ g} \cdot \frac{1}{0.100\ L} = 0.060\ M$

For each mole of $AgCH_3CO_2$ that dissolves, we get one mol of Ag^+ and one mol of $CH_3CO_2^-$. Since $[Ag^+] = [CH_3CO_2^-] = 0.060\ M$, $K_{sp} = (0.060)(0.060) = 3.6 \times 10^{-3}$.

20. For the process in which barium fluoride dissolves, we can write:

$BaF_2\ (s) \rightleftharpoons Ba^{+2}\ (aq) + 2\ F^-\ (aq)$

For each mole of BaF_2 that dissolves, we get one mol of Ba^{+2} and two mol of F^-.

Since $[F^-]$ at equilibrium = 0.0150 M, then $[Ba^{+2}] = 1/2 \cdot 0.0150\ M$

and $K_{sp} = [Ba^{+2}][F^-]^2 = [0.00750][0.0150]^2 = 1.69 \times 10^{-6}$.

22. For lead(II) hydroxide, the Ksp expression is $K_{sp} = [Pb^{+2}][OH^-]^2$.

Since we know the pH, we can calculate the $[OH^-]$.

pH = 8.92 and pOH = 14.00-8.92 = 5.08 so $[OH^-] = 8.3 \times 10^{-6}$.

For each mole of $Pb(OH)_2$ that dissolves, we get one mol of Pb^{+2} and two mol of OH^-.

Since $[OH^-]$ at equilibrium = 8.3×10^{-6} M, then $[Pb^{+2}] = 1/2 \cdot 8.3 \times 10^{-6}\ M$

$K_{sp} = [Pb^{+2}][OH^-]^2 = [4.2 \times 10^{-6}][8.3 \times 10^{-6}]^2 = 2.9 \times 10^{-16}$

Estimating Salt Solubility from K_{sp}

24. a. The K_{sp} for AgCN is 1.2×10^{-16}. The equation for AgCN dissolving is:

$$AgCN\ (s) \rightleftharpoons Ag^+\ (aq) + CN^-\ (aq)$$

From the equation we see that $[Ag^+] = [CN^-]$.

The K_{sp} expression for AgCN is: $K_{sp} = [Ag^+][CN^-] = 1.2 \times 10^{-16}$

$$K_{sp} = [Ag^+]^2 = [CN^-]^2 = 1.2 \times 10^{-16}$$

$$[Ag^+] = [CN^-] = 1.1 \times 10^{-8}\ M$$

These concentrations tell us that 1.1×10^{-8} mol/L of AgCN dissolve.

b. Solubility of AgCN in g/L:

$$1.1 \times 10^{-8} \frac{mol\ AgCN}{L} \cdot \frac{134\ g\ AgCN}{1\ mol\ AgCN} = 1.5 \times 10^{-6} \frac{g\ AgCN}{L}$$

26. $K_{sp} = [Ra^{2+}][SO_4^{2-}] = 4.2 \times 10^{-11}$ so $[Ra^{2+}] = [SO_4^{2-}] = 6.5 \times 10^{-6}$ M

$RaSO_4$ will dissolve to the extent of 6.5×10^{-6} mol/L

Express this as grams in 100. mL (or 0.1 L)

$$\frac{6.5 \times 10^{-6} \text{ mol } RaSO_4}{1 \text{ L}} \cdot \frac{0.1 \text{ L}}{1} \cdot \frac{322 \text{ g } RaSO_4}{1 \text{ mol } RaSO_4} = 2.09 \times 10^{-4} \text{ g}$$

Expressed as milligrams: 0.209 mg or 0.21 mg (2 sf) of $RaSO_4$ will dissolve!

28. The K_{sp} for $MgF_2 = 6.4 \times 10^{-9}$

$K_{sp} = [Mg^{2+}][F^-]^2 = 6.4 \times 10^{-9}$

if a mol/L of MgF_2 dissolve, $[Mg^{2+}] = a$ and $[F^-] = 2a$

$(a)(2a)^2 = 4a^3 = 6.4 \times 10^{-9}$ and $a = 1.17 \times 10^{-3}$

a. The molar solubility is then 1.2×10^{-3} M (2 sf)

b. Solubility in g/L

$$1.2 \times 10^{-3} \frac{\text{mol } MgF_2}{\text{L}} \cdot \frac{62.3 \text{ g } MgF_2}{1 \text{ mol } MgF_2} = 7.3 \times 10^{-2} \text{ g } MgF_2 \text{/L}$$

30. Compound in each pair that is **more soluble**.

	K_{sp}	
a. $PbCl_2$	1.7×10^{-5}	more soluble
$PbBr_2$	6.3×10^{-6}	
b. HgS	3.0×10^{-53}	
FeS	4.9×10^{-18}	more soluble
c. BiI_3	8.1×10^{-19}	more soluble
$Bi(OH)_3$	3.2×10^{-40}	
d. $Fe(OH)_2$	7.9×10^{-15}	more soluble
$Zn(OH)_2$	4.5×10^{-17}	

32. Compounds in order of increasing solubility in H_2O:

Compound	K_{sp}
$BaCO_3$	8.1×10^{-9}
Ag_2CO_3	8.1×10^{-12}
Na_2CO_3	not listed: Na_2CO_3 is very soluble in water and is certainly the most soluble salt of the 3.

To determine the relative solubilities find the molar solubilities.

For $BaCO_3$: $K_{sp} = [Ba^{2+}][CO_3^{2-}] = (x)(x) = 8.1 \times 10^{-9}$ and $x = 9.0 \times 10^{-5}$

The molar solubility of $BaCO_3$ is then 9.0×10^{-5} M.

and for Ag_2CO_3: $K_{sp} = [Ag^+]^2[CO_3^{2-}] = (2x)^2(x) = 8.1 \times 10^{-12}$

$$4x^3 = 8.1 \times 10^{-12}$$

and $x = 1.3 \times 10^{-4}$

The molar solubility of Ag_2CO_3 is 1.3×10^{-4} M.

In order of increasing solubility: $BaCO_3 < Ag_2CO_3 < Na_2CO_3$

34. The concentration corresponding to 5.0×10^{-3} g $NiCO_3$ in 1.0 L of water:

$$\frac{5.0 \times 10^{-3} \text{ g NiCO}_3}{1} \cdot \frac{1 \text{ mol NiCO}_3}{119 \text{ g NiCO}_3} = 4.2 \times 10^{-5} \text{ M}$$

This means that 4.2×10^{-5}mol/L would dissolve, providing 4.2×10^{-5}mol/L of Ni^{2+} and 4.2×10^{-5} mol/L of CO_3^{2-}

Q_c for this solution would be $[Ni^{2+}][CO_3^{2-}] = (4.2 \times 10^{-5})^2 = 1.8 \times 10^{-9}$

Since $Q_c < K_{sp}$ for $NiCO_3$ (6.6×10^{-9}), all the solid dissolves.

Precipitations

36. Given the equation for $PbCl_2$ dissolving in water:

$PbCl_2$ (s) $\rightleftharpoons$ Pb^{+2} (aq) + 2 Cl^- we can write the K_{sp} expression :

$K_{sp} = [Pb^{+2}][Cl^-]^2 = 1.7 \times 10^{-5}$

Substituting the ion concentrations into the Ksp expression we get:

$Q_c = [Pb^{+2}][Cl^-]^2 = (0.0012)(0.010)^2 = 1.2 \times 10^{-7}$

Since Q_c is less than K_{sp} , **no $PbCl_2$ precipitates.**

38. If $Zn(OH)_2$ is to precipitate, the reaction quotient (Q_c) must exceed the K_{sp} for the salt. 4.0 mg of NaOH in 10. mL corresponds to a concentration of:

$$[OH^-] = \frac{4.0 \times 10^{-3} \text{ g NaOH}}{0.0100 \text{ L}} \cdot \frac{1 \text{ mol NaOH}}{40.0 \text{ g NaOH}} = 0.01 \text{ M}$$

The value of Q_c is :

$$[Zn^{2+}][OH^-]^2 = (1.6 \times 10^{-4})(1.0 \times 10^{-2})^2 = 1.6 \times 10^{-8}$$

The value of Q_c is greater than the K_{sp} for the salt (4.5×10^{-17}), so **$Zn(OH)_2$ precipitates.**

40. To prevent CaF_2 from precipitating, Q_c for the salt **must not exceed** the K_{sp} for the salt. The $K_{sp} = [Ca^{+2}][F^-]^2 = [0.0020][F^-]^2 = 3.9 \times 10^{-11}$; solving for $[F^-]$ we get:

    $[F^-]^2 = \frac{3.9 \times 10^{-11}}{0.0020} = 1.95 \times 10^{-8}$ and $[F^-] = 1.4 \times 10^{-4}$ M

    So long as the fluoride ion concentration does not exceed 1.4×10^{-4} M, no CaF_2 precipitates.

42. Calculate the concentration of magnesium ion and hydroxide ion after mixing--but before reaction.

$$0.10 \text{ M } Mg^{2+} \cdot \frac{75.0 \text{ mL}}{(75.0 + 25.0 \text{ mL})} = 7.5 \times 10^{-2} \text{ M } Mg^{2+}$$

$$0.010 \text{ M } OH^- \cdot \frac{25.0 \text{ mL}}{(75.0 + 25.0 \text{ mL})} = 2.5 \times 10^{-3} \text{ M } OH^-$$

$Q_c = [Mg^{2+}][OH^-]^2 = (7.5 \times 10^{-2})(2.5 \times 10^{-3})^2 = 4.7 \times 10^{-7}$

and since $Q_c > K_{sp}$ (1.5×10^{-11}) **$Mg(OH)_2$ precipitates.**

44. Calculate the concentration of lead ion and chloride ion after mixing--but before reaction.

$$0.0010 \text{ M } Pb^{2+} \cdot \frac{10. \text{ mL}}{(10. + 5.0 \text{ mL})} = 6.7 \times 10^{-4} \text{ M } Pb^{2+}$$

$$0.015 \text{ M } Cl^- \cdot \frac{5.0 \text{ mL}}{(10. + 5.0 \text{ mL})} = 5.0 \times 10^{-3} \text{ M } Cl^-$$

$Q_c = [Pb^{2+}][Cl^-]^2 = (6.7 \times 10^{-4})(5.0 \times 10^{-3})^2 = 1.7 \times 10^{-8}$

and since $Q_c < K_{sp}$ (1.7×10^{-5}) **$PbCl_2$ does not precipitate.**

46. These metal sulfates are 1:1 salts. The K_{sp} expression has the general form:

$$K_{sp} = [M^{2+}][SO_4^{2-}]$$

To determine the $[SO_4^{2-}]$ necessary to begin precipitation, we can divide the equation by the metal ion concentration to obtain:

$$\frac{K_{sp}}{[M^{2+}]} = [SO_4^{2-}]$$

The concentration of the three metal ions under consideration are each 0.10 M. Substitution of the appropriate K_{sp} for the sulfates and 0.10 M for the metal ion concentration yields the sulfate ion concentrations in the table below. As the soluble sulfate is added to the metal ion solution, the sulfate ion concentration increases from zero molarity. The lowest sulfate ion concentration is reached first, with higher concentrations reached later. The order of precipitation is listed in the last column of the table below.

Compound	K_{sp}	Maximum $[SO_4^{2-}]$	Order of Precipitation
$BaSO_4$	1.1×10^{-10}	1.1×10^{-9}	1
$SrSO_4$	2.8×10^{-7}	2.8×10^{-6}	3
$PbSO_4$	1.8×10^{-8}	1.8×10^{-7}	2

48. The K_{sp} expressions for the hydroxides of Fe^{3+}, Pb^{2+}, and Al^{3+} are:

$[Fe^{3+}][OH^-]^3 = 6.3 \times 10^{-38}$ $[Pb^{2+}][OH^-]^2 = 2.8 \times 10^{-16}$

$$[Al^{3+}][OH^-]^3 = 1.9 \times 10^{-33}$$

The solution in question contains the cations each in 0.1 M concentration. Substituting this value for the metal ion concentrations, and solving for the $[OH^-]$ yields:

$Fe(OH)_3$: $[OH^-] = (6.3 \times 10^{-37})^{1/3} = 8.6 \times 10^{-13}$ M

$Pb(OH)_2$: $[OH^-] = (2.8 \times 10^{-15})^{1/2} = 5.3 \times 10^{-8}$ M

$Al(OH)_3$: $[OH^-] = (1.9 \times 10^{-32})^{1/3} = 2.7 \times 10^{-11}$ M

The salts would precipitate in the order: $Fe(OH)_3$, $Al(OH)_3$, $Pb(OH)_2$.

Common Ion Effect

50. The equilibrium for AgSCN dissolving is: $AgSCN\ (s) \rightleftharpoons Ag^+\ (aq) + SCN^-\ (aq)$.
As x mol/L of AgSCN dissolve in pure water, x mol/L of Ag^+ and x mol/L of SCN^- are produced.

$K_{sp} = [Ag^+][SCN^-] = x^2 = 1.0 \times 10^{-12}$ and $x = 1.0 \times 10^{-6}$ M

So 1.0×10^{-6} mol AgSCN/L dissolve in pure water.

The equilibrium for AgSCN dissolving in NaSCN (0.010 M) is like that above. Equimolar amounts of Ag^+ and SCN^- ions are produced as the solid dissolves. However the $[SCN^-]$ is augmented by the soluble NaSCN.

$$K_{sp} = [Ag^+][SCN^-] = (x)(x + 0.010) = 1.0 \times 10^{-12}$$

We can simplify the expression by assuming that $x + 0.010 \approx 0.010$.

$$(x)(0.010) = 1.0 \times 10^{-12} \quad \text{and } x = 1.0 \times 10^{-10} \text{ M}$$

The solubility of AgSCN in 0.010 M NaSCN is 1.0×10^{-10} M—reduced by four orders of magnitude from its solubility in pure water. LeChatelier strikes again!

52. The K_{sp} expression for Ag_3PO_4 is: $K_{sp} = [Ag^+]^3[PO_4^{3-}] = 1.3 \times 10^{-20}$

a. The stoichiometry of the compound is such that when the solid dissolves,

$$[Ag^+] = 3 \bullet [PO_4^{3-}]$$

Substituting into the K_{sp} expression gives $(3 \bullet [PO_4^{3-}])^3[PO_4^{3-}] = 1.3 \times 10^{-20}$

$$\text{or } 27 \bullet [PO_4^{3-}]^4 = 1.3 \times 10^{-20} \text{ and } [PO_4^{3-}] = 4.7 \times 10^{-6} \text{ M}$$

The molar amount of solid that dissolves will be equal to the concentration of phosphate ion. Expressing this concentration in mg/mL:

$$\frac{4.7 \times 10^{-6} \text{ mol}}{1 \text{ L}} \bullet \frac{419 \text{ g } Ag_3PO_4}{1 \text{ mol}} \bullet \frac{1000 \text{ mg}}{1 \text{ g}} \bullet \frac{1 \text{ L}}{1000 \text{ mL}} = 2.0 \times 10^{-3} \text{ mg/mL}$$

b. The solubility of the salt in 0.020 M $AgNO_3$:

While the ratio of silver and phosphate ions from the solid dissolving remains the same, the equilibrium concentration of Ag^+ will be increased by the 0.020 M Ag^+ ion. If we represent that molar solubility of Ag_3PO_4 as x, the concentrations at equilibrium will be:

$$[PO_4^{3-}] = x \quad \text{and } [Ag^+] = 3x + 0.020$$

Substituting the values into the K_{sp} expression:

$$[Ag^+]^3[PO_4^{3-}] = 1.3 \times 10^{-20} = (3x + 0.020)^3x = 1.3 \times 10^{-20}$$

The amount of silver phosphate which dissolves is small, hence let's simplify the expression to: $(0.020)^3x = 1.3 \times 10^{-20}$ and solving for x:

$x = 1.6 \times 10^{-15}$ M, and expressing this concentration in mg/mL:

$$\frac{1.6 \times 10^{-15} \text{ mol}}{1 \text{ L}} \bullet \frac{419 \text{ g } Ag_3PO_4}{1 \text{ mol}} \bullet \frac{1000 \text{ mg}}{1 \text{ g}} \bullet \frac{1 \text{ L}}{1000 \text{ mL}} = 6.8 \times 10^{-13} \text{ mg/mL}$$

Separations

54. a. To determine the maximum concentration of oxalate ion before the Mg^{2+} salt begins to precipitate, substitute the concentration of Mg^{2+} into the K_{sp} expression:

$K_{sp} = [Mg^{2+}][C_2O_4^{2-}] = 8.6 \times 10^{-5}$ and for a solution in which

$[Mg^{2+}] = 0.020$ M, the maximum $[C_2O_4^{2-}] = \dfrac{8.6 \times 10^{-5}}{0.020} = 4.3 \times 10^{-3}$ M.

b. When the magnesium salt just begins to precipitate, the $[C_2O_4^{2-}] = 4.3 \times 10^{-3}$ M. At that point the $[Ca^{2+}]$ would be:

$$K_{sp} = [Ca^{2+}][C_2O_4^{2-}] = 2.3 \times 10^{-9}$$

and the $[Ca^{2+}]$ would be $\dfrac{2.3 \times 10^{-9}}{4.3 \times 10^{-3}} = 5.3 \times 10^{-7}$ M.

56. The respective K_{sp}'s for the compounds are: $PbI_2 = 8.7 \times 10^{-9}$; $PbCO_3 = 1.5 \times 10^{-13}$

a. The K_{sp} expressions for these are:

$K_{sp} = [Pb^{2+}][I^-]^2 = 8.7 \times 10^{-9}$ and $K_{sp} = [Pb^{2+}][CO_3^{2-}] = 1.5 \times 10^{-13}$

Substituting the concentrations of iodide and carbonate ions, and solving for $[Pb^{2+}]$:

$[Pb^{2+}] = \dfrac{8.7 \times 10^{-9}}{(0.10)^2}$ $\qquad$ $[Pb^{2+}] = \dfrac{1.5 \times 10^{-13}}{0.10}$

$[Pb^{2+}] = 8.7 \times 10^{-7}$ M $\qquad$ $[Pb^{2+}] = 1.5 \times 10^{-12}$ M

Since the lead ion concentration will grow from 0 M, the lead ion concentration reached first is 1.5×10^{-12} M, so **$PbCO_3$ will begin to precipitate first.**

b. When PbI_2 begins to precipitate, $[Pb^{2+}] = 8.7 \times 10^{-7}$ M (as calculated above) and for a solution that is saturated in $PbCO_3$ (remember that $PbCO_3$ has been precipitating), we can write: $[Pb^{2+}][CO_3^{2-}] = 1.5 \times 10^{-13}$

Solving for $[CO_3^{2-}] = \dfrac{1.5 \times 10^{-13}}{8.7 \times 10^{-7}} = 1.7 \times 10^{-7}$ M

58. Separate the following pairs of ions:

a. **Ba^{2+} and Na^{+}** : Since most sodium salts are soluble, it is simple to find a barium salt which is not soluble--e.g. the sulfate. Addition of dilute sulfuric acid should provide a source of SO_4^{2-} ions in sufficient quantity to precipitate the barium ions.

b. **Bi^{3+} and Cd^{2+}**: The hydroxide ion will serve as an effective reagent for selective precipitation of the two ions with the less soluble $Bi(OH)_3$ ($K_{sp} = 3.2 \times 10^{-40}$) precipitating well before the cadmium salt ($K_{sp} = 1.2 \times 10^{-14}$).

Simultaneous Equilibria and Complex Ions

60. Show that the equation: $AuCl(s) + 2\ CN^-(aq) \rightleftharpoons Au(CN)_2^-(aq) + Cl^-(aq)$ is the sum of two equations.

AuCl dissolving:	$AuCl(s) \rightleftharpoons Au^+(aq) + Cl^-(aq)$	$K_{sp} = 2.0 \times 10^{-13}$
$Au(CN)_2^-$ formation:	$Au^+(aq) + 2\ CN^-(aq) \rightleftharpoons Au(CN)_2^-(aq)$	$K_f = 2.0 \times 10^{38}$
The net equation	$AuCl(s) + 2\ CN^-(aq) \rightleftharpoons Au(CN)_2^-(aq) + Cl^-$	$K_{net} = K_{sp} \bullet K_f$

$K_{net} = 2.0 \times 10^{-13} \bullet 2.0 \times 10^{38} = 4.0 \times 10^{25}$

62. The equation $AgCl(s) + I^-(aq) \rightleftharpoons AgI(aq) + Cl^-$ can be obtained by adding two equations:

1. $AgCl(s) \rightleftharpoons Ag^+(aq) + Cl^-(aq)$	$K_{sp1} = 1.8 \times 10^{-10}$
2. $Ag^+(aq) + I^-(aq) \rightleftharpoons AgI(s)$	$\frac{1}{K_{sp2}} = 6.7 \times 10^{15}$

The net equation: $AgCl(s) + I^-(aq) \rightleftharpoons AgI(aq) + Cl^-$

$$K_{net} = K_{sp1} \bullet \frac{1}{K_{sp2}} = \frac{1.8 \times 10^{-10}}{6.7 \times 10^{15}} = 1.2 \times 10^{6}$$

The equilibrium lies to the right. This indicates that **AgI will form** if I^- is added to a saturated solution of AgCl.

64. Will 5.0 mL of 2.5 M NH_3 dissolve 1.0×10^{-4} mol of AgBr ?

The reaction that must occur for AgBr to dissolve is:

$$AgBr(s) + 2\ NH_3(aq) \rightleftharpoons [Ag(NH_3)_2]^+(aq) + Br^-(aq)$$

The reaction is the sum of (1) AgBr dissolving and (2) $[Ag(NH_3)_2]^+$ forming from Ag^+ and NH_3. Accordingly,

$$K_{overall} = K_{sp} \cdot K_f$$

$$K_{overall} = (8.3 \times 10^{-13})(1.6 \times 10^7) = 5.3 \times 10^{-6}$$

$$K_{overall} = \frac{\{[Ag(NH_3)_2]^+\}[Br^-]}{[NH_3]^2} = 5.3 \times 10^{-6}$$

Calculate the $[NH_3]$ necessary to dissolve the AgBr.

1. If the AgBr dissolves, the equilibrium amount of Br^- will be 1.0×10^{-4} M, and the complex will be 1.0×10^{-4} M. Substituting into the $K_{overall}$ expression:

$$K_{overall} = \frac{\{[Ag(NH_3)_2]^+\}[Br^-]}{[NH_3]^2} = 5.3 \times 10^{-6}$$

$$\frac{(1.0 \times 10^{-4})(1.0 \times 10^{-4})}{[NH_3]^2} = 5.3 \times 10^{-6}$$

$$[NH_3] = 0.043 \text{ M}$$

2. The total ammonia necessary is the ammonia to form the complex $(2 \cdot 0.020) = 4.0 \times 10^{-2}$ M NH_3 **and** the ammonia to increase the NH_3 concentration to 0.043 M.

$$[NH_3] = 4.0 \times 10^{-2} + 0.043 = 0.083 \text{ M.}$$

The concentration of NH_3 (5.0 mL of 2.5 M NH_3) diluted to 1.0 L is 0.012 M. We see that this concentration of NH_3 **would not be sufficient to dissolve the AgBr.**

Solubility and pH

66. Soluble in strong acid: $Ba(OH)_2$, $BaCO_3$

For $Ba(OH)_2$ the addition of HCl for example would result in the acid/base reaction to form water, dissolving $Ba(OH)_2$.

$$2\,HCl\,(aq) + Ba(OH)_2(s) \rightarrow 2\,H_2O\,(\ell) + BaCl_2\,(aq)$$

For $BaCO_3$, HCl would react to form CO_2 and H_2O, causing the carbonate salt to dissolve.

$$2\,HCl\,(aq) + BaCO_3(s) \rightarrow H_2O\,(\ell) + CO_2\,(g) + BaCl_2\,(aq)$$

$BaSO_4$ does not react with HCl to an appreciable extent (K<1), so the salt does not dissolve.

68. BiI_3 is not expected to be soluble in a strong acid.

For BiI_3 to be soluble the reaction:

$$BiI_3(s) + 3\ HCl\ (aq) \rightleftharpoons 3\ HI\ (aq) + BiCl_3\ (aq)$$

would have to occur to an appreciable extent. Note that this reaction would take one strong acid (HCl) and form another strong acid (HI). This does not occur to an appreciable extent ($K<1$), so BiI_3 does not dissolve in strong acid.

$BaCO_3$ is expected to be soluble (See question 66 for an explanation).

$BiPO_4$ is expected to be soluble, since the PO_4^{3-} is the conjugate base of a weak acid. For the salt to be soluble, the reaction:

$$BiPO_4(s) + 3\ HCl\ (aq) \rightleftharpoons 3\ H_3PO_4\ (aq) + BiCl_3\ (aq)$$

has to occur to an appreciable extent. Since PO_4^{3-} is the conjugate base of a weak acid, the equation does occur to an appreciable extent ($K>1$) and $BiPO_4(s)$ dissolves.

General Questions

70. a. The equation for the solid dissolving in water is:

$Li_3PO_4\ (s) \rightleftharpoons 3\ Li^+\ (aq) + PO_4^{3-}\ (aq)$ and the K_{sp} expression would be:

$$K_{sp} = [\ Li^+]^3[PO_4^{3-}]$$

Given that the $[PO_4^{3-}] = 3.3 \times 10^{-3}$ M, we know (from the stoichiometry of the equation above) that $[\ Li^+] = 3 \bullet 3.3 \times 10^{-3}$ or 9.9×10^{-3} M

Substituting these concentrations into the Ksp expression we find:

$$K_{sp} = [\ Li^+]^3[PO_4^{3-}] = [9.9 \times 10^{-3}]^3[3.3 \times 10^{-3}] = 3.2 \times 10^{-9}$$

b. Since PO_4^{3-} is the conjugate base of a weak acid, the solid is expected to dissolve to a greater extent upon adding aqueous HCl (i.e. **become more soluble). See question 68 for additional explanation.**

72. a. To precipitate BaF_2, the ion product constant (right side of Ksp expression) will have to exceed the Ksp for the salt.

The Ksp expression is: $K_{sp} = [Ba^{2+}][F^-]^2 = 1.7 \times 10^{-6}$

and substituting the ion concentrations into that expression:

$$(0.015)[F^-]^2 = 1.7 \times 10^{-6} \text{ so } [F^-]^2 = \frac{1.7 \times 10^{-6}}{1.5 \times 10^{-2}} = 1.1 \times 10^{-4} \text{ and}$$

$$[F^-] = 0.011\ M$$

b. Mass of NaF to just begin precipitation (achieve a $[F^-]$ = 0.011 M)?

$$\frac{0.011 \text{ mol } F^-}{1 \text{ L}} \bullet \frac{0.100 \text{ L}}{1} \bullet \frac{1 \text{ mol NaF}}{1 \text{ mol } F^-} \bullet \frac{41.99 \text{ g NaF}}{1 \text{ mol NaF}} \bullet \frac{1000 \text{ mg NaF}}{1 \text{ g NaF}} = 45 \text{ mg NaF}$$

(Using the "unrounded" $[F^-] = 1.06 \times 10^{-2}$ M in part a.)

c. To determine the $[Ba^{2+}]$ remaining in solution after the addition of 0.50 g of NaF, ask two questions:

1. "What fluoride ion concentration would be generated by the addition of this mass of NaF?" and

2. "What barium ion concentration could be in equilibrium with this fluoride ion concentration ?"

Question 1:

$$0.50 \text{ g NaF} \bullet \frac{1 \text{ mol NaF}}{41.99 \text{ g NaF}} \bullet \frac{1 \text{ mol } F^-}{1 \text{ mol NaF}} \bullet \frac{1}{0.100\text{L}} = 0.12 \text{ M } F^-$$

Question 2:

$K_{sp} = [Ba^{2+}][F^-]^2 = 1.7 \times 10^{-6}$ and the $[F^-] = 0.12$ M so

$= [Ba^{2+}][0.12]^2 = 1.7 \times 10^{-6}$, and solving for $[Ba^{2+}]$,$[Ba^{2+}] = 1.2 \times 10^{-4}$ M

74. Determine if $Q_c > K_{sp}$ (for AgCl):

$K_{sp} = [Ag^+][Cl^-] = 1.8 \times 10^{-10}$ $Q_c = (1.0 \times 10^{-5})(2.0 \times 10^{-4}) = 2.0 \times 10^{-9}$

Since Q_c exceeds K_{sp}, **AgCl will precipitate.** (You are found out!)

76. Predict the order in which the Ag^+, Bi^{3+}, and Pb^{2+} iodides precipitate.

compound	Ksp
AgI	1.5×10^{-16}
BiI_3	8.1×10^{-19}
PbI_2	8.7×10^{-9}

Since these salts are 1:1 (AgI), 1:2 (PbI_2), and 1:3 (BiI_3) salts, it will be necessary to solve the Ksp expression for $[I^-]$. Substituting 0.10 M for each of the metal ions we get:

$Ksp = [Ag^+][I^-] = 1.5 \times 10^{-16}$ $\quad [I^-] = \frac{1.5 \times 10^{-16}}{0.10} = 1.5 \times 10^{-15}$ M

$Ksp = [Bi^{+3}][I^-]^3 = 8.1 \times 10^{-19}$ $\quad [I^-]^3 = \frac{8.1 \times 10^{-19}}{0.10} = 8.1 \times 10^{-18}$

and taking the cube root of both sides, $[I^-] = 2.0 \times 10^{-6}$ M

$Ksp = [Pb^{+2}][I^-]^2 = 8.7 \times 10^{-9}$ $\quad [I^-]^2 = \frac{8.7 \times 10^{-9}}{0.10} = 8.7 \times 10^{-8}$

and taking the square root of both sides, $[I^-] = 2.9 \times 10^{-4}$ M

Since the iodide ion concentration begins at 0 M and increases as the sodium iodide is added, the salts will precipitate in the order: AgI then BiI_3 then PbI_2.

78. The equation for $Zn(OH)_2$ dissolving is : $Zn(OH)_2\ (s) \rightleftharpoons Zn^{2+}\ (aq) + 2\ OH^-\ (aq)$

If a saturated solution of the base has a pH = 8.65 then pOH = 5.35 and $[OH^-] = 4.5 \times 10^{-6}$ M. The stoichiometry of the solid says that for two hydroxyl ions, one zinc ion is produced. So the concentration of $Zn^{2+} = 1/2 \bullet (4.5 \times 10^{-6}) = 2.2 \times 10^{-6}$.

The K_{sp} is then: $K_{sp} = [Zn^{2+}][OH^-]^2 = (2.2 \times 10^{-6})(4.5 \times 10^{-6})^2 = 4.5 \times 10^{-17}$.

80. The equation for $Mg(OH)_2$ dissolving is : $Mg(OH)_2(s) \rightleftharpoons Mg^{2+}(aq) + 2\ OH^-(aq)$

If a saturated solution of the base has a pH = 10.49 then pOH = 3.51 and $[OH^-] = 3.1 \times 10^{-4}$ M. The stoichiometry of the solid says that for two hydroxyl ions, one magnesium ion is produced. So the concentration of $Mg^{2+} = 1/2(3.1 \times 10^{-4}) = 1.5 \times 10^{-4}$.
The Ksp is then: $[Mg^{2+}][OH^-]^2 = (1.5 \times 10^{-4})(3.1 \times 10^{-4})^2 = 1.5 \times 10^{-11}$

82. Calculate the initial concentrations:

$$\frac{2.00\ g\ AgNO_3}{0.0500\ L} \bullet \frac{1\ mol\ AgNO_3}{169.9\ g\ AgNO_3} = 0.235\ M\ AgNO_3$$

$$\frac{3.00\ g\ K_2CrO_4}{0.0500\ L} \bullet \frac{1\ mol\ K_2CrO_4}{194.2\ g\ K_2CrO_4} = 0.309\ M\ K_2CrO_4$$

Since NO_3^- and K^+ ions don't form precipitates, we can calculate their final concentrations easily.

$$[NO_3^-] = 0.235 \frac{mol\ AgNO_3}{L} \bullet \frac{1\ mol\ NO_3^-}{1\ mol\ AgNO_3} = 0.235\ M\ NO_3^-$$

$$[K^+] = 0.235 \frac{mol\ K_2CrO_4}{L} \bullet \frac{2\ mol\ K^+}{1\ mol\ K_2CrO_4} = 0.618\ M\ K^+$$

The **initial concentration of Ag^+** is 0.235 M and of CrO_4^{2-} is 0.309 M. Since the K_{sp} for Ag_2CrO_4 is 9.0×10^{-12}, Ag_2CrO_4 will certainly precipitate. ($Q_c > K$)

$$Q_c = [Ag^+]^2[CrO_4^{2-}]$$
$$= (0.235)^2(0.309) = 1.7 \times 10^{-2}$$

The precipitation will remove 2 Ag^+ and 1 CrO_4^{2-} for each formula unit.

What amount of each will remain?

If "all" the Ag^+ precipitates, we would consume: 1.18×10^{-2} mol Ag^+
(2.00 g of $AgNO_3 \rightarrow 1.18 \times 10^{-2}$ mol).

This would consume $1/2 \times 1.18 \times 10^{-2}$ mol CrO_4^{2-} (5.88×10^{-3} mol), leaving ($1.54 \times 10^{-2} - 5.88 \times 10^{-3}$) 9.53×10^{-3} mol CrO_4^{2-} with a concentration of

$$\frac{9.53 \times 10^{-3} \text{ mol } CrO_4^{2-}}{0.050 \text{ L}} \text{ or } 0.191 \text{ M.}$$

The silver present in solution would then be controlled by the solubility of Ag_2CrO_4:

$$[Ag^+]^2[CrO_4^{2-}] = 9.0 \times 10^{-12}$$
$$[Ag^+]^2 = \frac{9.0 \times 10^{-12}}{0.191}$$
$$[Ag^+] = 6.9 \times 10^{-6} \text{ M}$$

84. a. The equilibrium expression for the equation may be written:

$$K = \frac{\{[Ag(S_2O_3)_2]^{3-}\}[Br^-]}{[S_2O_3^{2-}]^2}$$

NOTE: We have used { } to indicate concentration when complex ions are present.

The K_{sp} expression for AgBr is written: $K_{sp} = [Ag^+][Br^-] = 3.3 \times 10^{-13}$

The K_f for $[Ag(S_2O_3)_2]^{3-}$ may be written: $K_f = \dfrac{\{[Ag(S_2O_3)_2]^{3-}\}}{[Ag^+][S_2O_3^{2-}]^2} = 2.0 \times 10^{13}$

$$K_{overall} = K_{sp} \cdot K_f = \frac{\{[Ag(S_2O_3)_2]^{3-}\}[Br^-]}{[S_2O_3^{2-}]^2} = (3.3 \times 10^{-13})(2.0 \times 10^{13}) = 6.6$$

b. 1.0 g AgBr in 1.0 L corresponds to $\dfrac{1.0 \text{ g AgBr}}{1 \text{ L}} \cdot \dfrac{1 \text{ mol AgBr}}{188 \text{ g AgBr}} = 5.3 \times 10^{-3}$ M

Note that the concentration, 5.3×10^{-3} M, corresponds to the concentration of both silver complex and bromide ion. Note also that for each mole of silver complex, two moles of thiosulfate ion are consumed. The amount of $S_2O_3^{2-}$ consumed is then ($2 \cdot 5.3 \times 10^{-3}$) or 1.1×10^{-2} M.

Substituting into the $K_{overall}$ expression gives:

$$\frac{\{[Ag(S_2O_3)_2]^{3-}\}[Br^-]}{[S_2O_3^{2-}]^2} = \frac{(5.3 \times 10^{-3})(5.3 \times 10^{-3})}{[S_2O_3^{2-}]^2} = 6.6$$

and $[S_2O_3^{2-}]^2 = 4.3 \times 10^{-6}$ and $[S_2O_3^{2-}] = 2.1 \times 10^{-3}$ M

Note that this concentration is the equilibrium concentration for $S_2O_3^{2-}$.
(the result of $[S_2O_3^{2-}]$ initial - $[S_2O_3^{2-}]$ change)

2.1×10^{-3} M = $[S_2O_3^{2-}]$ initial - 1.1×10^{-2} M
1.3×10^{-2} M = $[S_2O_3^{2-}]$ initial.

This corresponds to 1.3×10^{-2} moles of $Na_2S_2O_3$.

The mass is: $1.3 \times 10^{-2} \text{ mol } Na_2S_2O_3 \bullet \frac{158 \text{ g } Na_2S_2O_3}{1 \text{ mol } Na_2S_2O_3} = 2.0 \text{ g } Na_2S_2O_3$

Conceptual Questions

86. Determining the more soluble of two substances with the same general composition (dipositive cation and dipositive anion) is done by noting that the substance with the greater K_{sp} is the more soluble. Aragonite ($K_{sp} = 6.0 \times 10^{-9}$) is slightly more soluble than calcite ($K_{sp} = 3.8 \times 10^{-9}$).

88. The solubility of $Ni(OH)_2$ would be affected by any equilibrium that affects either Ni^{2+} or OH^-. The equilibrium expression for $Ni(OH)_2$ dissolving is:

$$Ni(OH)_2(s) \rightleftharpoons Ni^{2+}(aq) + 2\ OH^-(aq)$$

If we reduce $[OH^-]$ by adding acid—decreasing the pH, the solubility of $Ni(OH)_2$ would increase as (according to LeChatelier's Principle) the equilibrium adjusts to account for the decrease in OH^-. If we add base—increasing the $[OH^-]$—and simultaneously increasing the pH, the equilibrium would shift to the left, increasing the amount of solid—and reducing the solubility.

90. To determine the relative value of Ksp for mercury(I) chloride:

Assume first that the equation is HgCl (s) $\rightleftharpoons Hg^+$ (aq) + Cl^- (aq)

The Ksp expression would be $K_{sp} = [Hg^+][Cl^-]$.

Let's assume for our discussion that the student determined the $[Cl^-]$ to be 1×10^{-3} M. Then the Ksp would be : $K_{sp} = [Hg^+][Cl^-] = [1 \times 10^{-3}][1 \times 10^{-3}] = 1 \times 10^{-6}$

Now let's look at the "correct" reaction:

$$Hg_2Cl_2\ (s) \rightleftharpoons Hg_2^{+2}\ (aq) + 2\ Cl^-\ (aq)$$

The Ksp expression would be : $K_{sp} = [Hg_2^{+2}][Cl^-]^2$.

Once again, let's assume that $[Cl^-]$ is 1×10^{-3} M. The stoichiometry of this reaction tells us that the concentration of the mercury(I) dimer would be **half** that of the chloride ion, or 5×10^{-4} M. Substituting into the Ksp expression:

$$K_{sp} = [Hg_2^{+2}][Cl^-]^2 = [5 \times 10^{-4}][1 \times 10^{-3}]^2 = 5 \times 10^{-10}$$

The calculated value would be **too high** (1×10^{-6} vs 5×10^{-10} in our example).

92. For the reaction : $BaCO_3\ (s) + CO_2\ (g) + H_2O\ (\ell) \rightleftharpoons Ba^{2+}(aq) + 2\ HCO_3^-(aq)$
    a. Solubility of $BaCO_3$ affected by pressure of CO_2: LeChatelier's principle tells us that adding CO_2 (**increasing pressure**) to the equilibrium mixture **will shift the equilibrium to the right**, resulting in **increased solubility of $BaCO_3$.**
    b. Solubility of $BaCO_3$ affected by pH decrease: Decreasing pH (**increasing $[H_3O^+]$**) will consume additional HCO_3^- (and produce CO_2 and H_2O). The reduction in the amount of HCO_3^- will cause an **equilibrium shift to the right** —to attempt to "re-build" the amount of HCO_3^- present—**increasing the amount of $BaCO_3$ that dissolves.**

Challenging Questions

94. Organize the compounds: $Al(OH)_3$, AgSCN, AgCl, $Fe(OH)_3$, and Ag_3PO_4 in order of decreasing solubility in water:

Information gained from observation (a):

- The fact that AgCl dissolves when SCN^- is added, lets us know that AgSCN is **less soluble than AgCl.**
- The K_{sp} for AgSCN is 1.0×10^{-12}. Since each formula unit of AgSCN dissolves to give 1 Ag^+ and 1SCN^- ion the K_{sp} expression = $[Ag^+][SCN^-] = 1.0 \times 10^{-12}$. So the molar solubility is equal to $(1.0 \times 10^{-12})^{½}$ or 1.0×10^{-6} M.

(At this point our order of solubility is then AgCl > AgSCN)

Information gained from observation (b):

- Write the equation for $Al(OH)_3$ and $Fe(OH)_3$ dissolving: In general, the expression is:

  $$M(OH)_3\ (s) \rightleftharpoons M^{3+}\ (aq)\ + 3\ OH^-\ (aq)$$

  Given that K_{sp} for $Al(OH)_3 = 1.9 \times 10^{-33}$. Let the solubility of the salt be x M. In the saturated solution there are x M Al^{3+} and 3x M OH^-. Substituting into the K_{sp} expression we get $K_{sp} = [Al^{+3}][OH^-]^3 = (x)(3x)^3 = 1.9 \times 10^{-33}$. Solving for x gives 2.9×10^{-9} M. Now we can compare the solubility of $Al(OH)_3$ with the solubility of AgSCN-1.0×10^{-6} M. (At this point our order of solubility is AgCl > AgSCN> $Al(OH)_3$)

- A saturated solution of $Fe(OH)_3$ has a lower pH than a saturated solution of $Al(OH)_3$. From the general equation shown above this means that the $[OH^-]$ is less for the iron salt—and that **less** $Fe(OH)_3$ dissolves than $Al(OH)_3$.

  (At this point our order of solubility is AgCl > AgSCN> $Al(OH)_3$ > $Fe(OH)_3$.)

Information gained from observation (c):

- Examine the solubility expression for AgCl: $AgCl\ (s) \rightleftharpoons Ag^+\ (aq) + Cl^-\ (aq)$. If the molar solubility of AgCl is x M, the concentration of Ag^+ in the saturated solution is **also x M**. {The solubility of AgCl = $[Ag^+]$}.

  Examine the equation for silver phosphate dissolving:

  $$Ag_3PO_4\ (s) \rightleftharpoons 3\ Ag^+(aq) + PO_4^{-3}\ (aq)$$

  If x mol/L of Ag_3PO_4 dissolve, then we get 3x M Ag^+. Another way of saying this is that the $[Ag^+] = 3 \cdot$ solubility of Ag_3PO_4.

  To keep things straight, let's designate the $[Ag^+]$ for Ag_3PO_4 as $[Ag^+]_p$ and $[Ag^+]$ for AgCl as $[Ag^+]_c$

  Observation c tells us $[Ag^+]_p = 4 \cdot [Ag^+]_c$ If we substitute the solubilities of the salts:

  $$3 \cdot \text{solubility of } Ag_3PO_4 = 4 \cdot \text{solubility of AgCl and dividing by 4 we obtain:}$$

  $$\frac{3}{4} \cdot \text{solubility of } Ag_3PO_4 = \text{solubility of AgCl}$$

  This tells us that the solubility of Ag_3PO_4 is **greater than** the solubility of AgCl .

Finally we can write:

Our order of solubility is Ag_3PO_4 > AgCl > AgSCN> $Al(OH)_3$ > $Fe(OH)_3$.

96. The equilibrium for ZnS dissolving in acid is a bit complicated.

There is the K_{sp} : $ZnS \rightleftharpoons Zn^{2+} + S^{2-}$, and the resulting hydrolytic equilibrium of S^{2-}:

$$S^{2-} + H_2O \rightleftharpoons HS^- + OH^-$$

which provides an overall equation:

$$ZnS + 2\,H_3O^+ \rightleftharpoons Zn^{2+} + H_2S + 2\,H_2O \qquad K = 2 \times 10^{-4}$$

So $\dfrac{[Zn^{2+}][H_2S]}{[H_3O^+]^2} = 2 \times 10^{-4}$ and with pH = 1.50 $[H_3O^+] = 3.2 \times 10^{-2}$

$$\frac{[Zn^{2+}](0.1)}{(3.2 \times 10^{-2})^2} = 2 \times 10^{-4} \text{ and } [Zn^{2+}] = 2 \times 10^{-6}\text{ M.}$$

Summary Question

98. a. $AlCl_3(aq) + H_3PO_4(aq) \rightarrow AlPO_4(s) + 3\,HCl(aq)$

b. $152\text{ g }AlCl_3 \cdot \dfrac{1\text{ mol }AlCl_3}{133.3\text{ g}AlCl_3} = 1.14\text{ mol }AlCl_3$

$0.750\text{ M }H_3PO_4 \cdot 3.0\text{ L} = 2.3\text{ mol }H_3PO_4$

Examining the moles-available and moles-required ratios, we note that $AlCl_3$ is the limiting reagent, so the amount of $AlPO_4$ obtainable is:

$$1.14\text{ mol }AlCl_3 \cdot \frac{1\text{ mol }AlPO_4}{1\text{ mol }AlCl_3} \cdot \frac{122.0\text{ }AlPO_4}{1\text{ mol }AlPO_4} = 139\text{ g }AlPO_4$$

c. $\dfrac{25.0\text{ g }AlPO_4}{1\text{ L}} \cdot \dfrac{1\text{ mol }AlPO_4}{122.0\text{ g }AlPO_4} = 0.205\text{ M }AlPO_4$

Determine if this is a saturated solution:

$K_{sp} = [Al^{3+}][PO_4^{3-}] = 1.3 \times 10^{-20}$ Taking the square root of both sides gives:

$$[Al^{3+}] = [PO_4^{3-}] = 1.1 \times 10^{-10}\text{ M}$$

Since 25.0 g of $AlPO_4$ corresponds to a greater concentration(0.205 M $AlPO_4$), the solution is saturated--only 1.1×10^{-10} mol/L $AlPO_4$ will dissolve and

$$[Al^{3+}] = [PO_4^{3-}] = 1.1 \times 10^{-10}\text{ M}$$

d. Since the phosphate ion is capable of reacting with H_3O^+ to form other ions (HPO_4^{2-}, $H_2PO_4^-$), addition of HCl will shift the phosphate equilibrium toward formation of these protonated anions--reducing $[PO_4^{3-}]$.

The reduction of $[PO_4^{3-}]$ will **increase the amount of** $AlPO_4$ which dissolves.

e. Concentrations after mixing:

$$[Al^{3+}] = 2.5 \times 10^{-3}\ M\ Al^{3+} \cdot \frac{1.50\ L}{(1.50\ L + 2.50\ L)} = 9.4 \times 10^{-4}\ M$$

$$[PO_4^{3-}] = 3.5 \times 10^{-2}\ M\ PO_4^{3-} \cdot \frac{2.50\ L}{(1.50\ L + 2.50\ L)} = 2.2 \times 10^{-2}\ M$$

$$Q_c = [Al^{3+}][PO_4^{3-}] = (9.4 \times 10^{-4})(2.2 \times 10^{-2}) = 2.1 \times 10^{-5}$$

Since $Q_c > K_{sp}$ (1.3×10^{-20}), **$AlPO_4$ precipitates.**

The Al^{3+} is the limiting reagent and the maximum amount of $AlPO_4$ that can form is :

$$\frac{9.4 \times 10^{-4}\ mol\ Al^{3+}}{L} \cdot \frac{4.0\ L}{1} \cdot \frac{1\ mol\ AlPO_4}{1\ mol\ Al^{3+}} \cdot \frac{122\ g\ AlPO_4}{1\ mol\ AlPO_4} = 0.46\ g\ AlPO_4$$

Chapter 20:
Principles of Reactivity: Entropy and Free Energy

NUMERICAL AND OTHER QUESTIONS

Entropy

6. The sample with the higher entropy:
   a. A piece of silicon containing a trace of some other atoms — The entropy of a pure substance is lower than an a sample of a mixture
   b. Liquid water at 0° C — The entropy of a liquid is greater than the entropy of the solid form of the substance.
   c. I_2 vapor — The entropy of the gaseous form of a substance will be greater than the entropy of the solid form.

8. Compound with the higher entropy:
   a. $AlCl_3$ (s) - Entropy increases with molecular complexity.
   b. CH_3CH_2I (ℓ) - Entropy increases with molecular complexity.
   c. NH_4Cl (aq) - Entropy of solutions is greater than that of solids.

10. Entropy changes:
   a. C (diamond) $\rightarrow$ C (graphite)

$$\Delta S^\circ = 5.140 \frac{J}{K \cdot mol}(1\ mol) - 2.311 \frac{J}{K \cdot mol}(1\ mol) = +3.363 \frac{J}{K}$$

The increase in entropy reflects the greater order of diamond over graphite.

   b. Na (g) $\rightarrow$ Na (s)

$$\Delta S^\circ = 51.21 \frac{J}{K \cdot mol}(1\ mol) - 153.112 \frac{J}{K \cdot mol}(1\ mol) = -102.50 \frac{J}{K}$$

The lower entropy of the solid state is evidenced by the negative sign.

   c. Br_2 (ℓ) $\rightarrow$ Br_2 (g)

$$\Delta S^\circ = 245.463 \frac{J}{K \cdot mol}(1\ mol) - 152.2 \frac{J}{K \cdot mol}(1\ mol) = 93.3 \frac{J}{K}$$

The increase in entropy is expected with the transition to the disordered state of a gas.

12. $\Delta S°$ for the transition of C_2H_5OH (ℓ) → C_2H_5OH (g)

$$\Delta S° = \frac{\Delta H_{vap}}{T} = \frac{39.3 \times 10^3 \text{ J}}{351.2 \text{ K}} = 112 \frac{\text{J}}{\text{K}}$$

Reactions and Entropy Change

14. For the reaction: 3 C (graphite) + 4 H_2 (g) → C_3H_8 (g)

$$\Delta S° = 1 \cdot S° \, C_3H_8 - [3 \cdot S° \, C \text{ (graphite)} + 4 \cdot S° \, H_2 \text{ (g)}]$$

$$= (1 \text{ mol})(269.9 \frac{\text{J}}{\text{K•mol}}) - [(3 \text{ mol})(5.740 \frac{\text{J}}{\text{K•mol}}) + (4 \text{ mol})(130.684 \frac{\text{J}}{\text{K•mol}})]$$

$$= -270.1 \frac{\text{J}}{\text{K}}$$

16. Calculate the standard molar entropy change for each substance from its elements:

a. H_2 (g) + ½ O_2 (g) → H_2O (ℓ)

$$\Delta S° = (1 \text{ mol})(69.91 \frac{\text{J}}{\text{K•mol}}) - [(1 \text{ mol})(130.684 \frac{\text{J}}{\text{K•mol}}) + (\frac{1}{2} \text{ mol})(205.138 \frac{\text{J}}{\text{K•mol}})] = -163.34 \frac{\text{J}}{\text{K}}$$

b. Mg(s) + O_2(g) + H_2(g) → $Mg(OH)_2$(s)

$$\Delta S° = (1 \text{ mol})(63.18 \frac{\text{J}}{\text{K•mol}}) - [(1 \text{ mol})(32.68 \frac{\text{J}}{\text{K•mol}}) + (1 \text{ mol})(205.138 \frac{\text{J}}{\text{K•mol}}) + (1 \text{ mol})(130.684 \frac{\text{J}}{\text{K•mol}})] = -305.32 \frac{\text{J}}{\text{K}}$$

c. Pb (s) + Cl_2 (g) → $PbCl_2$ (s)

$$\Delta S° = 1 \cdot S° \, PbCl_2 \text{ (s)} - [1 \cdot S° \, Pb \text{ (s)} + 1 \cdot S° \, Cl_2 \text{ (g)}]$$

$$\Delta S° = (1 \text{ mol})(136.0 \text{ J/K} \cdot \text{mol}) - [(1 \text{ mol})(64.8 \frac{\text{J}}{\text{K•mol}}) + (1 \text{ mol})(223.1 \frac{\text{J}}{\text{K•mol}})] = -151.9 \frac{\text{J}}{\text{K}}$$

18. Calculate the standard molar entropy change for each of the following reactions:

a. Mg(s) + 2 H_2O(ℓ) → $Mg(OH)_2$(aq) + H_2 (g)

$$\Delta S° = [1 \cdot S° \, Mg(OH)_2\text{(aq)} + 1 \cdot S° \, H_2 \text{ (g)}] - [1 \cdot S° \, Mg\text{(s)} + 2 \cdot S° \, H_2O(\ell)]$$

$$\Delta S^\circ = [(1\text{ mol})(63.18\ \tfrac{J}{K\bullet mol}) + (1\text{ mol})(130.684\ \tfrac{J}{K\bullet mol})] - [(1\text{ mol})(32.68\ \tfrac{J}{K\bullet mol}) + (2\text{ mol})(69.91\ \tfrac{J}{K\bullet mol})] = 21.36\ \tfrac{J}{K}$$

b. $Na_2CO_3\ (s) + 2\ HCl\ (aq) \rightarrow 2\ NaCl\ (aq) + H_2O\ (\ell) + CO_2\ (g)$

$$\Delta S^\circ = [2 \bullet S^\circ\ NaCl\ (aq) + 1 \bullet S^\circ\ H_2O\ (\ell) + 1 \bullet S^\circ\ CO_2\ (g)] - [1 \bullet S^\circ\ Na_2CO_3\ (s) + 2 \bullet S^\circ\ HCl\ (aq)]$$

$$\Delta S^\circ = [(2\text{ mol})(115.5\ \tfrac{J}{K\bullet mol}) + (1\text{ mol})(69.91\ \tfrac{J}{K\bullet mol}) + (1\text{ mol})(213.74\ \tfrac{J}{K\bullet mol})] - [(1\text{ mol})(134.98\ \tfrac{J}{K\bullet mol}) + (2\text{ mol})(56.5\ \tfrac{J}{K\bullet mol})] = +266.7\ \tfrac{J}{K}$$

20. Sign of ΔH for $C_6H_6\ (\ell) + 3\ H_2\ (g) \rightarrow C_6H_{12}\ (\ell)$

$$\Delta H^\circ_{rxn} = [1 \bullet \Delta H^\circ_f\ C_6H_{12}\ (\ell)] - [1 \bullet \Delta H^\circ_f\ C_6H_6\ (\ell) + 1 \bullet \Delta H^\circ_f\ H_2\ (g)]$$

$$\Delta H^\circ_{rxn} = [(1\text{ mol})(-156.4\ \tfrac{kJ}{mol})] - [(1\text{ mol})(+49.0\ \tfrac{kJ}{mol}) + (3\text{ mol})(0\ \tfrac{kJ}{mol})]$$

$= -205.4$ kJ This change will release energy.

This reaction is *product-favored* or spontaneous (at room temperature).This is reasonable since cyclohexane is stable with respect to H_2 and O_2 (ΔH°_f is negative), Benzene is **not** stable with respect to H_2 and O_2 (ΔH°_f is positive). The **change in entropy** however will be **negative** (since the total number of moles of gas decreases as the reaction proceeds), which does **not favor products** from a spontaneity standpoint. **However** the large amount of energy released ($\Delta H^\circ rxn = -205.4$ kJ) indicates that the **reaction will be product-favored.**

22. Using Table 20.2 classify each of the reactions:

a. $\Delta H_{system} = -$, $\Delta S_{system} = -$ Product-favored at lower T

b. $\Delta H_{system} = +$, $\Delta S_{system} = -$ Not product-favored

24. Calculate ΔH°_{rxn} and ΔS°_{rxn} for the combustion of ethane:

	C_2H_6 (g)	+ 7/2 O_2 (g)	→	2 CO_2(g)	+ 3 H_2O (g)
ΔH°_f (kJ/mol)	-84.68	0		-393.509	-241.818
S° (J/K•mol)	+229.60	+205.138		+213.74	+188.825

$\Delta H°_{rxn}$ = $[2 \bullet \Delta H°_f\ CO_2(g) + 3 \bullet \Delta H°_f\ H_2O\ (g)] - [1 \bullet \Delta H°_f\ C_2H_6\ (g) + 7/2 \bullet \Delta H°_f\ O_2\ (g)]$

$= [(2\ mol)(-\ 393.509\ \frac{kJ}{mol}) + (3\ mol)(-\ 241.818\ \frac{kJ}{mol})] -$

$[(1\ mol)(-\ 84.68\ kJ/mol) + 0]$

$= -\ 1427.79\ kJ$

$\Delta S°_{rxn} = [2 \bullet S°\ CO_2(g) + 3 \bullet S°\ H_2O\ (g)] - [1 \bullet S°\ C_2H_6\ (g) + 7/2 \bullet S°\ O_2\ (g)]$

$= [2\ mol)(213.74\ \frac{J}{K \bullet mol}) + (3\ mol)(188.825\ \frac{J}{K \bullet mol})] -$

$[(1\ mol)(229.60\ \frac{J}{K \bullet mol}) + 7/2\ mol)(205.1381\ \frac{J}{K \bullet mol})]$

$= 46.37\ J/K$

$\Delta S°_{surroundings} = \frac{-\ \Delta H_{rxn}}{T}$ (Assuming we're at 298.15 K)

$= \frac{1427.79\ kJ}{298.15\ K}(1000\ J/kJ) = +4788.8\ \frac{J}{K}$

so $\Delta S°_{system} + \Delta S°_{surroundings} = 46.37\ \frac{J}{K} + 4788.8\ \frac{J}{K}$

$= 4835.2\ \frac{J}{K}$

Since $\Delta H°$ = - and $\Delta S°$ = +, the process is **product-favored.**
This calculation is consistent with our expectations. We know that hydrocarbons burn completely (in the presence of sufficient oxygen) to produce carbon dioxide and water.

Free Energy

26. Calculate $\Delta G°_{rxn}$ for

a.	$Cu(s)$ +	$Cl_2(g)$ →	$CuCl_2(s)$
S°(J/K•mol)	33.150	223.066	108.07
ΔH°f (kJ/mol)	0	0	-220.1

$\Delta H°_{rxn} = \Delta H°_f\ CuCl_2(s) - [\Delta H°_f\ Cu(s) + \Delta H°_f\ Cl_2(g)]$

$= [(1\ mol)(-220.1\ \frac{kJ}{mol})] - [(1\ mol)(0) + (1\ mol)(0)]$

$= -220.1\ kJ$

$$\Delta S^\circ_{rxn} = [\,(1\text{mol})(108.07\ \frac{J}{K\bullet mol})] - [(1\ \text{mol})(33.150\ \frac{J}{K\bullet mol}) + (1\ \text{mol})(223.066\ \frac{J}{K\bullet mol})]$$

$$= -148.15\ J/K$$

$$\Delta G^\circ rxn = \Delta H^\circ_{rxn} - T\Delta S^\circ_{rxn}$$

$$= -220.1\ kJ - (298.15\ K)(-148.15\ J/K)(\frac{1.000kJ}{1000K})$$

$$= -175.9\ kJ$$

b. $NH_3\ (g) + HCl\ (g) \rightarrow NH_4Cl\ (s)$

	NH_3 (g)	HCl (g)	NH_4Cl (s)
S° (J/K•mol)	192.45	186.91	94.6
ΔH°_f (kJ/mol)	- 46.11	- 92.31	- 314.43

$$\Delta S^\circ_{rxn} = 1 \bullet S^\circ\ NH_4Cl\ (s) - [1 \bullet S^\circ\ NH_3\ (g) + 1 \bullet S^\circ\ HCl\ (g)]$$

$$= (1\ \text{mol})(94.6\ J/K \bullet mol) - [(1\ \text{mol})(192.45\ J/K \bullet mol) + (1\ \text{mol})(186.91\ J/K \bullet mol)]$$

$$= -284.8\ J/K$$

$$\Delta H^\circ_{rxn} = 1 \bullet \Delta H^\circ_f\ NH_4Cl\ (s) - [1 \bullet \Delta H^\circ_f\ NH_3\ (g) + 1 \bullet \Delta H^\circ_f\ HCl\ (g)]$$

$$= (1\ \text{mol})(-314.43\ kJ/mol) - [(1\ \text{mol})(-46.11\ kJ/mol) + (1\ \text{mol})(-92.31\ kJ/mol)]$$

$$= -176.01\ kJ$$

$$\Delta G^\circ_{rxn} = \Delta H^\circ_f - T\,\Delta S^\circ_{rxn}$$

$$= -176.01\ kJ - (298.15\ K)(-284.76\ J/K)(\frac{1.000\ kJ}{1000\ J})$$

$$= -176.01\ kJ + 84.90\ kJ$$

$$= -91.10\ kJ$$

Part a corresponds to formation of one mole of a substance from its elements, each in their standard state--ΔG°_f. The value obtained for ΔG° agrees well with that in Appendix L. (-175.9 kJ vs -175.7 kJ).The values for ΔG° for **both equations** are negative, indicating that they **are product-favored**. Both reactions are enthalpy driven ($\Delta H < 0$).

28. Calculate the molar free energies of formation for:

a. CS_2 (g) The reaction is: C (graphite) + 2 S (s,rhombic) $\rightarrow$ CS_2 (g)

$$\Delta H^\circ_f = (1\ \text{mol})(117.36\ \frac{kJ}{mol}) - [0 + 0] = 117.36\ kJ$$

$$\Delta S^\circ = (1\text{ mol})(237.84\,\frac{\text{J}}{\text{K•mol}}) - [(1\text{ mol})(5.740\,\frac{\text{J}}{\text{K•mol}}) + (2\text{ mol})(31.80\,\frac{\text{J}}{\text{K•mol}})] = 168.50\text{ J/K}$$

$$\Delta G^\circ_f = \Delta H^\circ_f - T\Delta S^\circ$$
$$= (1\text{ mol})(117.36\,\frac{\text{kJ}}{\text{mol}}) - (298.15\text{ K})(168.50\,\frac{\text{J}}{\text{K}})(\frac{1.000\text{ kJ}}{1000.\text{ J}})$$
$$= 67.12\text{ kJ/mol}$$ Appendix value: 67.12 kJ/mol

Reaction is **not product-favored.**

b. $N_2H_4\,(\ell)$ The reaction is: $N_2(g) + 2\,H_2(g) \rightarrow N_2H_4\,(\ell)$

$$\Delta H^\circ_f = (1\text{ mol})(50.63\,\frac{\text{kJ}}{\text{mol}}) - [0 + 0] = 50.63\text{ kJ}$$

$$\Delta S^\circ = (1\text{ mol})(121.21\,\frac{\text{J}}{\text{K•mol}}) - [(1\text{ mol})(191.61\,\frac{\text{J}}{\text{K•mol}}) + (2\text{ mol})(130.684\,\frac{\text{J}}{\text{K•mol}})] = -331.77\text{ J/K}$$

$$\Delta G^\circ_f = \Delta H^\circ_f - T\Delta S^\circ$$
$$= (1\text{ mol})(50.63\text{ kJ/mol}) - (298.15\text{ K})(-331.77\text{ J/K})(\frac{1.000\text{ kJ}}{1000.\text{ J}})$$
$$= 149.55\text{ kJ/mol}$$ Appendix value: 149.34 kJ/mol

Reaction is **not product-favored.**

c. $COCl_2\,(g)$ The reaction is: C (graphite) + $\frac{1}{2}\,O_2\,(g) + Cl_2\,(g) \rightarrow COCl_2\,(g)$

$$\Delta H^\circ_f = (1\text{ mol})(-218.8\,\frac{\text{kJ}}{\text{mol}}) - [0 + 0 + 0] = -218.8\text{ kJ}$$

$$\Delta S^\circ = (1\text{ mol})(283.53\,\frac{\text{J}}{\text{K•mol}}) - [(1\text{ mol})(5.740\,\frac{\text{J}}{\text{K•mol}}) + (\tfrac{1}{2}\text{ mol})(205.138\,\frac{\text{J}}{\text{K•mol}}) + (1\text{ mol})(223.066\,\frac{\text{J}}{\text{K•mol}})]$$
$$= -47.85\text{ J/K}$$

$$\Delta G^\circ_f = \Delta H^\circ_f - T\Delta S^\circ$$
$$= (1\text{ mol})(-218.8\text{ kJ/mol}) - (298.15\text{ K})(-47.85\text{ J/K})(\frac{1.000\text{ kJ}}{1000.\text{ J}})$$
$$= -204.5\text{ kJ/mol}$$ Appendix value: - 204.6 kJ/mol

The formation of $COCl_2$ is predicted to be **product-favored.**

30. Reaction of N_2H_4 (ℓ): $N_2H_4\ (\ell) + O_2\ (g) \rightarrow 2\ H_2O\ (\ell) + N_2\ (g)$

ΔG°_{rxn} for reaction of 1.00 mole of N_2H_4 (ℓ):

$$\Delta G^\circ_{rxn} = [2 \bullet \Delta G^\circ_f H_2O\ (\ell) + 1 \bullet \Delta G^\circ_f N_2\ (g)] - [1 \bullet \Delta G^\circ_f N_2H_4\ (\ell) + 1 \bullet \Delta G^\circ_f O_2\ (g)]$$
$$= [(2\text{ mol})(-237.129\ \frac{kJ}{mol}) + 0] - [(1\text{ mol})(149.34\ \frac{kJ}{mol}) + 0] = -623.60\text{ kJ}$$

Free energy change for oxidation of 1.00×10^3 g of N_2H_4 (ℓ):

$$1.00 \times 10^3\text{ g of } N_2H_4 \bullet \frac{1\text{ mol } N_2H_4}{32.0\text{g } N_2H_4} \bullet \frac{-623.60\text{ kJ}}{1\text{ mol } N_2H_4} = -1.95 \times 10^4\text{ kJ}$$

32. Calculate ΔG°_{rxn} for the following reactions:

a. $Ca\ (s) + Cl_2\ (g) \rightarrow CaCl_2\ (s)$

$$\Delta G^\circ_{rxn} = (1\text{ mol})(-748.1\ \frac{kJ}{mol}) - [0 + 0]$$
$$= -748.1\text{ kJ}$$ **product-favored**

b. $2\ HgO\ (s) \rightarrow 2\ Hg\ (l) + O_2(g)$

$$\Delta G^\circ_{rxn} = [0 + 0] - (2\text{ mol})(-58.539\ \frac{kJ}{mol})$$
$$= 117.08\text{ kJ}$$ **reactant-favored**

c. $NH_3\ (g) + 2\ O_2\ (g) \rightarrow HNO_3\ (\ell) + H_2O\ (\ell)$

$$\Delta G^\circ_{rxn} = [(1\text{ mol})(-80.71\ \frac{kJ}{mol}) + (1\text{ mol})(-237.129\ \frac{kJ}{mol})] - [(1\text{ mol})(-16.45\ \frac{kJ}{mol}) + 0\ ]$$
$$= -301.39\text{ kJ}$$ **product-favored**

34. Value for ΔG°_f for $BaCO_3(s)$:

$$\Delta G^\circ_{rxn} = [\Delta G^\circ_f\ BaO(s) + \Delta G^\circ_f\ CO_2(g)] - [\Delta G^\circ_f\ BaCO_3(s)]$$
$$+218.1\text{ kJ} = [(1\text{ mol})(-525.1\ \frac{kJ}{mol}) + (1\text{ mol})(-394.359\ \frac{kJ}{mol})] - \Delta G^\circ_f\ BaCO_3(s)$$
$$+218.1\text{ kJ} = -919.5\text{ kJ} - \Delta G^\circ_f\ BaCO_3(s)$$
$$-1137.6\text{ kJ/mol} = +\Delta G^\circ_f\ BaCO_3(s)$$

36. Determine $\Delta H°$, $\Delta S°$, and $\Delta G°$ for: C_8H_{16} (g) + H_2 (g) → C_8H_{18} (g)

	C_8H_{16} (g)	H_2 (g)	C_8H_{18} (g)
$\Delta H°_f$ (kJ/mol)	- 81.4	0	- 208.6
S° (J/K • mol)	462.5	130.684	466.7

$$\Delta H°_{rxn} = [(1\ mol)(-208.6\frac{kJ}{mol})] - [(1\ mol)(-81.4\frac{kJ}{mol}) + 0]$$
$$= -127.2\ kJ$$

$$\Delta S°_{rxn} = [(1\ mol)(466.7\frac{J}{K\bullet mol})] - [(1\ mol)(462.5\frac{J}{K\bullet mol}) + (1\ mol)(130.684\frac{J}{K\bullet mol})]$$
$$= -126.5\ J/K$$

$$\Delta G°_{rxn} = \Delta H°_f - T\,\Delta S°_{rxn}$$
$$= -127.2\ kJ - (298.15\ K)(-126.5\ J/K)(\frac{1.000\ kJ}{1000\ J})$$
$$= -89.5\ kJ$$

The negative value for $\Delta G°_{rxn}$ indicates that the reaction is **product-favored** under standard conditions.

Thermodynamics and Equilibrium Constants

38. Calculate K_p for the reaction:

$$\frac{1}{2} N_2(g) + \frac{1}{2} O_2(g) \rightarrow NO\ (g) \qquad \Delta G°_f = +86.55\ kJ/mol\ NO$$

$$\Delta G°_{rxn} = -RT \ln K_p$$
$$86.55 \times 10^3\ J/mol = -(8.314\frac{J}{K\bullet mol})(298.15\ K) \ln K_p$$
$$-34.9 = \ln K_p$$
$$7 \times 10^{-16} = K_p$$

Note that the + value of $\Delta G°_f$ results in a value of K_p which is small—**reactants are favored.** A negative value would result in a large K_p -- a process in which the products were favored.

40. a. $\Delta G°_{rxn} = \Delta G°_f\ C_2H_6\ (g) - [\Delta G°_f\ C_2H_4\ (g) + \Delta G°_f\ H_2(g)]$

$$= (1\ mol)(-32.82\frac{kJ}{mol}) - [(1\ mol)(68.15\frac{kJ}{mol}) + (1\ mol)(0)]$$
$$= -100.97\ kJ$$

The reaction is predicted to be product-favored.

b. $\Delta G°rxn = -RT \ln K_p$

$-100.97 \times 10^3 \frac{J}{mol} = -(8.314 \frac{J}{K \cdot mol})(298.15\ K) \ln K_p$

$40.754 = \ln K_p$

$5.00 \times 10^{17} = K_p$

Note that a **large negative value for $\Delta G°$** results in a **large K_p**.

General Questions

42. For the reaction of sodium with water:

$Na\ (s) + H_2O(\ell) \rightarrow NaOH(aq) + ½\ H_2\ (g)$

Predict signs for $\Delta H°$ and $\Delta S°$:

This one seems easy !! The reaction of sodium with water **gives off heat**, and the heat frequently ignites the hydrogen gas that is concomitantly evolved. **$\Delta H°$= -**. Regarding entropy, the system changes from one with a solid (low entropy) and a liquid (higher entropy) to a solution (*frequently* higher entropy than liquid) and a gas (high entropy). So we would predict that the **entropy would increase, i.e. $\Delta S°$ = +**.

Now for the calculation:

$\Delta H°_{rxn} = [1 \cdot \Delta H°_f\ NaOH(aq) + ½ \cdot \Delta H°_f\ H_2(g)] - [1 \cdot \Delta H°_f\ Na(s) + 1 \cdot \Delta H°_f\ H_2O(\ell)]$

$= [(1\ mol)(-470.114 \frac{kJ}{mol}) + (½\ mol)(0)] - [(1\ mol)(0) + (1\ mol)(-285.830 \frac{kJ}{mol})]$

$= -184.284\ kJ$

$\Delta S°_{rxn} = [1 \cdot S°\ NaOH(aq) + ½ \cdot S°\ H_2(g)] - [1 \cdot S°\ Na(s) + 1 \cdot S°\ H_2O(\ell)]$

$= [(1\ mol)(48.1 \frac{J}{K \cdot mol}) + (½\ mol)(130.684 \frac{J}{K \cdot mol})] -$

$[(1\ mol)(51.21 \frac{J}{K \cdot mol}) + (1\ mol)(69.91 \frac{J}{K \cdot mol})]$

$= -7.7 \frac{J}{K \cdot mol}$

As expected, the $\Delta H°_{rxn}$ for the reaction is negative! The surprise comes in the calculation for $\Delta S°_{rxn}$. While we anticipate the sign to be positive, we find a **slightly negative** number—reflecting the order (hence a decrease in entropy) that can occur as solutions form.

44. For the reaction : $C_6H_{12}O_6\ (aq) \rightarrow 2\ C_2H_5OH\ (\ell) + 2\ CO_2\ (g)$

$S°\ (\frac{J}{K\cdot mol})$	289	160.7	213.74
$\Delta H°(\frac{kJ}{mol})$	-1260.0	-277.69	-393.509

$$\Delta H°_{rxn} = [2 \cdot \Delta H°_f\ C_2H_5OH\ (\ell) + 2 \cdot \Delta H°_f\ CO_2\ (g)] - [1 \cdot \Delta H°_f\ C_6H_{12}O_6\ (aq)]$$

$$\Delta H°_{rxn} = [(2\ mol)(-277.69 \frac{kJ}{mol}) + (2\ mol)(-393.509 \frac{kJ}{mol})] - [(1\ mol)(-1260.0 \frac{kJ}{mol})]$$

$$= -82.4\ kJ$$

$$\Delta S°_{rxn} = [2 \cdot S°\ C_2H_5OH\ (\ell) + 2 \cdot S°\ CO_2\ (g)] - [1 \cdot S°\ C_6H_{12}O_6\ (aq)]$$

$$\Delta S°rxn = [(2\ mol)(160.7 \frac{J}{K\cdot mol}) + (2\ mol)(213.74 \frac{J}{K\cdot mol})] - [(1\ mol)(289 \frac{J}{K\cdot mol})]$$

$$= 460.\ J/K$$

$$\Delta G°_{rxn} = \Delta H°_f - T\,\Delta S°_{rxn}$$

$$= -82.4\ kJ - (298.15\ K)(460. \frac{J}{K})(\frac{1.000\ kJ}{1000.\ J})$$

$= -219.4\ kJ$ The reaction is **product-favored.**

46. What is $\Delta G°$ for the equilibrium: $N_2O_4\ (g) \rightleftharpoons 2\ NO_2\ (g)$ given $Kp = 0.14$ at 25 °C

$$\Delta G° = -RT \ln K = -(8.314 \frac{J}{K\cdot mol})(298.15\ K) \ln (0.14)$$

$$= 4.87 \times 10^3 \frac{J}{mol} \text{ or } 4.87\ kJ/mol$$

The calculated value from $\Delta G°$ is:

$$\Delta G°rxn = 2 \cdot \Delta G°f\ NO_2\ (g) - 1 \cdot \Delta G°f\ N_2O_4\ (g)$$

$$= (2\ mol)(51.31 \frac{kJ}{mol}) - (1\ mol)(97.89 \frac{kJ}{mol})$$

$$= 4.73\ kJ/mol$$

The values are quite comparable.

48. What is $\Delta G°_{rxn}$ for the equilibrium: butane (g) $\rightleftharpoons$ isobutane (g) given K = 2.5 at 25 °C

$$\Delta G°_{rxn} = -RT \ln K = -(8.314 \frac{J}{K\cdot mol})(298.15\ K) \ln (2.5)$$

$$= -2.3 \times 10^3 \frac{J}{mol} \text{ or } -2.3\ kJ/mol$$

50. The key phrase needed to answer the question:"What is the sign......" is "Iodine dissolves readily...." . This phrase tell us that ΔG° **is negative.**

Enthalpy-driven processes are exothermic. The "neutrality" of the ΔH for this reaction tells us that the process is NOT enthalpy-driven. Since the iodine goes from the solid state to the "solution" state, we anticipate an **increase in entropy**, and would therefore state that the **process is entropy-driven**.

52. For the reaction: $2\ SO_3\ (g) \rightarrow 2\ SO_2\ (g) + O_2\ (g)$

$$\Delta H^\circ_{rxn} = [2 \bullet \Delta H^\circ_f\ SO_2\ (g) + 1 \bullet \Delta H^\circ_f\ O_2\ (g)] - [2 \bullet \Delta H^\circ_f\ SO_3\ (g)]$$
$$= [(2\ mol)(-\ 296.830\ \frac{kJ}{mol}) + 0] - [(2\ mol)(-395.72\ \frac{kJ}{mol})] = 197.78\ kJ$$

$$\Delta S^\circ_{rxn} = [2 \bullet S^\circ\ SO_2\ (g) + 1 \bullet S^\circ\ O_2\ (g)] - [2 \bullet S^\circ\ SO_3\ (g)]$$
$$= [(2\ mol)(248.22\ \frac{J}{K \bullet mol}) + (1\ mol)(205.138\ \frac{J}{K \bullet mol})] - [(2\ mol)(256.76\ \frac{J}{K \bullet mol})]$$
$$= 188.06\ J/K$$

a. Is the reaction product-favored at 25 °C ?

$$\Delta G^\circ_{rxn} = \Delta H^\circ_f - T\,\Delta S^\circ_{rxn}$$
$$= 197.78\ kJ - (298.15\ K)(188.06\ \frac{J}{K})(\frac{1.000\ kJ}{1000\ J})$$
$$= 141.73\ kJ$$

The reaction is **reactant-favored.**

b. The reaction can become product-favored if there is some T at which $\Delta G^\circ_{rxn} < 0$.

To see if such a T is feasible, let's set $\Delta G^\circ_{rxn} = 0$ and solve for T!

Remember that the units of energy must be the same, so let's convert units of J (for the entropy term) into units of kJ.

$$\Delta G^\circ_{rxn} = \Delta H^\circ_f - T\,\Delta S^\circ_{rxn}$$
$$0 = 197.78\ kJ - T(0.18806\ \frac{kJ}{K})$$
$$T = \frac{197.78\ kJ}{0.18806\ \frac{kJ}{K}} = 1051.7\ K \quad \text{or } (1051.7 - 273.2) = 778.5\ ^\circ C$$

c. The equilibrium constant for the reaction at 1500 °C

Since we know that $\Delta G^\circ_{rxn} = \Delta H^\circ_f - T\,\Delta S^\circ_{rxn} = -\ RT \ln K$,

we can solve for K if we know ΔG°_{rxn} at 1500 °C

$$\Delta G^\circ_{rxn} = \Delta H^\circ_f - T\,\Delta S^\circ_{rxn} = -\ RT \ln K$$

$$= 197.78 \text{ kJ} - (1773 \text{ K})(188.06 \text{ J/K})(\frac{1 \text{ kJ}}{1000 \text{ J}})$$

$$-135.7 \text{ kJ} = -8.314 \frac{\text{J}}{\text{K} \cdot \text{mol}}(\frac{1 \text{ kJ}}{1000 \text{ J}})(1773 \text{ K})\ln K$$

$$K = 9.9 \times 10^3 \text{ (2 sf)}$$

54. Reaction: $H_2S\,(g) + 2\,O_2\,(g) \rightarrow H_2SO_4\,(\ell)$

	H_2S (g)	$2\,O_2$ (g)	H_2SO_4 (ℓ)
$\Delta H°_f$ (kJ/mol)	-20.63	0	- 813.989
S° (J/K • mol)	205.79	205.138	156.904

$$\Delta H°_{rxn} = [(1 \text{ mol})(-813.989 \frac{\text{kJ}}{\text{mol}})] - [(1 \text{ mol})(-20.63 \frac{\text{kJ}}{\text{mol}}) + 0]$$

$$= -793.36 \text{ kJ}$$

$$\Delta S°_{rxn} = [(1 \text{ mol})(156.904 \frac{\text{J}}{\text{K•mol}})] - [(1 \text{ mol})(205.79 \frac{\text{J}}{\text{K•mol}}) + (2 \text{ mol})(205.138 \frac{\text{J}}{\text{K•mol}})]$$

$$= -459.16 \text{ J/K}$$

$$\Delta G°_{rxn} = \Delta H°_f - T\,\Delta S°_{rxn}$$

$$= -793.36 \text{ kJ} - (298.15 \text{ K})(-459.16 \frac{\text{J}}{\text{K}})(\frac{1.000 \text{ kJ}}{1000 \text{ J}})$$

$$= -656.46 \text{ kJ}$$

The reaction is **product-favored** at 25 °C ($\Delta G° < 0$) and **enthalpy-driven** ($\Delta H°_{rxn} < 0$).

Conceptual Questions

56. Calculate ΔS° for: $HCl(g) + H_2O(\ell) \rightarrow HCl(aq)$

	HCl(g)	$H_2O(\ell)$	HCl(aq)
S° ($\frac{\text{J}}{\text{K•mol}}$)	186.908	69.91	56.5

$$\Delta S° = (1 \text{ mol})(56.5 \frac{\text{J}}{\text{K•mol}}) - [(1 \text{ mol})(186.908 \frac{\text{J}}{\text{K•mol}}) + (1 \text{ mol})(69.91 \frac{\text{J}}{\text{K•mol}})]$$

$$= -200.3 \text{ J/K}$$

We anticipate that the **entropy should decrease** as the gas dissolves in water. This expectation is born out by the calculated ΔS° .

58. Calculate the $\Delta G°$ for the transition of S_8 (rhombic) $\rightarrow$ S_8 (monoclinic)

a. At 80 °C $\Delta G°_{rxn} = \Delta H° - T\,\Delta S°$

$$\Delta G°_{rxn} = 3.213 \text{ kJ} - (353 \text{ K})(0.0087 \frac{\text{kJ}}{\text{K}})$$

$$= 0.1419 \text{ kJ} \quad \text{or } +100 \text{ J} \qquad (80\text{ °C has 1 sf})$$

At 100 °C $\Delta G°_{rxn} = \Delta H° - T\,\Delta S°$

$$\Delta G°_{rxn} = 3.213 \text{ kJ} - (373 \text{ K})(0.0087 \frac{\text{kJ}}{\text{K}})$$

$$= -0.0321 \text{ kJ or } -30 \text{ J} \qquad (100\text{ °C has 1 sf})$$

The rhombic form of sulfur is the more stable at lower temperature, while the monoclinic form is the more stable at higher temperature. The transition to monoclinic form is product-favored at temperatures above 100 degrees C.

b. The temperature at which $\Delta G°_{rxn} = 0$:

$$\Delta G°_{rxn} = 3.213 \text{ kJ} - (T)(0.0087 \frac{\text{kJ}}{\text{K}})$$

$$0 = 3.213 \text{ kJ} - (T)(0.0087 \frac{\text{kJ}}{\text{K}})$$

$$T = \frac{3.213 \text{ kJ}}{0.0087 \frac{\text{kJ}}{\text{K}}} = 370 \text{ K} = 96 \text{ °C}$$

Challenging Problems

60. a. $\Delta G°_{rxn}$ and K_p for steps :

Step 1: $\Delta G°_{rxn} = [\Delta G°_f\, Ti(s) + \Delta G°_f\, CO_2(g)] - [\Delta G°_f\, TiO_2(s) + \Delta G°_f\, C(s)]$

$$= [(1 \text{ mol})(0) + (1\text{mol})(-394.359 \frac{\text{kJ}}{\text{mol}})] - [(1 \text{ mol})(-884.5 \frac{\text{kJ}}{\text{mol}}) + (1 \text{ mol})(0)]$$

$$= +490.1 \text{ kJ}$$

With $\Delta G°_{rxn}$ we can calculate K:

$$\Delta G° = -RT\ln K_p$$

$$+490.1 \times 10^3 \text{ J} = -(8.314 \frac{\text{J}}{\text{K·mol}})(298.15 \text{ K}) \ln K_p$$

$$-197.8 = \ln K_{p1} \quad \text{or} \quad e^{-197.8}$$

$$1 \times 10^{-86} = K_{p1}$$

Step 2: ΔG°_{rxn} = $[\Delta G^\circ_f\ TiCl_4(\ell)] - [\Delta G^\circ_f\ Ti(s) + 2\ \Delta G^\circ_f\ Cl_2(g)]$

$= [(1\ mol)(-737.2\ \frac{kJ}{mol})] - [(1\ mol)(0) + (2\ mol)(0)]$

$= -737.2\ kJ$

Now that we have ΔG°_{rxn} we can calculate K:

$$\Delta G^\circ = -RT\ln K_p$$
$$-737.2 \times 10^3\ J = -(8.314\ \frac{J}{K \cdot mol})(298.15\ K)\ \ln K_{p2}$$
$$297.5 = \ln K_{p2}$$
$$1 \times 10^{129} = K_{p2}$$

b. Overall : ΔG°_{rxn} = $[\Delta G^\circ_f\ TiCl_4(\ell) + \Delta G^\circ_f\ CO_2(g)] -$
$[\Delta G^\circ_f\ TiO_2(s) + \Delta G^\circ_f\ C(s) + 2\ \Delta G^\circ_f\ Cl_2(g)]$

$= [(1\ mol)(-737.2\ \frac{kJ}{mol}) + (1\ mol)(-394.359\ \frac{kJ}{mol})] -$
$[(1\ mol)(-884.5\ \frac{kJ}{mol}) + (1\ mol)(0) + (2\ mol)(0)]$

$= -247.1\ kJ$

Note that $\Delta G^\circ_{overall} = \Delta G^\circ_1 + \Delta G^\circ_2 = (+490.1\ kJ + -737.2\ kJ) = -247.1\ kJ$

Overall

$$\Delta G^\circ = -RT\ln K_p$$
$$-247.1 \times 10^3\ J = -(8.314\ \frac{J}{K \cdot mol})(298.15\ K)\ \ln K_p$$
$$99.7 = \ln K_p \quad \text{or} \quad e^{99.7}$$
$$2 \times 10^{43} = K_p \quad \text{or} \quad K_{p1} \cdot K_{p2}$$

c. The overall negative free energy indicates the overall process is product-favored, even though the entropy is decreasing, so **the process is enthalpy driven.**

62. For the reaction:

	$4\ Ag(s)$	+ $O_2(g)$	→	$2\ Ag_2O(s)$
$S^\circ\ (\frac{J}{K \cdot mol})$	42.55	205.138		121.3
$\Delta H^\circ\ (\frac{kJ}{mol})$	0	0		-31.05

a. Calculate $\Delta H^\circ rxn$, ΔS° , and ΔG°_{rxn} :

$\Delta H^\circ rxn = [(2\ mol)(-31.05\ \frac{kJ}{mol})] - [(4\ mol)(0) + (1\ mol)(0)] = -62.10\ kJ$

$$\Delta S^\circ = [(2\text{ mol})(121.3\ \frac{J}{K\cdot mol})] - [(4\text{ mol})(42.55\ \frac{J}{K\cdot mol}) + (1\text{ mol})(205.138\ \frac{J}{K\cdot mol})]$$

$$= -132.7\text{ J/K}$$

$$\Delta G^\circ_{rxn} = (-62.10\text{ kJ}) - (298.15\text{ K})(-132.7\text{ J/K})(\frac{1.000\text{kJ}}{1000\text{ J}})$$

$$= -22.52\text{ kJ}$$

b. To calculate P_{O_2} we'll need K_p.

$$\Delta G^\circ_{rxn} = -RT\ln K_p$$

$$-22.52 \times 10^3\text{ J} = -(8.314\ \frac{J}{K\cdot mol})(298.15\text{ K})\ln K_p$$

$$+9.08 = \ln K_p$$

$$K_p = \frac{1}{P_{O_2}}$$ (Remember solids are omitted from equilibrium expressions.)

$$\frac{1}{P_{O_2}} = 8.79 \times 10^3 \quad \text{and} \quad P_{O_2} = 1.1 \times 10^{-4}\text{ atm}$$

c. T at which $P_{O_2} = 1.00$ atm

If $P_{O_2} = 1.00$ atm, then $K_p = \frac{1}{1.00\text{ atm}} = 1.00$

Let's calculate the ΔG° at which $K_p = 1.00$

$$\Delta G^\circ_{rxn} = -RT\ln K_p$$

The difficulty is that if T is not 25 °C, we need to calculate ΔG°, $(\Delta H - T\Delta S)$, so

$$\Delta H^\circ - T\Delta S^\circ = -RT\ln(1.00)$$

or $\Delta H^\circ = -RT\ln(1.00) + T\Delta S^\circ$

or $\frac{\Delta H^\circ}{T} = -R\ln(1.00) + \Delta S^\circ$

Using ΔH° and ΔS° data (from part a)

$$\frac{-62.10\text{ kJ}}{T} = -(8.314\ \frac{J}{K\cdot mol})(0) + -132.7\text{ J/K}$$

or $\frac{-62.10\text{ kJ}}{T} = -132.7\text{ J/K}$

and solving for T : 468 K or 195 °C.

Summary Questions

64. a. Calculate $\Delta G°$ for the reaction of benzene with water to form phenol and hydrogen:

$$C_6H_6\,(\ell) + H_2O\,(\ell) \rightarrow C_6H_5OH\,(s) + H_2\,(g)$$

	$C_6H_6\,(\ell)$	$H_2O\,(\ell)$	$C_6H_5OH\,(s)$	$H_2\,(g)$
$S°\,(\frac{J}{K\bullet mol})$	172.8	69.91	144.0	130.684
$\Delta H°(\frac{kJ}{mol})$	49.03	-285.830	-165.1	0

$$\Delta H°_{rxn} = [1 \bullet \Delta H°_f\, C_6H_5OH\,(s) + 1\bullet \Delta H°_f\, H_2\,(g)] - [1\bullet \Delta H°_f\, C_6H_6\,(\ell) + 1\bullet \Delta H°_f\, H_2O\,(\ell)]$$

$$\Delta H°_{rxn} = [(1\text{ mol})(-165.1\frac{kJ}{mol}) + (1\text{ mol})(0)] - [(1\text{ mol})(49.03\frac{kJ}{mol}) + (1\text{ mol})(-285.830\frac{kJ}{mol})]$$

$$= 71.7\text{ kJ}$$

$$\Delta S°_{rxn} = [1 \bullet S°\, C_6H_5OH\,(s) + 1\bullet S°\, H_2\,(g)] - [1\bullet S°\, C_6H_6\,(\ell) + 1\bullet S°\, H_2O\,(\ell)]$$

$$\Delta S°rxn = [(1\text{ mol})(144.0\frac{J}{K\bullet mol}) + (1\text{ mol})(130.684\frac{J}{K\bullet mol})] - [(1\text{ mol})(172.8\frac{J}{K\bullet mol}) + (1\text{ mol})(69.91\frac{J}{K\bullet mol})]$$

$$= 32.0\frac{J}{K}$$

$$\Delta G°_{rxn} = \Delta H°_f - T\,\Delta S°_{rxn}$$

$$= 71.7\text{ kJ} - (298.15\text{ K})(32.0\frac{J}{K})(\frac{1.000\text{ kJ}}{1000.\text{ J}})$$

$$= 62.2\text{ kJ}$$

The reaction is **not product-favored** under standard conditions at 25 °C ($\Delta G° > 0$)

b. What is the C-O-H bond angle in phenol?

With four groups on the O (2 atoms and 2 lone pairs) the electronic group geometry is tetrahedral, and the molecular geometry is bent. (The angle is approximately 109°.)

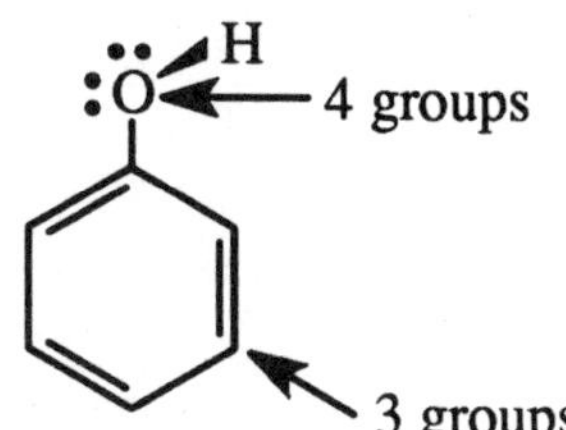

c. What is the hybridization of the C atoms in phenol? ....the O atom ?

The C atoms are sp^2 hybridized (three groups attached-- bond angle 120°)

The O atom is sp^3 hybridized (four groups attached -- bond angle 109°)

d. What is the pH of a 0.012 M solution of phenol?

The Ka for phenol = 1.3×10^{-10}. (From Appendix)

The equilibrium for the weak acid in water can be written:

$$C_6H_5OH + H_2O \rightleftharpoons C_6H_5O^- + H_3O^+$$

$$Ka = \frac{[C_6H_5O^-][H_3O^+]}{[C_6H_5OH]} = 1.3 \times 10^{-10}$$

Since phenol is a monoprotic acid, the concentrations of the phenoxide ion and hydronium ion are equal. We can rewrite the Ka expression as:

$$Ka = \frac{[H_3O^+]^2}{[0.012]} = 1.3 \times 10^{-10}$$

and $[H_3O^+]^2 = (0.012)(1.3 \times 10^{-10}) = 1.56 \times 10^{-12}$

$pH = -\log(1.2 \times 10^{-6}) = 5.90$

Chapter 21
Principles of Reactivity: Electron Transfer Reactions

NUMERICAL AND OTHER QUESTIONS

Balancing Equations for Redox Reactions

12. Balance the following:

	reactant is	overall process is
a. $Cr (s) \rightarrow Cr^{3+} (aq) + 3 e^-$	reducing agent	oxidation
b. $AsH_3 (g) \rightarrow As (s) + 3 H^+ (aq) + 3 e^-$	reducing agent	oxidation
c. $VO_3^- (aq) + 6 H^+ (aq) + 3 e^- \rightarrow V^{2+} (aq) + 3 H_2O (\ell)$	oxidizing agent	reduction

Note: e^- are used to balance charge; H^+ balances only H atoms; H_2O balances both H and O atoms.

14. Balance the following (in acidic solution)

	reactant is	overall process is
a. $Cr_2O_7^{2-} (aq) + 14 H^+ (aq) + 6 e^- \rightarrow 2 Cr^{3+} (aq) + 7 H_2O (\ell)$	oxidizing agent	reduction
b. $CH_3CHO (aq) + H_2O (\ell) \rightarrow CH_3CO_2H (aq) + 2 H^+ (aq) + 2 e^-$	reducing agent	oxidation
c. $Bi^{3+} (aq) + 3 H_2O (\ell) \rightarrow HBiO_3 (aq) + 5 H^+ (aq) + 2 e^-$	reducing agent	oxidation

16. Balance the following (in basic solution)

	reactant is	overall process is
a. $Sn (s) + 4 OH^- (aq) \rightarrow Sn(OH)_4^{2-} (aq) + 2 e^-$	reducing agent	oxidation
b. $MnO_4^- (aq) + 2 H_2O (\ell) + 3 e^- \rightarrow MnO_2 (s) + 4 OH^- (aq)$	oxidizing agent	reduction
c. $ClO^- (aq) + H_2O (\ell) + 2 e^- \rightarrow Cl^- (aq) + 2 OH^- (aq)$	oxidizing agent	reduction

18. Balancing redox equations in neutral or acidic solutions may be accomplished in several steps. They are:

1. Separating the equation into two equations which represent reduction and oxidation
2. Balancing mass of elements (other than H or O)
3. Balancing mass of O by adding H_2O
4. Balancing mass of H by adding H^+
5. Balancing charge by adding electrons
6. Balancing electron gain (in the reduction half-equation) with electron loss (in the oxidation half-equation)
7. Combining the two half equations

For the parts (a-c) of this problem, each step will be identified with a number corresponding to the list above. In addition, the physical states of all species will be omitted in all but the final step. While this omission is <u>not generally recommended</u>, it should increase the clarity of the steps involved. In addition when a step leaves a half equation unchanged from the previous step, we have omitted the half equation.

a. $Cl_2\ (aq) + Br^-\ (aq) \rightarrow Br_2\ (aq) + Cl^-\ (aq)$

1.	$Cl_2 \rightarrow Cl^-$	$Br^- \rightarrow Br_2$
2.	$Cl_2 \rightarrow 2\ Cl^-$	$2\ Br^- \rightarrow Br_2$
5.	$Cl_2 + 2e^- \rightarrow 2\ Cl^-$	$2\ Br^- \rightarrow Br_2 + 2e^-$
7.	$Cl_2\ (aq) + 2\ Br^-\ (aq) \rightarrow 2\ Cl^-\ (aq) + Br_2\ (aq)$	

b. $Sn\ (s) + H^+\ (aq) \rightarrow Sn^{2+}\ (aq) + H_2\ (g)$

1.	$Sn \rightarrow Sn^{2+}$	$H^+ \rightarrow H_2$
3.		
4.		$2\ H^+ \rightarrow H_2$
5.	$Sn \rightarrow Sn^{2+} + 2e^-$	$2\ H^+ + 2e^- \rightarrow H_2$
7.	$Sn\ (s) + 2\ H^+\ (aq) \rightarrow Sn^{2+}\ (aq) + H_2\ (g)$	

c. $Zn\ (s) + VO_2^+\ (aq) \rightarrow Zn^{2+}\ (aq) + V^{3+}\ (aq)$

1.	$Zn \rightarrow Zn^{2+}$	$VO_2^+ \rightarrow V^{2+}$
3.		$VO_2^+ \rightarrow V^{2+} + 2\ H_2O$

4. $VO_2^+ + 4\,H^+ \rightarrow V^{2+} + 2\,H_2O$
5. $Zn \rightarrow Zn^{2+} + 2\,e^-$ $VO_2^+ + 4\,H^+ + 3\,e^- \rightarrow V^{2+} + 2\,H_2O$
6. $3\,Zn \rightarrow 3\,Zn^{2+} + 6\,e^-$ $2\,VO_2^+ + 8\,H^+ + 6\,e^- \rightarrow 2\,V^{2+} + 4\,H_2O$
7. $3\,Zn\,(s) + 2\,VO_2^+\,(aq) + 8\,H^+\,(aq) \rightarrow 3\,Zn^{2+}\,(aq) + 2\,V^{2+}\,(aq) + 4\,H_2O\,(\ell)$

20. See problem 18 for explanation of step numbers.

a. $Ag^+\,(aq) + HCHO\,(aq) \rightarrow Ag\,(s) + HCO_2H\,(aq)$

1. $Ag^+ \rightarrow Ag$ $HCHO \rightarrow HCO_2H$
3. $H_2O + HCHO \rightarrow HCO_2H$
4. $H_2O + HCHO \rightarrow HCO_2H + 2\,H^+$
5. $Ag^+ + 1\,e^- \rightarrow Ag$ $H_2O + HCHO \rightarrow HCO_2H + 2\,H^+ + 2\,e^-$

6. $2\,Ag^+ + 2e^- \rightarrow 2\,Ag$
7. $2\,Ag^+\,(aq) + H_2O\,(\ell) + HCHO\,(aq) \rightarrow 2\,Ag\,(s) + HCO_2H\,(aq) + 2\,H^+\,(aq)$

b. $H_2S\,(aq) + Cr_2O_7^{2-}\,(aq) \rightarrow S\,(s) + Cr^{3+}\,(aq)$

1. $H_2S \rightarrow S$ $Cr_2O_7^{2-} \rightarrow Cr^{3+}$
2. $Cr_2O_7^{2-} \rightarrow 2\,Cr^{3+}$
3. $Cr_2O_7^{2-} \rightarrow 2\,Cr^{3+} + 7\,H_2O$
4. $H_2S \rightarrow S + 2\,H^+$ $14\,H^+ + Cr_2O_7^{2-} \rightarrow 2\,Cr^{3+} + 7\,H_2O$
5. $H_2S \rightarrow S + 2\,H^+ + 2e^-$ $14\,H^+ + Cr_2O_7^{2-} + 6\,e^- \rightarrow 2\,Cr^{3+} + 7\,H_2O$

6. $3\,H_2S \rightarrow 3\,S + 6\,H^+ + 6e^-$
7. $3\,H_2S\,(aq) + 8\,H^+\,(aq) + Cr_2O_7^{2-}\,(aq) \rightarrow 3\,S\,(s) + 2\,Cr^{3+}\,(aq) + 7\,H_2O\,(\ell)$

Note the removal of 6 H^+ (aq) from both sides of the equation.

c. $H_2C_2O_4\,(aq) + MnO_4^-\,(aq) \rightarrow CO_2\,(g) + Mn^{2+}\,(aq)$

1. $H_2C_2O_4 \rightarrow CO_2$ $MnO_4^- \rightarrow Mn^{2+}$
2. $H_2C_2O_4 \rightarrow 2\,CO_2$
3. $MnO_4^- \rightarrow Mn^{2+} + 4\,H_2O$
4. $H_2C_2O_4 \rightarrow 2\,CO_2 + 2\,H^+$ $8\,H^+ + MnO_4^- \rightarrow Mn^{2+} + 4\,H_2O$

5. $H_2C_2O_4 \rightarrow 2\,CO_2 + 2\,H^+ + 2\,e^-$ $8\,H^+ + MnO_4^- + 5\,e^- \rightarrow Mn^{2+} + 4\,H_2O$

6. $5\,H_2C_2O_4 \rightarrow 10\,CO_2 + 10\,H^+ + 10\,e^-$ $16\,H^+ + 2\,MnO_4^- + 10\,e^- \rightarrow 2\,Mn^{2+} + 8\,H_2O$

7. $5\,H_2C_2O_4 + 6\,H^+ (aq) + 2\,MnO_4^- (aq) \rightarrow 2\,Mn^{2+} (aq) + 10\,CO_2 (g) + 8\,H_2O (\ell)$

22. One method for balancing redox reactions in basic solution is quite similar to the procedure in question 18 <u>with two exceptions:</u>

3. Balance mass of O by adding OH^-. Use twice as many OH^- as you need oxygens.
4. Balance H by adding H_2O.

Once again to clarify the steps, states of matter will be noted only in the final step. When no change is made from one step to the next, the equation will be written only for the first step.

a. $Zn (s) + ClO^- (aq) \rightarrow Zn(OH)_2 (s) + Cl^- (aq)$

1. $Zn \rightarrow Zn(OH)_2$ $ClO^- \rightarrow Cl^-$
3. $Zn + 2\,OH^- \rightarrow Zn(OH)_2$ $ClO^- \rightarrow Cl^- + 2\,OH^-$
4. $H_2O + ClO^- \rightarrow Cl^- + 2\,OH^-$
5. $Zn + 2\,OH^- \rightarrow Zn(OH)_2 + 2\,e^-$ $2\,e^- + H_2O + ClO^- \rightarrow Cl^- + 2\,OH^-$
7. $Zn (s) + H_2O (\ell) + ClO^- (aq) \rightarrow Zn(OH)_2 (s) + Cl^- (aq)$

b. $ClO^- (aq) + CrO_2^- (aq) \rightarrow Cl^- (aq) + CrO_4^{2-} (aq)$

1. $ClO^- \rightarrow Cl^-$ $CrO_2^- \rightarrow CrO_4^{2-}$
3. $ClO^- \rightarrow Cl^- + 2\,OH^-$ $4\,OH^- + CrO_2^- \rightarrow CrO_4^{2-}$
4. $H_2O + ClO^- \rightarrow Cl^- + 2\,OH^-$ $4\,OH^- + CrO_2^- \rightarrow CrO_4^{2-} + 2\,H_2O$
5. $H_2O + ClO^- + 2\,e^- \rightarrow Cl^- + 2\,OH^-$ $4\,OH^- + CrO_2^- \rightarrow CrO_4^{2-} + 2\,H_2O + 3\,e^-$
6. $3\,H_2O + 3\,ClO^- + 6\,e^- \rightarrow 3\,Cl^- + 6\,OH^-$ $8\,OH^- + 2\,CrO_2^- \rightarrow 2\,CrO_4^{2-} + 4\,H_2O + 6\,e^-$
7. $3\,ClO^- (aq) + 2\,OH^- (aq) + 2\,CrO_2^- (aq) \rightarrow 3\,Cl^- (aq) + 2\,CrO_4^{2-} (aq) + H_2O(\ell)$

c. $Br_2 (\ell) \rightarrow Br^- (aq) + BrO_3^- (aq)$

1.	$Br_2 \rightarrow Br^-$	$Br_2 \rightarrow BrO_3^-$
2	$Br_2 \rightarrow 2\,Br^-$	$Br_2 \rightarrow 2\,BrO_3^-$
3.		$12\,OH^- + Br_2 \rightarrow 2\,BrO_3^-$
4.		$12\,OH^- + Br_2 \rightarrow 2\,BrO_3^- + 6\,H_2O$
5.	$2e^- + Br_2 \rightarrow 2\,Br^-$	$12\,OH^- + Br_2 \rightarrow 2\,BrO_3^- + 6\,H_2O + 10\,e^-$

6. $10\,e^- + 5\,Br_2 \rightarrow 10\,Br^-$
7. $6\,Br_2 (\ell) + 12\,OH^- (aq) \rightarrow 10\,Br^- (aq) + 2\,BrO_3^- (aq) + 6\,H_2O\ (\ell)$

Note that all coefficients in Step 7 are divisible by two. The overall balanced equation is then: $3\,Br_2 (\ell) + 6\,OH^- (aq) \rightarrow 5\,Br^- (aq) + BrO_3^- (aq) + 3\,H_2O\ (\ell)$

Electrochemical Cells and Cell Potentials

24.

a. Half equations	b. Processes	c. Compartment
$Cr(s) \rightarrow Cr^{3+}(aq) + 3\,e^-$	oxidation	anode
$Fe^{2+}(aq) + 2\,e^- \rightarrow Fe(s)$	reduction	cathode

26. The copper electrode is found to be the external anode (+) and the tin electrode the external cathode (-). The copper electrode is therefore the **internal cathode** and the tin electrode the **internal anode.**

The half-reactions occurring in the half-cells are:

	Processes:	Compartment
$Cu^{2+} (aq) + 2\,e^- \rightarrow Cu\,(s)$	reduction	cathode
$Sn\,(s) \rightarrow Sn^{2+} (aq) + 2\,e^-$	oxidation	anode

28. For a cell with E° = +2.91, the ΔG° is:

$$\Delta G^\circ = -nFE^\circ$$
$$= -2\text{ mole} \cdot 9.65 \times 10^4 \frac{\text{J}}{\text{volt}\cdot\text{mol}} \cdot 2.91\text{ volt} \cdot \frac{1.000\text{ kJ}}{1000\text{ J}}$$
$$= -561\text{ kJ}$$

30. Calculate E° and decide if each reaction is product-favored:

a. $2\ I^- (aq) + Zn^{2+} (aq) \rightarrow I_2 (s) + Zn (s)$

	process	potential
$2\ I^- (aq) \rightarrow I_2 (s) + 2\ e^-$	oxidation	- 0.535 V
$Zn^{2+} (aq) + 2\ e^- \rightarrow Zn (s)$	reduction	- 0.763 V
	Process is **not** product-favored. $E°_{net}$ =	- 1.298 V

b. $Zn (s) + Ni^{2+} (aq) \rightarrow Zn^{2+} (aq) + Ni (s)$

	process	potential
$Ni^{2+} (aq) + 2\ e^- \rightarrow Ni (s)$	reduction	- 0.25 V
$Zn (s) \rightarrow Zn^{2+} (aq) + 2\ e^-$	oxidation	+ 0.763 V
	Process **is** product-favored. $E°_{net}$ =	+ 0.51 V

c. $Cu (s) + Cl_2 (g) \rightarrow 2\ Cl^- (aq) + Cu^{2+} (aq)$

	process	potential
$Cl_2 (g) + 2\ e^- \rightarrow 2\ Cl^- (aq)$	reduction	+ 1.360 V
$Cu (s) \rightarrow Cu^{2+} (aq) + 2\ e^-$	oxidation	- 0.337 V
	Process **is** product-favored. $E°_{net}$ =	+ 1.023 V

32. a. $Sn^{2+} (aq) + 2\ Ag (s) \rightarrow Sn (s) + 2\ Ag^+ (aq)$

Sn^{2+} is reduced (- 0.14 V); Ag is oxidized (- 0.7994 V)

E° = (- 0.14 - 0.7994) = - 0.94 V **not** product-favored

b. $Zn (s) + Sn^{4+} (aq) \rightarrow Sn^{2+} (aq) + Zn^{2+} (aq)$

Sn^{4+} is reduced (+ 0.15 V); Zn is oxidized (+ 0.763 V)

E° = (+ 0.15 + 0.763) = + 0.91 V **is** product-favored

c. $I_2 (aq) + 2\ Br^- (aq) \rightarrow 2\ I^- (aq) + Br_2 (\ell)$

I_2 is reduced (+0.535 V) ; Br^- is oxidized (- 1.08 V)

E° = (+ 0.535 - 1.08) = - 0.55 V **not** product-favored

34. For the half-reactions listed:

a. Weakest oxidizing agent: V^{2+}

The more positive the E°, the better the oxidizing ability of the specie.

b. Strongest oxidizing agent: Cl_2

c. Strongest reducing agent: V

The more negative the E°, the better the reducing ability of a substance.

d. Weakest reducing agent: Cl^-

e. Pb (s) **cannot** reduce V^{2+} (aq).

Since the reduction potential of lead is less negative than that of V^{2+}, lead cannot reduce V^{2+}.

f. I^- can reduce Cl_2 to Cl^- .

The less positive the value of E°, the better the reducing ability of a substance, The E° for Cl_2 is greater than that for I_2, hence I^- can reduce Cl_2 to Cl^- .

g. Pb(s) can reduce Cl_2, and I_2

The comment from part c applies. The reduction potential of Pb is more negative than that for Cl_2 or I_2, making Pb capable of reducing either of these substances.

36. a. Maximum positive standard potential:

Cl_2 (g) + V (s) → 2 Cl^- (aq) + V^{2+} (aq) E° = +2.54 V

b.

	potential
I_2 (g) + 2 e^- → 2 I^- (aq)	+ 0.535 V
V (s) → V^{2+} (aq) + 2 e^-	+ 1.18 V
$E°_{net}$ =	+ 1.72 V

The product-favored reaction is: I_2 (g) + V (s) → 2 I^- (aq) + V^{2+} (aq)

38. The standard reduction potentials for the two species are:

Zn^{2+} (aq) + 2 e^- → Zn (s) E° = - 0.763 V

Ag^+ (aq) + 1 e^- → Ag (s) E° = + 0.7994 V

a. The product-favored reaction would be:

2 Ag^+ (aq) + Zn (s) → 2 Ag (s) + Zn^{2+} (aq)

The potential for the cell would be: E° = (+ 0.7994 + 0.763) = + 1.56 V

b. Silver is the cathode and zinc is the anode.

c. The diagram for the cell:

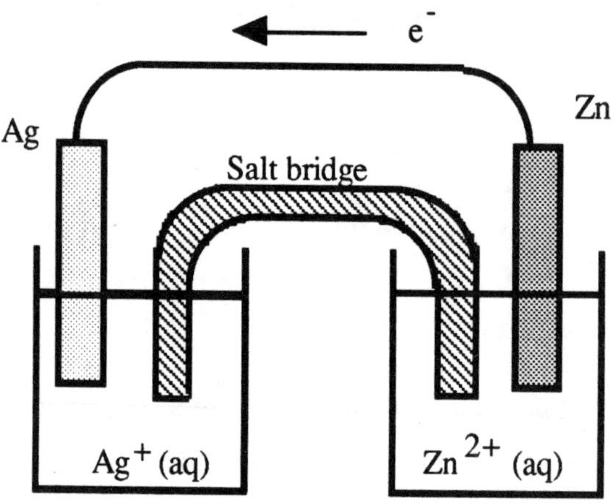

d. A strip of silver would serve as the cathode.

e. Electrons would flow from the zinc electrode to the silver electrode.

f. Nitrate ions would flow from the salt bridge to the Zn compartment (as electrons leave the external Zn electrode, an increasing positive charge would develop).

Cells Under Nonstandard Conditions, E and K

40. The Nernst equation for this reaction can be written:

$$E_{net} = E°_{net} - \frac{0.0257}{n} \ln \frac{[Fe^{2+}]^2}{[Fe^{3+}]^2[I^-]^2}$$

Substituting appropriate values:

$$E_{net} = (0.771 - 0.535) - \frac{0.0257}{2} \ln \frac{(0.1)^2}{(0.1)^2(0.1)^2}$$

$$= 0.236 - (0.01285 \bullet \ln (100))$$

$$= 0.177 \text{ V}$$

Note that the potential of the cell, E_{net}, is **decreased** from that of the standard potential of the cell.

42. From Appendix M:

$Ag^+ (aq) + 1 e^- \rightarrow Ag (s)$ $E° = +0.7994$ V

$Fe^{3+} (aq) + 1 e^- \rightarrow Fe^{2+} (aq)$ $E° = +0.771$ V

a. The reaction for the cell operation:

$$Ag^+ (aq) + Fe^{2+} (aq) \rightarrow Ag (s) + Fe^{3+} (aq)$$

b. $E°_{net} = (+0.7994 - 0.771) = +0.028$ V

c. **Assume** that the reaction is the same under the non-standard conditions as under the standard conditions. Q can be calculated:

$$Q = \frac{[Fe^{3+}]}{[Fe^{2+}][Ag^{+}]} = \frac{(1.0\ M)}{(1.0\ M)(0.1\ M)} = 10$$

Substitution into the Nernst equation yields:

$$E_{net} = E^{\circ}_{net} - \frac{0.0257}{n} \ln Q$$

$$= +0.028\ V - \frac{0.0257}{1} \ln (10) = -0.031\ V$$

The cell reaction—contrary to our assumption—is **not** the same as under standard conditions. The net reaction is now:

$$Ag\ (s) + Fe^{3+}\ (aq) \rightarrow Ag^{+}\ (aq) + Fe^{2+}\ (aq)$$

44. To calculate the equilibrium constants for the reactions, we use the Nernst equation expressed as follows:

$$E_{net} = E^{\circ}_{net} - \frac{0.0257}{n} \ln K$$

At equilibrium $E_{net} = 0$. E°_{net} must be calculated.

		potential
a.	$2\ I^{-}\ (aq) \rightarrow I_2\ (aq) + 2\ e^{-}$	- 0.535 V
	$Fe^{3+}\ (aq) + 1\ e^{-} \rightarrow Fe^{2+}\ (aq)$	+ 0.771 V
	E°_{net}	= + 0.236 V

The number of electrons in the balanced overall equation (n) is 2.

Substituting we get:

$$0 = 0.236\ V - \frac{0.0257}{2} \ln K$$

$$\frac{(2)(0.236)}{(0.0257)} = \ln K = 18.4 \text{ and } K = 9.5 \times 10^{7}$$

		potential
b.	$I_2\ (aq) + 2\ e^{-} \rightarrow 2\ I^{-}\ (aq)$	+ 0.535 V
	$2\ Br^{-}\ (aq) \rightarrow Br_2\ (aq) + 2\ e^{-}$	- 1.08 V
	E°_{net}	= - 0.55 V

The number of electrons in the balanced overall equation (n) is 2.

Substituting we get:

$$0 = -0.55\ V - \frac{0.0257}{2} \ln K$$

$$\frac{(2)(-0.55)}{(0.0257)} = \ln K = -42.41 \text{ and } K = 4 \times 10^{-19}$$

Electrolysis, Electrical Energy, and Power

46. $1.00\text{ g Au}^{3+} \cdot \frac{1\text{ mol Au}^{3+}}{197.0\text{ g Au}^{3+}} \cdot \frac{3\text{ mol e}^-}{1\text{ mol Au}^{3+}} \cdot \frac{96500\text{ C}}{1\text{ mol e}^-} \cdot \frac{1\text{ A} \cdot \text{s}}{1\text{ C}} \cdot \frac{1}{2.00\text{ A}} =$

735 s (or 12.2 min)

48. Solutions to problems of this sort are best solved by beginning with a factor containing the desired units. Connecting this factor to data provided usually gives a direct path to the answer.

units desired

↓

$$\frac{58.69\text{ g Ni}}{1\text{ mol Ni}} \cdot \frac{1\text{ mol Ni}}{2\text{ mol e}^-} \cdot \frac{1\text{ mol e}^-}{9.65 \times 10^4\text{ C}} \cdot \frac{1\text{ C}}{1\text{ amp} \cdot \text{s}}$$

$$\cdot \frac{2.50\text{ amps}}{1} \cdot \frac{3600\text{ s}}{\text{hr}} \cdot \frac{2.00\text{ hr}}{1} = 5.47\text{ g Ni}$$

The second factor ($\frac{1\text{ mol Ni}}{2\text{ mol e}^-}$) is arrived at by looking at the reduction half-reaction:

$$Ni^{2+}\text{ (aq)} + 2\text{ e}^- \rightarrow \text{Ni(s)}$$

All other factors are either data or common unity factors (e.g. $\frac{3600\text{ s}}{1\text{ hr}}$).

50. Current flowing if 0.052 g Ag are deposited in 450 s:

$$\frac{1\text{ amp} \cdot \text{s}}{1\text{ C}} \cdot \frac{1}{450\text{ s}} \cdot \frac{9.65 \times 10^4\text{ C}}{1\text{ mol e}^-} \cdot \frac{1\text{ mol e}^-}{1\text{ mol Ag}}$$

$$\cdot \frac{1\text{ mol Ag}}{108\text{ g Ag}} \cdot \frac{0.052\text{ g Ag}}{1} = 0.10\text{ amperes}$$

52. The mass of aluminum produced in 8.0 hr by a current of 1.0×10^5 amp:

$$\frac{26.98\text{ g Al}}{1\text{ mol Al}} \cdot \frac{1\text{mol Al}}{3\text{ mol e}^-} \cdot \frac{1\text{ mol e}^-}{9.65 \times 10^4\text{ C}} \cdot \frac{1\text{ C}}{1\text{ amp} \cdot \text{s}}$$

$$\cdot \frac{1.0 \times 10^5\text{ amp}}{1} \cdot \frac{3600\text{ s}}{1\text{ hr}} \cdot \frac{8.0\text{ hr}}{1} = 2.7 \times 10^5\text{ g Al}$$

54. The mass of lead consumed by a current of 1.0 amp for 50. hours:

During discharge of the lead storage battery, the anode reaction is

$$Pb\ (s) \rightarrow Pb^{2+}\ (aq) + 2e^-.$$

$$\frac{207\ g\ Pb}{1\ mol\ Pb} \cdot \frac{1\ mol\ Pb}{2\ mol\ e^-} \cdot \frac{1\ mol\ e^-}{9.65 \times 10^4\ C} \cdot \frac{1\ C}{1\ amp \cdot s}$$

$$\cdot \frac{1.0\ amp}{1} \cdot \frac{3600\ s}{1\ hr} \cdot \frac{50.\ hr}{1} = 190\ g\ Pb$$

56. $1.138 \times 10^{10}\ kg\ Cl_2 \cdot \frac{1 \times 10^3\ g\ Cl_2}{1.0\ kg\ Cl_2} \cdot \frac{1\ mol\ Cl_2}{70.906\ g\ Cl_2} \cdot \frac{2\ mol\ e^-}{1\ mol\ Cl_2} \cdot \frac{9.65 \times 10^4\ C}{1\ mol\ e^-}$

$$= 3.097 \times 10^{16}\ C$$

The power is then:

$$\frac{3.097 \times 10^{16}\ C}{1} \cdot \frac{4.6\ V}{1} \cdot \frac{1\ J}{1\ V \cdot C} \cdot \frac{1\ kwh}{3.6 \times 10^6\ J} = 4.0 \times 10^{10}\ kwh.$$

(2 sf)

The factor, $\frac{1\ kwh}{3.6 \times 10^6\ J}$, is derived from the fact that a kwh = 1000 watts • 1 hr

and the relationship that 1 J = 1 watt • s or 1 J/s = 1 watt

$$1\ kwh = 1000\ watts \cdot 1\ hr \cdot \frac{1 J/s}{1 watt} \cdot \frac{3600\ s}{1\ hr} = \frac{3.60 \times 10^6\ J}{1\ kwh}$$

See Example 21.15 for another example of this concept.

58. a. At the anode, oxidation occurs : $2\ Cl^-\ (aq) \rightarrow Cl_2\ (g) + 2\ e^-$.

The polarity is **positive**. Recall that electrolytic cells and batteries have opposite signs for the anode and cathode!

b. At the cathode, reduction occurs: $Ni^{2+}\ (aq) + 2\ e^- \rightarrow Ni\ (s)$

The polarity is **negative**. Recall that electrolytic cells and batteries have opposite signs for the anode and cathode!

c. Mass of Ni and Cl_2 produced by a current of 0.0250 amperes flowing for 1.25 hrs:

$$\frac{70.91\ g\ Cl_2}{1\ mol\ Cl_2} \cdot \frac{1\ mol\ Cl_2}{2\ mol\ e^-} \cdot \frac{1\ mol\ e^-}{9.65 \times 10^4\ C} \cdot$$

$$\frac{1\ C}{1\ amp \cdot s} \cdot \frac{0.0250\ amp}{1} \cdot \frac{3600\ s}{1\ hr} \cdot \frac{1.25\ hr}{1} = 0.0413\ g\ Cl_2$$

$$\frac{58.69 \text{ g Ni}}{1 \text{ mol Ni}} \bullet \frac{1 \text{ mol Ni}}{2 \text{ mol e}^-} \bullet \frac{1 \text{ mol e}^-}{9.65 \times 10^4 \text{ C}} \bullet$$

$$\frac{1 \text{ C}}{1 \text{ amp} \bullet \text{s}} \bullet \frac{0.0250 \text{ amp}}{1} \bullet \frac{3600 \text{ s}}{1 \text{ hr}} \bullet \frac{1.25 \text{ hr}}{1} = 0.0342 \text{ g Ni}$$

60. a. Electrolysis of KBr (aq):

Reduction of K^+	E° = - 2.925 V
Reduction of H_2O	= - 0.83 V

So H_2O will be reduced to H_2 at the cathode (in preference to elemental K).

Oxidation of Br^-	E° = - 1.066 V
Oxidation of H_2O	= - 1.23 V (+0.6 V overvoltage)

So Br^- will be oxidized to Br_2 at the anode.

b. Electrolysis of NaF (molten):

At the anode: $2 F^-(aq) \rightarrow F_2(g) + 2 e^-$

At the cathode: $Na^+(aq) + e^- \rightarrow Na(s)$

c. Electrolysis of NaF (aq):

Reduction of Na^+	E° = -2.71 V
Reduction of H_2O	= -0.83 V

So H_2O will be reduced to H_2 at the cathode.

Oxidation of F^-	E° = - 2.87 V
Oxidation of H_2O	= - 1.23 V (+0.6 V overvoltage)

So H_2O will be oxidized to O_2 at the anode.

General Questions

62. See detailed steps for balancing redox equations in Question 18 (p 314)

a. $3 Br_2(aq) + 6 e^- \rightarrow 6 Br^-(aq)$

$I^-(aq) + 3 H_2O(\ell) \rightarrow IO_3^-(aq) + 6 H^+(aq) + 6 e^-$

$I^-(aq) + 3 Br_2(aq) + 3 H_2O(\ell) \rightarrow IO_3^-(aq) + 6 Br^-(aq) + 6 H^+(aq)$

b. $5 U^{4+}(aq) + 10 H_2O(\ell) \rightarrow 5 UO_2^{2+}(aq) + 20 H^+(aq) + 10 e^-$

$2 MnO_4^-(aq) + 16 H^+(aq) + 10 e^- \rightarrow 2 Mn^{2+}(aq) + 8 H_2O(\ell)$

$5 U^{4+}(aq) + 2 H_2O(\ell) + 2 MnO_4^-(aq) \rightarrow 5 UO_2^{2+}(aq) + 2 Mn^{2+}(aq) + 4 H^+(aq)$

Note that upon the addition of the two half-equations, it is necessary to "subtract" 8 H_2O molecules and 16 H^+ ions from both sides.

c. $MnO_2(s) + 4\,H^+(aq) + 2\,e^- \rightarrow Mn^{2+}(aq) + 2\,H_2O(\ell)$

$2\,I^-(aq) \rightarrow I_2(s) + 2\,e^-$

$MnO_2(s) + 2\,I^-(aq) + 4\,H^+(aq) \rightarrow Mn^{2+}(aq) + I_2(s) + 2\,H_2O(\ell)$

64. The possible reductions are:

$Mg^{2+}(aq) + 2\,e^- \rightarrow Mg(s)$ -2.37 V

$Al^{3+}(aq) + 3\,e^- \rightarrow Al(s)$ -1.66 V

$Zn^{2+}(aq) + 2\,e^- \rightarrow Zn(s)$ -0.763 V

With the magnesium couple having the most negative E°, when combined with the chlorine half-cell, the largest possible positive $E^\circ net$ will result.

The balanced equation for the reaction would be:

$Mg(s) + Cl_2(g) \rightarrow 2\,Cl^-(aq) + Mg^{2+}(aq)$

and $E^\circ net = (+1.360 + 2.37) = +3.73$ V

66. a. The equilibrium constant for the lead storage battery : $E^\circ = 2.04$ V

K can be calculated using Equation 21.3: $\ln K = \frac{nE^\circ}{0.0257\text{ V}}$

$\ln K = \frac{2 \cdot 2.04\text{ V}}{0.0257\text{ V}} = 159$ and $K = 8.8 \times 10^{68}$

b. The net reaction for the lead storage battery is:

$Pb(s) + PbO_2(s) + 2\,H_2SO_4(aq) \rightarrow 2\,PbSO_4(s) + 2\,H_2O(\ell)$

From this the "equilibrium constant term" is : $\frac{1}{[H_2SO_4]^2}$

Substitution into the Nernst equation gives:

$$E_{net} = E^\circ net - \frac{0.0257}{n} \ln \frac{1}{[H_2SO_4]^2}$$

$$= 2.04\text{ V} - \frac{0.0257}{2} \ln \frac{1}{(6.00)^2}$$

$$= 2.04\text{ V} - \frac{0.0257}{2} \ln(0.0278)$$

$$= 2.04\text{ V} + 0.046 = 2.086\text{ V (or 2.09 V)}$$

68. a. Since the magnesium is more easily oxidized than the iron, the iron is "forced" to be reduced—a process that doesn't occur. This is sometimes referred to as cathodic protection.

b. Time to consume 5.0 kg of Mg

$$5.0 \times 10^3 \text{ g Mg} \cdot \frac{1 \text{ mol Mg}}{24.3 \text{ g Mg}} \cdot \frac{2 \text{ mol e}^-}{1 \text{ mol Mg}} \cdot \frac{9.65 \times 10^4 \text{ C}}{1 \text{ mol e}^-} \cdot \frac{1 \text{ A} \cdot \text{s}}{1 \text{ C}} \cdot$$

$$\frac{1}{0.030 \text{ A}} \cdot \frac{1 \text{ hr}}{3600 \text{ s}} \cdot \frac{1 \text{ day}}{24 \text{ hr}} \cdot \frac{1 \text{ yr}}{365.25 \text{ day}} = 42 \text{ yr}$$

70. a. Balanced equation: $2 \text{ Al(s)} + 3 \text{ Cl}_2\text{(g)} \rightarrow 2 \text{ Al}^{3+}\text{(aq)} + 6 \text{ Cl}^-\text{(aq)}$

b. The anode reaction: $\text{Al(s)} \rightarrow \text{Al}^{3+}\text{(aq)}$ polarity : negative
The cathode reaction: $\text{Cl}_2\text{(g)} \rightarrow 2 \text{ Cl}^-\text{(aq)}$ polarity : positive

c. $E°_{net} = E°_{chlorine} + E°_{aluminum}$

$= 1.358 + 1.66 = 3.02 \text{ V}$

d. $$E_{net} = E°_{net} - \frac{0.0257}{n} \ln \frac{[\text{Al}^{3+}]^2[\text{Cl}^-]^2}{[\text{Cl}_2]^3}$$

$$= 3.02 \text{ V} - \frac{0.0257}{6} \ln \frac{1}{[0.50]^3} = 3.02 - \frac{0.0257}{6} \ln(\frac{1}{0.125})$$

$$= 3.02 - 0.0089 = 3.01 \text{ V}$$

Note the **voltage decreases** if the pressure of chlorine is less than 1 atm.

e. Time to consume 30.0 g Al while delivering 0.75 amperes:

$$30.0 \text{ g Al} \cdot \frac{1 \text{ mol Al}}{26.98 \text{ g Al}} \cdot \frac{3 \text{ mol e}^-}{1\text{mol Al}} \cdot \frac{9.65 \times 10^4 \text{ C}}{1 \text{ mol e}^-} \cdot \frac{1 \text{ A} \cdot \text{s}}{1 \text{ C}} \cdot \frac{1}{0.75 \text{ A}} =$$

4.29×10^5 s or 119 hrs

72. a. The stoichiometry of the silver/zinc battery indicates a reaction of one mole each of silver oxide, zinc, and water. The mass of one mole of each of these three substances is:

1 mol Ag_2O	231.7 g
1 mol Zn	65.4 g
1 mol H_2O	18.0 g
	315.1 g

The energy associated with the battery is: $\frac{0.10 \text{ C}}{1 \text{ s}} \cdot \frac{1.59 \text{ V}}{1} = \frac{0.159 \text{ V} \cdot \text{C}}{\text{s}}$

Since 1 V • C = 1 Joule, this energy corresponds to 0.159 J/s. [1 watt = 1 J/s]

The energy/gram for the silver/zinc battery is: $\frac{0.159 \text{ J/s}}{315.1 \text{ g}} = 5.0 \times 10^{-4}$ watts/gram. (2 sf)

b. Performing the same calculations for the lead storage battery, using a stoichiometric amount for the overall battery reaction:

1 mol Pb	207.2 g
1 mol PbO_2	239.2 g
2 mol H_2SO_4	196.2 g
	642.6 g

The energy associated with the battery is: $\frac{0.10\ C}{1\ s} \bullet \frac{2.0\ V}{1} = \frac{0.20\ V \bullet C}{s}$

The energy/gram for the lead storage battery is: $\frac{0.20\ J/s}{642.6\ g} = 3.1 \times 10^{-4}$ watts/g

c. The silver/zinc battery produces more energy (and more power)/gram.

Conceptual Questions

74. a. Since A and C reduce H^+ to elemental hydrogen, they are both stronger reducing agents than H_2.

Reducing agent strength: $H_2 < A,C$

b. C reduces A, B, and D—hence it is the strongest reducing agent.

A, B, D < C and from part a. above: $B, D < H_2 < A < C$

c. D reduces B^{n+}, so D is a stronger reducing agent than B.

So $B < D < H_2 < A < C$

76. When carbon dioxide dissolves in water, acid(H_3O^+) is formed:

$$CO_2\ (g) + 2\ H_2O\ (\ell) \rightleftharpoons HCO_3^-\ (aq) + H_3O^+\ (aq)$$

Iron resists corrosion owing to the presence of a $Fe(OH)_2$ layer. When the acid—formed by the reaction of carbon dioxide in water—reacts with this hydroxide, metallic iron is exposed. The metallic iron can then oxidize to form "rust".

$$2\ H_3O^+\ (aq) + Fe(OH)_2\ (s) \rightarrow 4\ H_2O\ (\ell) + Fe^{2+}\ (aq)$$

Challenging Questions

78. a. $Ni^{2+}(aq) + Cd(s) \rightarrow Ni(s) + Cd^{2+}(aq)$

b. Ni^{2+} is reduced (from 2+ to 0) and functions as the oxidizing agent.
Cd is oxidized (from 0 to 2+) and functions as the reducing agent.

c. **Cd** is the **anode**, and **Ni** is the **cathode**.
Since Cd furnishes electrons to the external circuit, **Cd is negative**.

d. $E^\circ_{net} = E^\circ_{cathode} + E^\circ_{anode}$

$= -0.25 + 0.40 = +0.15\ V$

e. Electrons flow from the Cd electrode to the Ni electrode.

f. As Cd is oxidized, there will be a shortage of anions in the anode compartment. Similarly as Ni^{2+} is reduced, there will be an excess of anions in the cathode compartment. The NO_3^- will migrate toward the anode compartment (cadmium).

g. $E_{net} = E^\circ_{net} - \frac{0.0257}{n} \ln K$. At equilbrium, we know that $E_{net} = 0$.

Substituting 0 into the expression and rearranging the equation to solve for (lnK):

$$\frac{nE^\circ}{0.0257V} = \ln K = \frac{(2)(+0.15\ V)}{0.0257V} = 11.67 \quad \text{(12 to 2 sf)}$$

and $K = 1.2 \times 10^5$

h. For $[Cd^{2+}] = 0.010\ M$ and $[Ni^{2+}] = 1.0\ M$

$$E_{net} = E^\circ_{net} - \frac{0.0257}{2} \ln\frac{[Cd^{2+}]}{[Ni^{2+}]} = +0.15 - \frac{0.0257}{2} \ln\left(\frac{1 \times 10^{-2}}{1.0}\right)$$

$$E_{net} = +0.15 + 0.0592 = +0.21\ V$$

The reaction is still the reaction given in part a (E_{net} is positive).

i. Battery lifetime:

Which reactant is consumed first?

Recall that the spontaneous reaction **reduces Ni^{2+} and oxidizes Cd.**

$$1.0\ L \bullet \frac{1.0\ mol\ Ni^{2+}}{1.0\ L} = 1.0\ mol\ Ni^{2+}$$

$$50.0\ g\ Cd \bullet \frac{1\ mol\ Cd}{112.4 g\ Cd} = 0.445\ mol\ Cd$$ (Cd is the limiting reagent!)

$$0.445 \text{ mol Cd} \cdot \frac{2 \text{ mol e}^-}{1 \text{ mol Cd}} \cdot \frac{9.65 \times 10^4 \text{ C}}{1 \text{ mol e}^-} \cdot \frac{1 \text{ A} \cdot \text{s}}{1 \text{ C}} \cdot \frac{1}{0.050 \text{ A}}$$

$$= 1.7 \times 10^6 \text{ s or } 480 \text{ hrs}$$

80. To calculate the charge on the Rh^{x+} ion, we need to know two things:

1. How many moles of elemental rhodium are reduced?
2. How many moles of electrons caused that reduction?

Moles of rhodium: $\frac{0.038 \text{ g Rh}}{1} \cdot \frac{1 \text{ mol Rh}}{102.9 \text{ g Rh}} = 3.7 \times 10^{-4} \text{ mol Rh}$

Moles of electrons:

$$0.0100 \text{ amp} \cdot \frac{1 \text{ C}}{1 \text{ amp} \cdot \text{s}} \cdot \frac{3600 \text{ s}}{1 \text{ hr}} \cdot \frac{3 \text{ hr}}{1} \cdot \frac{1 \text{ mol e}^-}{9.65 \times 10^4 \text{ C}} = 1.1 \times 10^{-3} \text{ mol e}^-$$

Recall that our general reduction reactions are written: $M^{+x} + x\,e^- \rightarrow M$

If we know the number of $\frac{\text{moles of electrons}}{\text{mol of metal}}$, we know the charge on the cation, hence for the Rh^{x+} ion we have $\frac{1.1 \times 10^{-3} \text{ mol e}^-}{3.7 \times 10^{-4} \text{ mol Rh}} = 3.0 \frac{\text{mol e}^-}{\text{mol Rh}}$.

The ion is therefore the Rh^{3+} ion!

82. Since the reaction depends on the oxidation of elemental hydrogen to water (2 mol e^- per mol H_2), we must detemine the amount of H_2 present:

$$n = \frac{(200.\text{ atm})(1.0 \text{ L})}{(0.0821 \frac{\text{L}\cdot\text{atm}}{\text{K}\cdot\text{mol}})(298 \text{ K})} = 8.2 \text{ mol } H_2$$

The amount of time this can produce current:

$$8.2 \text{ mol } H_2 \cdot \frac{2 \text{ mol e}^-}{1 \text{ mol } H_2} \cdot \frac{9.65 \times 10^4 \text{ C}}{1 \text{ mol e}^-} \cdot \frac{1 \text{ A} \cdot \text{s}}{1 \text{ C}} \cdot \frac{1}{1.5 \text{ A}} = 1.1 \times 10^6 \text{ s}$$

(290 hrs)

Summary Question

84. a. Will NAD^+ oxidize iron (II) to iron (III) ?

To answer this question, construct a cell with NAD^+ being reduced and iron(II) oxidized:

$NAD^+ + H^+ + 2\,e^- \rightarrow NADH$	$E^\circ = -0.320$ V
$Fe^{2+} + 1\,e^- \rightarrow Fe^{3+}$	$E^\circ = -0.771$ V
	$E^\circ_{net} = -1.091$ V

The negative voltage for this cell indicates that **NAD^+ can not oxidize iron (II) to iron (III).**

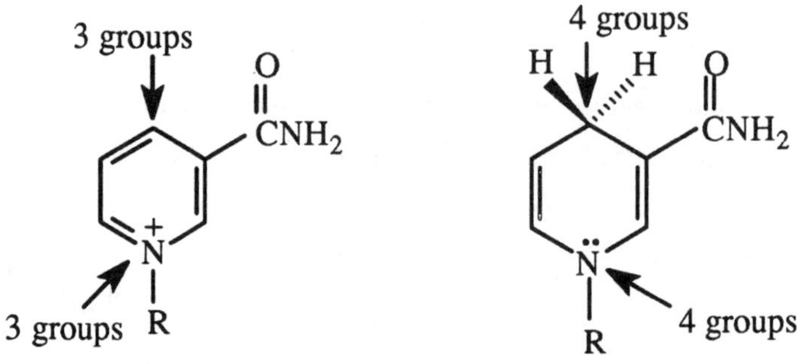

3 groups
3 groups R
$\overset{O}{\parallel}$ CNH_2

4 groups
H H O
CNH_2
R 4 groups

b. Changes in structure of the C_5N ring:

NAD^+ is an aromatic ring, and **is planar**; NADH has two atoms (C,N) that have four groups attached to them, and the **ring is non-planar.**

The electron pair geometry around the N atom in the ring?

The electron pair geometry around NAD^+ has **three groups** (two C,R) around it and has a trigonal planar geometry (bond angles of 120°). Around the N of the reduced specie, there are **four groups** (two C, R, a lone pair of e), and a tetrahedral electron pair geometry around it.

The electron pair geometry around the C atom *para* to the N atom?

The C atom para to the N (in NAD^+) has three groups attached to it (two C,H)— and has a trigonal planar geometry (bond angles of 120°)while the C atom (in NADH) has four groups (two C, and two H) around it and a tetrahedral geometry around it (109°angles)

Hybridizations around the C and N atoms change?

For both C and N, the initial hybridizations would be sp^2 (with bond angles of approximately 120°). In the reduced specie—with tetrahedral electron pair geometry (and 109° bond angles) around both atoms, the hybridization would be sp^3.

Chapter 22
The Chemistry of The Main Group Elements

Review Questions

6. Electron configurations for the elements in the second period:

Atomic #	Symbol	E' configuration	Valence electrons
3	Li	$1s^2\ 2s^1$	1
4	Be	$1s^2\ 2s^2$	2
5	B	$1s^2\ 2s^2\ 2p^1$	3
6	C	$1s^2\ 2s^2\ 2p^2$	4
7	N	$1s^2\ 2s^2\ 2p^3$	5
8	O	$1s^2\ 2s^2\ 2p^4$	6
9	F	$1s^2\ 2s^2\ 2p^5$	7
10	Ne	$1s^2\ 2s^2\ 2p^6$	8

8. [X] $ns^2\ np^2$ where X represents the noble (or inert) gas for the (n - 1) period and n represents the period in which the element is located.
   e.g. C(in period 2) has the configuration [He] $2s^2\ 2p^2$
   Group 4A elements typically share 4 electrons from other atoms to achieve a noble gas configuration, e.g. CH_4.

10. Four ions with an electron configuration that is the same as argon's: (18 electrons)
    S^{2-}, Cl^-, K^+, Ca^{2+}
    sulfide ion, chloride ion, potassium ion, calcium ion

12. The reaction of an alkali metal with chlorine: $2\ Li + Cl_2 \rightarrow 2\ LiCl$.
    Given the stability of M^+ cations and X^- anions compared to the metal and nonmetal, the **reaction will likely be exothermic**, and an ionic product will result. Reactions between elements in "widely separated groups" tend to form ionic products (metal + nonmetal) while covalent products frequently are formed between elements in "closer groups" (e.g. two nonmetals).

14. Predict: color, physical state, and water solubility for the product formed in question 12

color: white ---------------- colored compounds have "d" electrons

physical state: solid ------- ionic compounds are solid

water solubility: soluble ----------- monopositive cations & mononegative anions form water soluble compounds

16. 10 most **abundant elements in earth's crust** from lowest to greatest abundance:
Main group elements are identified with an asterisk (*)

Abundance	Element	Specie
0.6 %	Ti	$FeTiO_3$, ilmenite; TiO_2, rutile
0.9 %	H(*)	H_2O
1.9 %	Mg(*)	oxides & chlorides; e.g. $MgCl_2{\cdot}KCl{\cdot}6\,H_2O$, carnallite
2.4 %	K(*)	KCl
2.6 %	Na(*)	NaCl, $Na_2B_4O_7{\cdot}\,10\,H_2O$
3.4 %	Ca(*)	dolomite, $CaCO_3{\cdot}MgCO_3$; limestone, $CaCO_3$
4.7 %	Fe	oxide ores: Fe_2O_3 , Fe_3O_4
7.4 %	Al(*)	bauxite, $Al_2O_3{\cdot}x\,H_2O$
25.7 %	Si(*)	SiO_2 ; silicates, SiO_4^{4-}
49.5 %	O(*)	oxides of all types; water; $CaCO_3$

18. For the first 10 elements, those found free in the earth's crust? combined in the earth's crust?

Element	Atomic #	Free	Combined
H	1		√
He	2		
Li	3		√
Be	4		√
B	5		√
C	6	√	
N	7		√
O	8		√
F	9		√
Ne	10		

20. In general metal oxides (also commonly called base anhydrides) are more basic than nonmetal oxides (also known as acid anhydrides). We would predict the order to be (from least basic to most basic:

$$SO_3 < SiO_2 < Al_2O_3 < Na_2O$$

Al_2O_3 and SiO_2 dissolve <u>only very slightly</u>!

22. Balanced equations:

a. $2\ K(s) + I_2(g) \rightarrow 2\ KI(s)$

b. $Ba(s) + O_2(g) \rightarrow BaO(s)$

c. $16\ Al(s) + 3\ S_8(s) \rightarrow 8\ Al_2S_3(s)$

d. $Si(s) + 2\ Cl_2(g) \rightarrow SiCl_4(\ell)$

Hydrogen

24. $2\ K(s) + H_2(g) \rightarrow 2\ KH(s)$ (potassium hydride)

The compound will be ionic. One would anticipate the ionic bond to indicate a high melting point (i.e. **KH is a solid**). We anticipate the hydride to be a good reducing agent.

26. A balanced equation for the formation of H_2 from methane and water:

	$CH_4(g)$	+	$H_2O(g)$	$\rightarrow$	$3\ H_2(g)$	+	$CO(g)$
$S^\circ\ (\frac{J}{K\bullet mol})$	186.264		188.825		130.684		197.674
$\Delta H^\circ\ (\frac{kJ}{mol})$	-74.81		-241.818		0		-110.525

$$\Delta H^\circ = [(3\ mol)(0) + (1\ mol)(-110.525\ \frac{kJ}{mol})] - [(1\ mol)(-74.81\ \frac{kJ}{mol}) + (1\ mol)(-241.818\ \frac{kJ}{mol})]$$

$$= 206.10\ kJ$$

$$\Delta S^\circ = [(3\ mol)(130.684\ \frac{J}{K\bullet mol}) + (1\ mol)(197.674\ \frac{J}{K\bullet mol})] - [(1\ mol)(186.264\ \frac{J}{K\bullet mol}) + (1mol)(188.825\ \frac{J}{K\bullet mol})]$$

$$= 214.637\ J/K$$

$$\Delta G^\circ = 206.10 \text{ kJ} - (298 \text{ K})(214.637 \text{ J/K})(\frac{1 \text{ kJ}}{1000 \text{ J}})$$

$$= 142.14 \text{ kJ}$$

28. An experiment to identify a gas that is either hydrogen, nitrogen, or oxygen:

To differentiate between these three gases, one can use the chemical reactivity of the three gases. Using a piece of tubing, allow part of the gas to escape via a small orifice, near which a small flame is located. If the **gas is hydrogen**, the escaping gas will burn with a light blue flame—even after the match is withdrawn. If the **gas is oxygen**, the match flame will glow much more brightly—with the increased concentration of oxygen available to the combustion reaction of the match. If **the gas is nitrogen**, the flame may indeed by extinguished as the nitrogen from the orifice reduces the oxygen available to the flame.

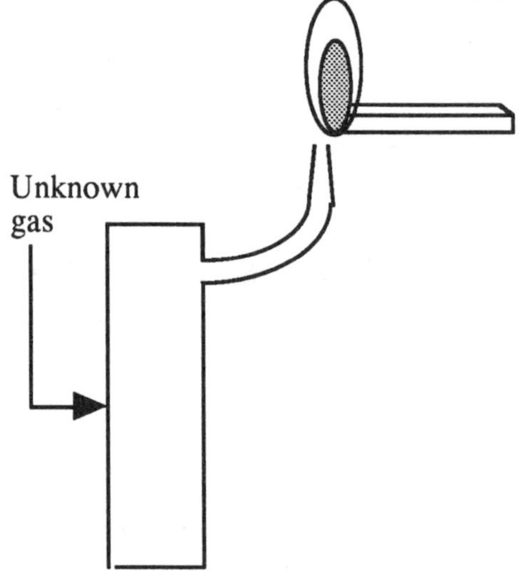

One can also use the physical property of the boiling point. If the gas is chilled, and the temperature monitored, one can differentiate the three gases. Elemental hydrogen boils at -253°C, oxygen at -183°C, and nitrogen at -196°C.

Alkali Metals

30. Equations for the reaction of sodium with the halogens:

$2 \text{ Na(s)} + F_2\text{(g)} \rightarrow 2 \text{ NaF(s)}$

$2 \text{ Na(s)} + Cl_2\text{(g)} \rightarrow 2 \text{ NaCl(s)}$

$2 \text{ Na(s)} + Br_2(\ell) \rightarrow 2 \text{ NaBr(s)}$

$2 \text{ Na(s)} + I_2\text{(s)} \rightarrow 2 \text{ NaI(s)}$

We anticipate that these salts are ionic, are good conductors (in the liquid state or solution), have high melting points (i.e.they are solid at room temperature), and are readily soluble in water.

32. The reaction of lithium, sodium, and potassium with oxygen:

$4\,Li(s) + O_2(g) \rightarrow 2\,Li_2O(s)$ oxides

$2\,Na(s) + O_2(g) \rightarrow Na_2O_2(s)$ peroxides

$K(s) + O_2(g) \rightarrow KO_2(s)$ superoxide

34. In the electrolysis of aqueous NaCl:

a. The balanced equation for the process: (While this is a reduction-oxidation reaction, it can be balanced easily by inspection:

$$2\,NaCl(aq) + 2\,H_2O(\ell) \rightarrow 2\,NaOH(aq) + Cl_2(g) + H_2(g)$$

b. Anticipated mass ratios:

$$\frac{1 \text{ mol } Cl_2}{2 \text{ mol NaOH}} = \frac{71 \text{ g } Cl_2}{80 \text{ g NaOH}} = 0.89 \text{ g } Cl_2/\text{g NaOH}$$

$$\text{Actual: } \frac{1.14 \times 10^{10} \text{ kg } Cl_2}{1.19 \times 10^{10} \text{ kg NaOH}} = 0.96 \frac{\text{kg } Cl_2}{\text{kg NaOH}}$$

The difference in ratios means that alternative methods of producing chlorine are used. One of these is the Kel-Chlor process which uses HCl, NOCl, and O_2. Other products are NO and H_2O.

Alkaline Earths

36. Balanced equations for the reaction of magnesium with nitrogen and oxygen:

$3\,Mg(s) + N_2(g) \rightarrow Mg_3N_2(s)$ $2\,Mg(s) + O_2(g) \rightarrow 2\,MgO(s)$

38. Uses of limestone:

agricultural: to furnish Ca^{2+} to plants and neutralize acidic soils

building: lime (CaO) is used in mortar and absorbs CO_2 to form $CaCO_3$

steel-making: $CaCO_3$ furnishes lime (CaO) in the basic oxygen process. The lime reacts with gangue (SiO_2) to form calcium silicate.

The balanced equation for the reaction of limestone with carbon dioxide in water:

$$CaCO_3(s) + H_2O(\ell) + CO_2(g) \rightleftharpoons Ca^{2+}(aq) + 2\,HCO_3^-(aq)$$

This reaction is important in the formation of "hard water" (not particularly a great happening for plumbing) and stalagmites and stalactites (aesthetically pleasing in caves).

40. The amount of SO_2 that could be removed by 1.2×10^3 kg of CaO by the reaction:

$$CaO\,(s) + SO_2\,(g) \rightarrow CaSO_3\,(s)$$

$$1.2 \times 10^6 \text{ g CaO} \cdot \frac{1 \text{ mol CaO}}{56.079 \text{ g CaO}} \cdot \frac{1 \text{ mol } SO_2}{1 \text{ mol CaO}} \cdot \frac{64.059 \text{ g } SO_2}{1 \text{ mol } SO_2} = 1.4 \times 10^6 \text{ g } SO_2$$

Aluminum

42. The equations for the reaction of aluminum with HCl, Cl_2 and O_2:

$$2\,Al(s) + 6\,HCl(aq) \rightarrow 2\,Al^{3+}(aq) + 6\,Cl^-(aq) + 3\,H_2(g)$$

$$2\,Al(s) + 3\,Cl_2(g) \rightarrow 2\,AlCl_3(s)$$

$$4\,Al(s) + 3\,O_2(g) \rightarrow 2\,Al_2O_3(s)$$

44. The equation for the reaction of aluminum dissolving in aqueous NaOH:

$$2\,Al\,(s) + 2\,NaOH\,(aq) + 6\,H_2O\,(\ell) \rightarrow 2\,NaAl(OH)_4\,(aq) + 3\,H_2\,(g)$$

Volume of H_2 (in mL) produced when 13.2 g of Al react:

$$13.2 \text{ g Al} \cdot \frac{1 \text{ mol Al}}{26.98 \text{ g Al}} \cdot \frac{3 \text{ mol } H_2}{2 \text{ mol Al}} = 0.734 \text{ mol } H_2$$

$$V = \frac{(0.734 \text{ mol } H_2)(0.082057 \frac{L \cdot atm}{K \cdot mol})(295.7 \text{ K})}{735 \text{ mm Hg} \cdot \frac{1 \text{ atm}}{760 \text{ mm Hg}}} = 18.4 \text{ L}$$

or 1.84×10^4 mL

46. The equation for the reaction of aluminum oxide with sulfuric acid:

$$Al_2O_3(s) + 3\,H_2SO_4(aq) \rightarrow Al_2(SO_4)_3\,(aq) + 3\,H_2O(\ell)$$

$$1.00 \times 10^3 \text{ g } Al_2(SO_4)_3 \cdot \frac{1 \text{ mol } Al_2(SO_4)_3}{342.1 \text{ g } Al_2(SO_4)_3} \cdot \frac{1 \text{ mol } Al_2O_3}{1 \text{ mol } Al_2(SO_4)_3} \cdot$$

$$\frac{102.1 \text{ g } Al_2O_3}{1 \text{ mol } Al_2O_3} \cdot \frac{1 \text{ kg}}{1 \times 10^3 \text{ g}} = 0.298 \text{ kg } Al_2O_3$$

$$1.00 \times 10^3 \text{ g } Al_2(SO_4)_3 \bullet \frac{1 \text{ mol } Al_2(SO_4)_3}{342.1 \text{ g } Al_2(SO_4)_3} \bullet \frac{3 \text{ mol } H_2SO_4}{1 \text{ mol } Al_2(SO_4)_3} \bullet$$

$$\frac{98.07 \text{ g } H_2SO_4}{1 \text{ mol } H_2SO_4} \bullet \frac{1 \text{ kg}}{1 \times 10^3 \text{ g}} = 0.860 \text{ kg } H_2SO_4$$

48. A Lewis electron dot structure for $AlCl_4^-$:

```
[      :Cl:       ]-
        |
[ :Cl — Al — Cl:  ]
        |
[      :Cl:       ]
```

The $AlCl_4^-$ will have a **tetrahedral geometry**. The aluminum atom would utilize **sp^3 hybridization.**

Silicon

50. The structure of SiO_2 is tetrahedral with the silicon atom being surrounded by four oxygen atoms. The geometry around C in CO_2 is linear with two groups attached to the C atom.

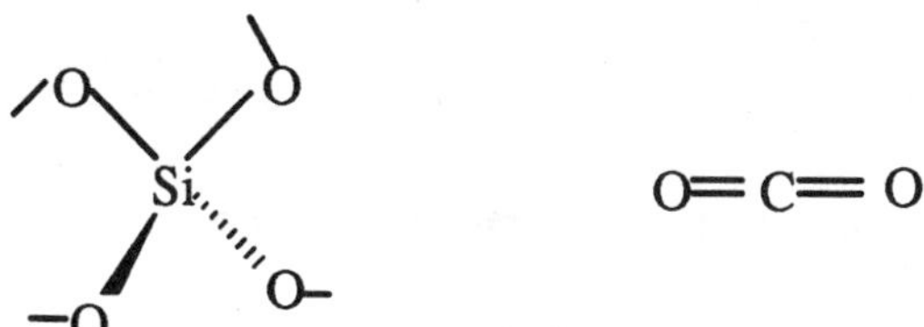

The energies of four Si - O single bonds is greater than two Si=O double bonds. CO_2, on the other hand, forms discrete C=O double bonds. The high melting point of SiO_2 is due to the **network structure** of the SiO_4 tetrahedron, while CO_2 exists as discrete nonpolar molecules—and as a gas at room temperature.

52. In the synthesis of dichlorodimethylsilane:

a. The balanced equation: $Si\ (s) + 2\ CH_3Cl\ (g) \rightarrow (CH_3)_2SiCl_2\ (\ell)$

b. Stoichiometric amount of CH_3Cl to react with 2.65 g silicon:

$$2.65 \text{ g Si} \bullet \frac{1 \text{ mol Si}}{28.09 \text{ g Si}} \bullet \frac{2 \text{ mol } CH_3Cl}{1 \text{ mol Si}} = 0.189 \text{ mol } CH_3Cl$$

$$P = \frac{(0.189 \text{ mol } CH_3Cl)(0.082057 \frac{L \bullet atm}{K \bullet mol})(297.7 \text{ K})}{5.60 \text{ L}} = 0.823 \text{ atm}$$

c. Mass of $(CH_3)_2SiCl_2$ produced assuming 100 % yield:

$$0.0943 \text{ mol Si} \bullet \frac{1 \text{ mol } (CH_3)_2SiCl_2}{1 \text{ mol Si}} \bullet \frac{129.1 \text{ g } (CH_3)_2SiCl_2}{1 \text{ mol } (CH_3)_2SiCl_2} = 12.2 \text{ g } (CH_3)_2SiCl_2$$

Nitrogen and Phosphorus

54. The enthalpy data from Appendix L are shown below:

compound:	NO(g)	NO_2(g)	N_2O(g)	N_2O_4(g)
$\Delta H°$ $(\frac{kJ}{mol})$	+90.25	+33.18	+82.05	+9.16

Note that the + signs for each of these oxides lets us know that **none** of these oxides is stable with respect to elemental nitrogen or oxygen. This can be seen if one writes the "general" equation for the formation of these oxides.

$$\frac{x}{2} N_2\,(g) + \frac{y}{2} O_2(g) \rightarrow N_xO_y$$

Writing the $\Delta H°$ expression (omitting the coefficients for clarity) for the reaction would give:

$$\Delta H°_{rxn} = \Delta H°_f\, N_xO_y - (\Delta H°_f\, N_2 + \Delta H°_f\, O_2)$$

Since the $\Delta H°_f$ for both elemental oxygen and elemental nitrogen are 0 $\frac{kJ}{mol}$, the $\Delta H°_{rxn}$ would be **endothermic** in the case of **all of these four oxides.**

56. Calculate $\Delta H°_{rxn}$ for the reaction: 2 NO(g) + O_2(g) → 2 NO_2(g)

	2 NO(g)	O_2(g)	2 NO_2(g)
$\Delta H°$ $(\frac{kJ}{mol})$	+90.25	0	+33.18

$$\Delta H°_{rxn} = [(2 \text{ mol})(+33.18 \frac{kJ}{mol})] - [(2 \text{ mol})(+90.25 \frac{kJ}{mol})] = -114.14 \text{ kJ}$$

The **reaction is exothermic.**

58. a. The reaction of hydrazine with dissolved oxygen:

$$N_2H_4\,(aq) + O_2\,(aq) \rightarrow N_2\,(g) + 2\,H_2O(\ell)$$

b. Mass of hydrazine to consume the oxygen in 3.00×10^4 L of water:

$$3.00 \times 10^4 \text{ L} \bullet \frac{3.08 \text{ cm}^3\, O_2}{0.100 \text{ L}} \bullet \frac{1 \text{ mol } O_2}{22400 \text{ cm}^3} \bullet \frac{1 \text{ mol } N_2H_4}{1 \text{ mol } O_2} \bullet \frac{32.05 \text{ g } N_2H_4}{1 \text{ mol } N_2H_4} = 1.32 \times 10^3 \text{ g } N_2H_4$$

↑ at STP

60. The half equations:

$$N_2H_5^+ (aq) \rightarrow N_2 (g) + 5 H^+ (aq) + 4 e^-$$

$$IO_3^- (aq) \rightarrow I_2 (s)$$

are balanced according to the procedure in Chapter 21 to give:

$$5 N_2H_5^+ (aq) + 4 IO_3^- (aq) \rightarrow 5 N_2 (g) + 2 I_2 (s) + 11 H_2O(\ell) + H_3O^+ (aq)$$

E° for the reaction is: E° for the hydrazine equation (oxidation) = + 0.23 V
E° for the iodate equation (reduction) = + 1.195 V
E°net = + 1.43 V

62. The dot structure for the azide ion: $[\,\ddot{\underset{..}{N}}{=}N{=}\ddot{\underset{..}{N}}\,]^-$

Note that with (5+5+5+1) 16 electrons, we could begin with single bonds (1 pair of electrons) between the nitrogen atoms. That structure would leave the "central" N atom with a deficiency of electrons. Double bonds between the three N atoms provide 8 electrons for **each**.

Oxygen and Sulfur

64. a. Allowable release of SO_2: (0.30 %)

$$1.80 \times 10^6 \text{ kg } H_2SO_4 \cdot \frac{1 \text{ mol } H_2SO_4}{98.07 \text{ g } H_2SO_4} \cdot \frac{1 \text{ mol } SO_2}{1 \text{ mol } H_2SO_4} \cdot \frac{64.06 \text{ g } SO_2}{1 \text{ mol } SO_2}$$

$$\cdot \frac{0.0030 \text{ kg } SO_2 \text{ released}}{1.00 \text{ kg } SO_2 \text{ produced}} = 3.5 \times 10^3 \text{ kg } SO_2$$

(2 sf)

b. Mass of $Ca(OH)_2$ to remove 3.5×10^3 kg SO_2:

$$3.5 \times 10^3 \text{ kg } SO_2 \cdot \frac{1 \text{ mol } SO_2}{64.06 \text{ g } SO_2} \cdot \frac{1 \text{ mol } Ca(OH)_2}{1 \text{ mol } SO_2} \cdot \frac{74.09 \text{ g } Ca(OH)_2}{1 \text{ mol } Ca(OH)_2}$$

$$= 4.1 \times 10^3 \text{ kg } Ca(OH)_2$$

66. The disulfide ion S_2^{2-} can be pictured as : $[:\ddot{\underset{..}{S}}:\ddot{\underset{..}{S}}:]^{2-}$

With (6+6+2) 14 electrons, a single pair of electrons between the S atoms, and three lone (or nonbonding) pairs on each of the atoms provides eight electrons for each of the atoms in this dianion.

Chlorine

68. Calculate the equivalent net cell potential for the oxidation:

	$6\,[Mn^{2+}(aq) + 4\,H_2O(\ell) \rightarrow MnO_4^-(aq) + 8\,H^+(aq) + 5\,e^-]$	-1.51 V
	$5\,[BrO_3^-(aq) + 6\,H^+(aq) + 6\,e^- \rightarrow Br^-(aq) + 3\,H_2O(\ell)]$	+1.44 V
net	$6\,Mn^{2+}(aq) + 9\,H_2O(\ell) + 5\,BrO_3^-(aq) \rightarrow 6\,MnO_4^-(aq) + 18\,H^+(aq) + 5\,Br^-(aq)$	-0.07 V

The negative net potential indicates that this **process doesn't favor products** with 1.0 M bromate ion.

70. The balanced equation for the reaction of Cl_2 with Br^-

	$2\,Br^-(aq) \rightarrow Br_2(\ell) + 2\,e^-$	-1.066 V
	$Cl_2(g) + 2\,e^- \rightarrow 2\,Cl^-(aq)$	+1.358 V
net:	$Cl_2(g) + 2\,Br^-(aq) \rightarrow 2\,Cl^-(aq) + Br_2(\ell)$	+ 0.292 V

- Note that bromide ions are losing electrons (and donating them to chlorine), causing chlorine to be reduced—so **bromide is the reducing agent** and chlorine, removing the electrons from bromide ions, is causing the bromide ions to be oxidized—so **chlorine is the oxidizing agent.**
- Note also that the voltage for the cell we "constructed" is positive, making this process a **product-favored** reaction.

General Questions

72. Describe the elements in the third period:

Atomic No.	Element	Type	Color	State
11	Sodium	metal	grey,shiny	solid
12	Magnesium	metal	grey,shiny	solid
13	Aluminum	metal	grey,shiny	solid
14	Silicon	metalloid	grey	solid

Atomic No.	Element	Type	Color	State
15	Phosphorus	nonmetal	red,white, black	solid
16	Sulfur	nonmetal	yellow	solid
17	Chlorine	nonmetal	yellow-green	gas
18	Argon	nonmetal	colorless	gas

74. Reactions of Na, Mg, Al, Si, P, S

a. Balanced equations of the elements with elemental chlorine:	b. Bonding in product:
$2\ Na\ (s) + Cl_2\ (g) \rightarrow 2\ NaCl\ (s)$	ionic
$Mg\ (s) + Cl_2\ (g) \rightarrow MgCl_2\ (s)$	ionic
$2\ Al\ (s) + 3\ Cl_2\ (g) \rightarrow 2\ AlCl_3\ (s)$	covalent
$Si\ (s) + 2\ Cl_2\ (g) \rightarrow 2\ SiCl_4\ (\ell)$	covalent
$P_4\ (s) + 10\ Cl_2\ (g) \rightarrow 4\ PCl_5\ (s)$ (excess Cl_2)	covalent
$S_8\ (s) + 16\ Cl_2\ (g) \rightarrow 8\ SCl_2(s)$	covalent

(Solid sulfur is best represented S_8 and solid phosphorus, P_4. Other compounds of chlorine with sulfur and phosphorus are also known.

c. Electron dot structures for the products; electron-pair geometry; molecular geometry

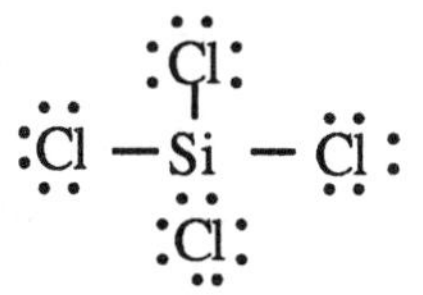

Electron dot structure	Electron-pair geometry	Molecular geometry
$SiCl_4$	tetrahedral	tetrahedral
PCl_5	trigonal bipyramidal	trigonal bipyramidal

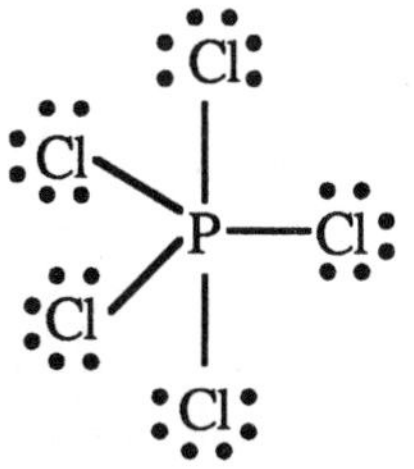

76. The products expected when molten LiH is electrolyzed:

Molten LiH contains lithium (Li^+) and hydride (H^-) ions.

At the cathode, lithium ions are reduced: $Li^+ + e^- \rightarrow Li\ (s)$

At the anode, hydride ions are oxidized: $2\ H^- \rightarrow H_2\ (g) + 2\ e^-$

78. Calculate ΔG° for the decomposition of the metal carbonates for Mg, Ca, and Ba

M ΔG° (kJ/mol)	$MCO_3(s)$	→	MO(s)	+	$CO_2(g)$
Mg	-1012.1		-569.43		-394.359
Ca	-1128.79		-604.03		-394.359
Ba	-1137.6		-525.1		-394.359

$\Delta G°_{rxn} = \Delta G°_f\ MO + \Delta G°_f\ CO_2 - \Delta G°_f\ MCO_3$

$MgCO_3 = (-569.43 \frac{kJ}{mol})(1\ mol) + (-394.359 \frac{kJ}{mol})(1\ mol) - (-1012.1 \frac{kJ}{mol})(1\ mol)$
$= 48.3\ kJ$

$CaCO_3 = (-604.03 \frac{kJ}{mol})(1\ mol) + (-394.359 \frac{kJ}{mol})(1\ mol) - (-1128.79 \frac{kJ}{mol})(1\ mol)$
$= 130.40\ kJ$

$BaCO_3 = (-525.1 \frac{kJ}{mol})(1\ mol) + (-394.359 \frac{kJ}{mol})(1\ mol) - (-1137.6 \frac{kJ}{mol})(1\ mol)$
$= 218.1\ kJ$

The relative tendency for decomposition diminishes in the order: $MgCO_3 > CaCO_3 > BaCO_3$

80. a. Since $\Delta G°_{rxn} < 0$ for the reaction to be product-favored, calculate the value for $\Delta G°_f$ MX that will make the $\Delta G°_{rxn}$ zero.

$\Delta G°_{rxn} = \Delta G°_f\ (MX_n) - n\ \Delta G°_f\ (HX)$ and for HCl $= \Delta G°_f\ (MX_n) - n(-95.3 \frac{kJ}{mol})$

so if n(-95.3 kJ) = $\Delta G°_f(MX_n)$ then ΔG° = 0 and if n(-95.3 kJ) > ΔG° (MX_n) then $\Delta G°_{rxn} < 0$.

b. Examine ΔG°MX values for metal:

metal:	Ba	Pb	Hg	Ti
$\Delta G°_{MX}$:	-810.4	-314.10	-178.6	-737.2
n:	2	2	2	4
n(-95.3):	-190.6	-190.6	-190.6	-381.2
	- 619.6	- 123.5	+ 12.0	- 356.0

For Barium, Lead, and Titanium, n (-95.3) > $\Delta G°(MX)$, and we would expect these reactions to be spontaneous.

82. a. The N-O bonds are the same length owing to the delocalization of a second pair of electrons between the two N-O bonds. A dot picture of this would show two reasonable structures (resonance hybrids). In essence there is a bond order of 1.5 for these two bonds; compared to a bond order of 1 for the N-OH bond.

b. The bond angle for the oxygens involved in the delocalized bond is only slightly larger than anticipated for the trigonal planar geometry (120°). The larger angle reflects the increased electron density (and repulsion) of this bond. The increased bond angle would result in a slightly smaller-than-ideal angle between the two "non-H" oxygens and the O-H bond. The bond angle N-O-H is only slightly less than anticipated for the tetrahedral orientation around the oxygen (two atoms and two lone pairs), a finding consistent with the two lone pairs of electrons on that oxygen.

c. The central **N atom has sp^2 hybridization.** The "unhybridized p" orbital on the N can participate in a π - type overlap with the orbitals on the two oxygen atoms, resulting in the pi bond.

Challenge Questions

84. Reaction scheme:

Clue

1. 1.00 g A (white solid) + heat → B (white solid) + gas (P = 209 mm; V = 450 mL; T = 298 K)

2. Gas (from 1) + $Ca(OH)_2$ (aq) → C (s) (white solid)

3. Aqueous solution of B is basic (turns red litmus paper blue)

4. B (aq) + HCl (aq) + heat → D (white solid)

5. Flame test for B: green flame

6. B (aq) + H_2SO_4 (aq) → E (white solid)

Clue 5 indicates that **B** is a barium salt.

Clue 2 suggests that the gas evolved in Clue 1 is CO_2, and that **C** would be $CaCO_3$.

Heating of carbonates liberates CO_2 (g).

Compound **B** is a metal oxide (Clue 3), and probably has the formula BaO.

Clues 3 and 5 suggest the oxide reacts with HCl and H_2SO_4 to form $BaCl_2$ (compound **D**) and $BaSO_4$ (compound **E**) respectively.

Since **B** is most likely BaO, compound **A** must be $BaCO_3$.

One gram of $BaCO_3$ (Molar mass 197) corresponds to 5.06×10^{-3} mol $BaCO_3$.
Compare this amount of substance to the amount of gas liberated when substance **A** is heated. Substitution of data from clue 1 yields:

$$n = \frac{(209 \text{ mm Hg})\ (0.450 \text{ L})}{(62.4 \frac{\text{L} \bullet \text{mm Hg}}{\text{K} \bullet \text{mol}})\ (298 \text{ K})} = 5.06 \times 10^{-3} \text{ mol gas}$$

This is the quantity of CO_2 anticipated from the thermal decomposition of $BaCO_3$.

86. Examine the enthalpy change for the reaction:

$$2\ N_2\ (g) + 5\ O_2\ (g) + 2\ H_2O\ (\ell) \rightarrow 4\ HNO_3\ (aq)$$

$$\begin{aligned} \Delta H^\circ_{rxn} &= [(4 \text{ mol})(-207.36 \tfrac{\text{kJ}}{\text{mol}})] - \\ &\quad [(2 \text{ mol})(0 \tfrac{\text{kJ}}{\text{mol}}) + (2 \text{ mol})(0 \tfrac{\text{kJ}}{\text{mol}}) + (2 \text{ mol})(-285.830 \tfrac{\text{kJ}}{\text{mol}})] \\ &= -257.78 \text{ kJ} \end{aligned}$$

The reaction is **exothermic** so it is a reasonable "first-guess" that this might be a way to "fix" nitrogen. The only way to be certain is to calculate the ΔG°_{rxn}.

$$\begin{aligned} \Delta G^\circ_{rxn} &= [(4 \text{ mol})(-111.25 \tfrac{\text{kJ}}{\text{mol}})] - \\ &\quad [(2 \text{ mol})(0 \tfrac{\text{kJ}}{\text{mol}}) + (2 \text{ mol})(0 \tfrac{\text{kJ}}{\text{mol}}) + (2 \text{ mol})(-237.129 \tfrac{\text{kJ}}{\text{mol}})] \\ &= 29.26 \text{ kJ} \end{aligned}$$

The positive value for ΔG°_{rxn} tells us that the **reaction is not likely at 25 °C**, and no amount of research will change the **sign of this process**, so research into this process would be very unproductive.

Summary Questions

88. a. Volume of seawater to obtain 1.00 kg Mg:

$$\frac{1.00 \times 10^3 \text{ g Mg}}{1} \bullet \frac{1 \text{ mol Mg}}{24.31 \text{ g Mg}} \bullet \frac{1 \text{ L seawater}}{0.050 \text{ mol Mg}} = 823 \text{ L seawater}$$

(820 L to 2 sf)

Mass of CaO to precipitate the magnesium:

The precipitation equation may be written as two steps:

1. $CaO(s) + H_2O(\ell) \rightarrow Ca(OH)_2(s)$
2. $Ca(OH)_2(s) + Mg^{2+}(aq) \rightarrow Ca^{2+}(aq) + Mg(OH)_2(s)$

The result of these processes is that 1 mole CaO precipitates 1 mole of Magnesium ions.

$$\frac{1.00 \times 10^3 \text{ g Mg}}{1} \bullet \frac{1 \text{ mol Mg}}{24.31 \text{ g Mg}} \bullet \frac{1 \text{ mol CaO}}{1 \text{ mol Mg}} \bullet \frac{56.08 \text{ g CaO}}{1 \text{ mol CaO}} = 2.3 \times 10^3 \text{ g CaO}$$

or 2.3 kg CaO

b. $MgCl_2\ (\ell) \xrightarrow{\text{electricity}} Mg\ (s) + Cl_2\ (g)$

Mass of Mg produced at the cathode:

$$1.2 \times 10^3 \text{ kg } MgCl_2 \bullet \frac{1 \text{ mol } MgCl_2}{95.211 \text{ g } MgCl_2} \bullet \frac{1 \text{ mol Mg}}{1 \text{ mol } MgCl_2} \bullet \frac{24.305 \text{ g Mg}}{1 \text{ mol Mg}} = 310 \text{ kg Mg}$$

Note the absence of a conversion of mass of $MgCl_2$ from kg to grams. Since the answer was requested in units of kg, any conversion to units of grams would have necessitated a conversion back to kg at the end of the calculation. The two conversion factors would cancel each other, and "leaving them out" causes no harm to the integrity of the reasoning (or the answer).

At the anode, chlorine is produced. [$2\ Cl^- \rightarrow Cl_2 + 2\ e^-$]

Mass of Cl_2 produced:

$$1.2 \times 10^3 \text{ kg } MgCl_2 \bullet \frac{1 \text{ mol } MgCl_2}{95.211 \text{ g } MgCl_2} \bullet \frac{1 \text{ mol } Cl_2}{1 \text{ mol } MgCl_2} \bullet \frac{70.906 \text{ g } Cl_2}{1 \text{ mol } Cl_2} = 890 \text{ kg } Cl_2$$

Faradays used in the process:

The reduction of magnesium requires 2 Faradays per mole: $Mg^{2+} + 2e^- \rightarrow Mg$

$$310 \text{ kg Mg} \cdot \frac{1.000 \times 10^3 \text{ g Mg}}{1.0 \text{ kg Mg}} \cdot \frac{1 \text{ mol Mg}}{24.305 \text{ g Mg}} \cdot \frac{2 \text{ F}}{1 \text{ mol Mg}} = 2.5 \times 10^4 \text{ F}$$

The oxidation of chlorine requires 2 Faradays per mole of chlorine: $2\ Cl^- \rightarrow Cl_2 + 2\ e^-$

$$890 \text{ kg } Cl_2 \cdot \frac{1.000 \times 10^3 \text{ g } Cl_2}{1.0 \text{ kg } Cl_2} \cdot \frac{1 \text{ mol } Cl_2}{70.906 \text{ g } Cl_2} \cdot \frac{2 \text{ F}}{1 \text{ mol } Cl_2} = 2.5 \times 10^4 \text{ F}$$

The total number of Faradays of electricity used in the process is 2.5×10^4 F.

c. Joules required per mole of magnesium:

$$\frac{8.4 \text{ kwh}}{1 \text{ lb Mg}} \cdot \frac{3.60 \times 10^6 \text{ J}}{1 \text{ kwh}} \cdot \frac{1 \text{ lb Mg}}{454 \text{ g Mg}} \cdot \frac{24.305 \text{ g Mg}}{1 \text{ mol Mg}} = 1.6 \times 10^6 \frac{\text{J}}{\text{mol Mg}}$$

The reaction, $MgCl_2$ (s) $\rightarrow$ Mg (s) + Cl_2 (g), represents the reverse of the formation of magnesium chloride from its elements (each in their standard states). From Appendix L, the ΔH for the process is +641.32 kJ/mol or 6.4×10^5 J/mol. The difference between this value and the value calculated above may be attributed to the energy required to melt the $MgCl_2$.

Chapter 23
The Transition Elements

Configurations and Physical Properties

14. Cr^{3+} 3d [↑][↑][↑][ ][ ] 4s [ ] paramagnetic

Cr^{6+} 3d [ ][ ][ ][ ][ ] 4s [ ]

16. Transition metal ions with the electron configuration
    a. [Ar] $3d^6$ Fe^{2+}, Co^{3+}
    b. [Ar] $3d^5$ Mn^{2+}, Fe^{3+}
    c. [Ar] $3d^{10}$ Cu^{1+}, Zn^{2+}
    d. [Ar] $3d^8$ Ni^{2+}, Cu^{3+} Cu^{3+} is not a common ion, owing to the ease of reduction of the ion to the 2+ or 1+ ion.

Metallurgy

18. Balance:
    a. $Cr_2O_3(s) + 2\ Al(s) \rightarrow Al_2O_3(s) + 2\ Cr(s)$
    b. $TiCl_4(\ell) + 2\ Mg(s) \rightarrow Ti(s) + 2\ MgCl_2(s)$
    c. $2\ [Ag(CN)_2]^-(aq) + Zn(s) \rightarrow 2\ Ag(s) + Zn^{2+}(aq) + 4\ CN^-(aq)$

20. Volume of 18.0 M H_2SO_4 required to react with 1.00 kg of ilmenite:

$$1.00 \times 10^3 \text{ g } FeTiO_3 \cdot \frac{1 \text{ mol } FeTiO_3}{151.7 \text{ g } FeTiO_3} \cdot \frac{3 \text{ mol } H_2SO_4}{1 \text{ mol } FeTiO_3} \cdot \frac{1 \text{ L}}{18.0 \text{ mol } H_2SO_4} =$$

1.10 L H_2SO_4

Mass of TiO_2 produced:

$$1.00 \times 10^3 \text{ g } FeTiO_3 \cdot \frac{1 \text{ mol } FeTiO_3}{151.7 \text{ g } FeTiO_3} \cdot \frac{1 \text{ mol } TiO_2}{1 \text{ mol } FeTiO_3} \cdot \frac{79.88 \text{ g } TiO_2}{1 \text{ mol } TiO_2} = 527 \text{ g } TiO_2$$

or 0.527 kg TiO_2

Liquids and Formulas of Complexes

22. Classify each of the following as monodentate or multidentate:

a. CH_3NH_2	monodentate (lone pair on N)
b. $CH_3C{\equiv}N$	monodentate (lone pair on N)
c. N_3^-	monodentate (lone pair on a N atom)
d. $C_2O_4^{2-}$	multidentate (lone pairs on terminal O atoms)
e. ethylenediamine	multiidentate (lone pairs on terminal N atoms)
f. Br^-	monodentate (lone pair on Br)
g. phenanthroline	multidentate (lone pairs on N atoms)

24.

	Compound	Metal	Oxidation Number
a.	$[Mn(NH_3)_6]SO_4$	Mn	+2

With NH_3 being neutral, and sulfate having a 2- charge, Mn must be +2.

b.	$K_3[Co(CN)_6]$	Co	+3

The complex ion has a -3 charge, CN^- has a -1 charge.

c.	$[Co(NH_3)_4Cl_2]Cl$	Co	+3

With NH_3 being neutral, and 3 Cl (each with a -1 charge), Co must be +3.

d.	$Mn(en)_2Cl_2$	Mn	+2

en is a neutral ligand. With 2 Cl (each with a -1 charge), Mn must be +2.

Naming

26. Formulas for:

a. dichlorobis(ethylenediamine)nickel(II)	$Ni(en)_2Cl_2$
b. potassium tetrachloroplatinate(II)	$K_2[PtCl_4]$
c. potassium dicyanocuprate(I)	$K[Cu(CN)_2]$
d. diaquatetraammineiron(II)	$[Fe(H_2O)_2(NH_3)_4]^{2+}$

28.

	Formula	Name
a.	$[Ni(C_2O_4)_2(H_2O)_2]^{2-}$	diaquabis(oxalato)nickelate(II) ion
b.	$[Co(en)_2Br_2]^+$	dibromobis(ethylenediamine)cobalt(III) ion
c.	$[Co(en)_2(NH_3)Cl]^{2+}$	amminechlorobis(ethylenediamine)cobalt(III) ion
d.	$Pt(NH_3)_2(C_2O_4)$	diammineoxalatoplatinum(II)

30. The name or formula for the ions or compounds shown below:

a. $[Fe(OH)(H_2O)_5]^{2+}$	Hydroxopentaaquairon(III) ion
b. $K_2[Ni(CN)_4]$	Potassium tetracyanonickelate(II)
c. $K[Cr(C_2O_4)_2(H_2O)_2]$	Potassium diaquabis(oxalato)chromate(III)
d. $(NH_4)_2[PtCl_4]$	Ammonium tetrachloroplatinate(IV)

Isomerization

32. Geometric Isomers of

a. $Fe(NH_3)_2Cl_2$

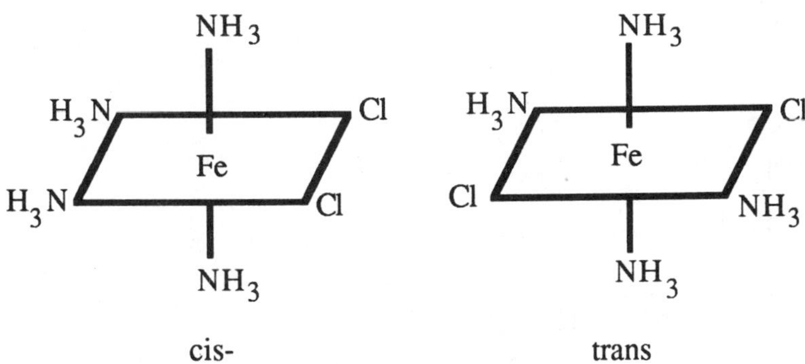

b. $Pt(NH_3)_2(NCS)(Br)$

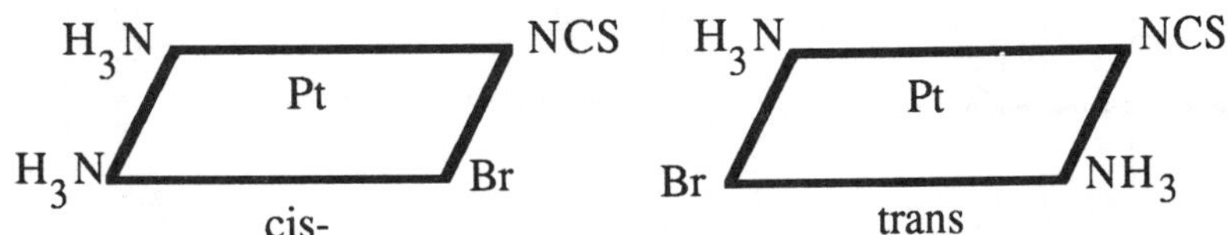

c. $Co(NH_3)_3(NO_2)_3$

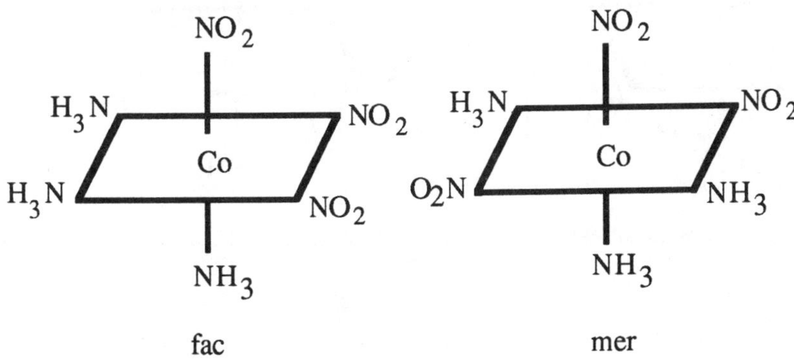

d. $[Co(en)Cl_4]^-$

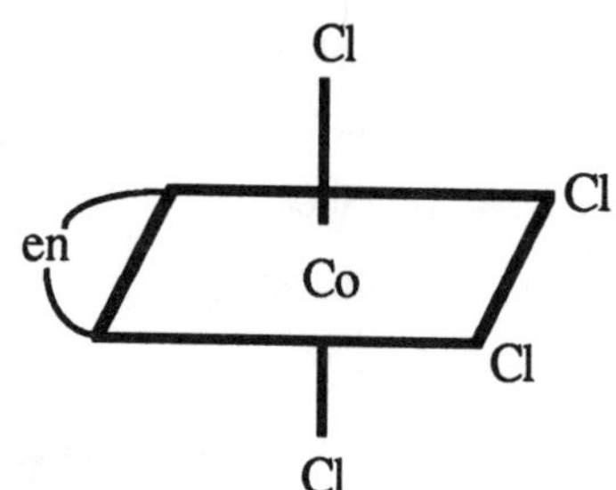

The ligand ethylenediamine is represented in these drawings by the symbolism :

= en

34. Molecules possess a chiral center:

a.	CH_2Cl_2	no
b.	$H_2N\underline{C}H(CH_3)CO_2H$	yes
c.	$Cl\underline{C}H(OH)CH_2Cl$	yes
d.	$CH_3CH_2CH{=}CHC_6H_5$	no

36. The four isomers of $[Co(en)(NH_3)_2(H_2O)Cl]^{2+}$:

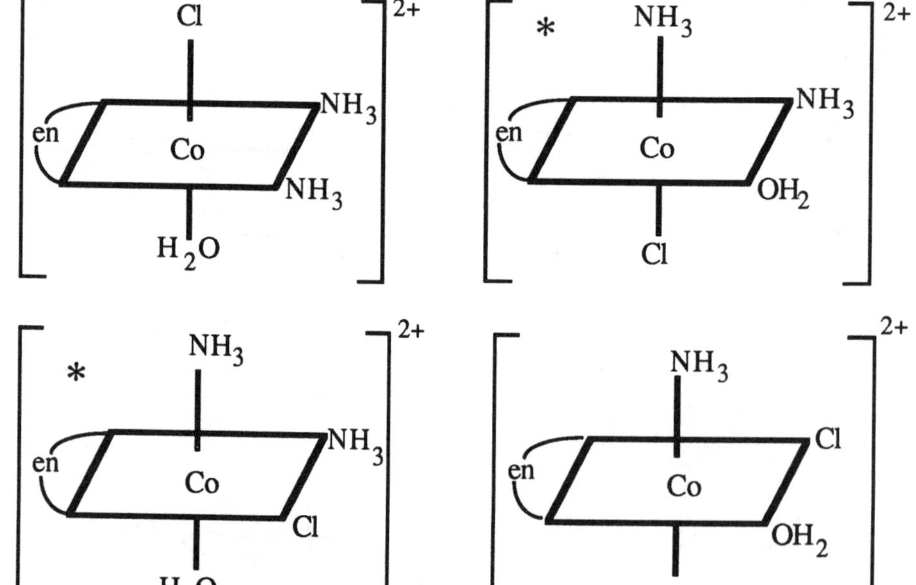

The two isomers marked with an asterisk (*) have nonsuperimposable mirror images.

Magnetism of Coordination Complexes

38. **Only d^4 through d^7 metal ions** can exhibit both high and low spin. With fewer than four electrons, all the electrons would occupy the three "lower energy" orbitals. Metals with d^8 through d^{10} could not place any electrons in one of the lower energy orbitals—since they would be full.

40. a. $[Fe(CN)_6]^{4-}$ Fe^{2+} has a d^6 configuration

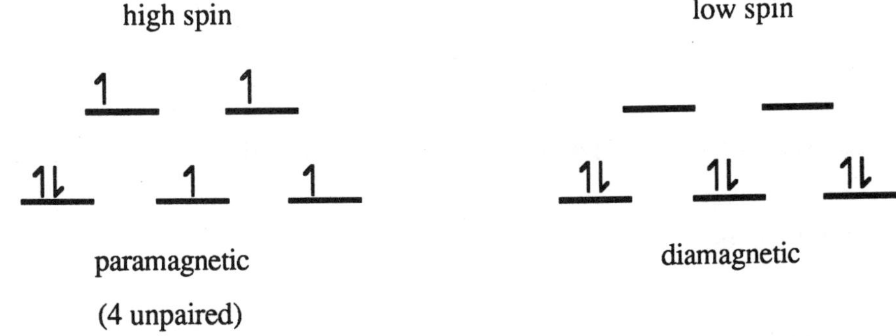

b. $[Co(NH_3)_6]^{3+}$ Co^{3+} has a d^6 configuration—like part a above.

c. $[Fe(H_2O)_6]^{3+}$ Fe^{3+} has a d^5 configuration

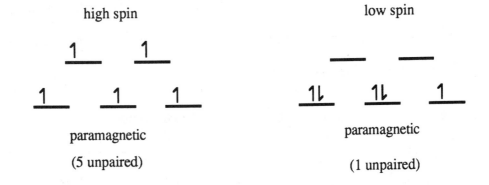

high spin

low spin

paramagnetic

(5 unpaired)

paramagnetic

(1 unpaired)

d. $[CrF_6]^{4-}$ Cr^{2+} has the d^4 configuration

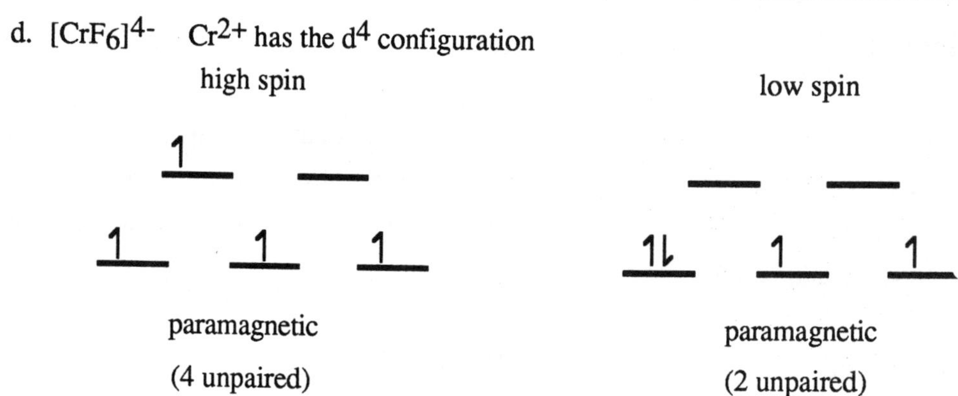

Color

42. For an ion to be violet, it has to be absorbing yellow-green light.

General Questions

44. For $[Fe(H_2O)_6]^{2+}$
    a. The coordination number of iron is 6. 6 monodentate ligands are attached
    b. The coordination geometry is octahedral. (Six groups attached to the central metal ion)
    c. The oxidation state of iron is 2+. The charge on the complex is 2+. Water is a neutral ligand.
    d. Fe^{2+} is a d^6 case. For a high spin core there are 4 unpaired electrons.(See question 40a)
    e. The complex would be paramagnetic.

46.

complex	spin	character	unpaired electrons
a. $[Fe(CN)_6]^{4-}$	low	diamagnetic	0
b. $[MnF_6]^{4-}$	high	paramagnetic	5
c. $[Cr(en)_3]^{3+}$	either	paramagnetic	3
d. $[Cu(phen)_3]^{2+}$	either	paramagnetic	1

Electron configuration:

a. Fe^{2+} ______ ______

↑↓ ↑↓ ↑↓

c. Cr^{3+} ______ ______

↑ ↑ ↑

b. Mn^{2+} ↑ ↑

↑ ↑ ↑

d. Cu^{2+} ↑↓ ↑

↑↓ ↑↓ ↑↓

48. The square planar complex $Pt(NH_3)_2(C_2O_4)$:

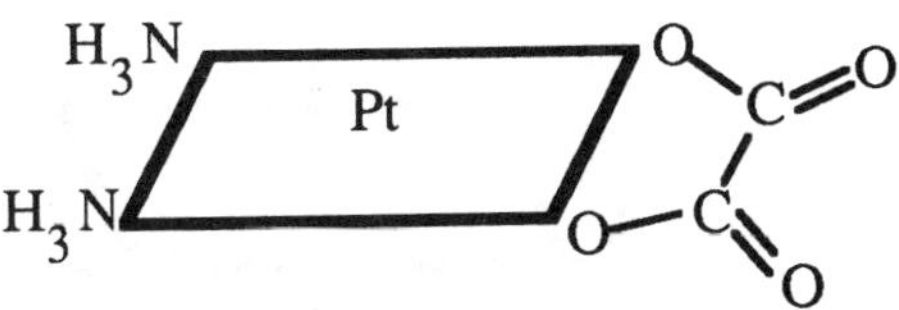

Since the complex is neutral, the ammine ligands are neutral and the oxalate ion is (2-), the charge on platinum is 2+. The compound is named: diammineoxalatoplatinum(II).

50. Formula of a complex containing a Co^{3+} ion, two ethylenediamine molecules, one water molecule, and one chloride ion: $[Co(en)_2(H_2O)Cl]^{2+}$

The ethylenediamine molecules and the water molecule are neutral, so with a (3+) and (1-) charge, the **net charge on the ion is 2+.**

52. Chiral center (In this question, the charges on the ions are omitted for clarity.)

a. [<u>Fe</u>(en)$_3$]$^{2+}$ yes

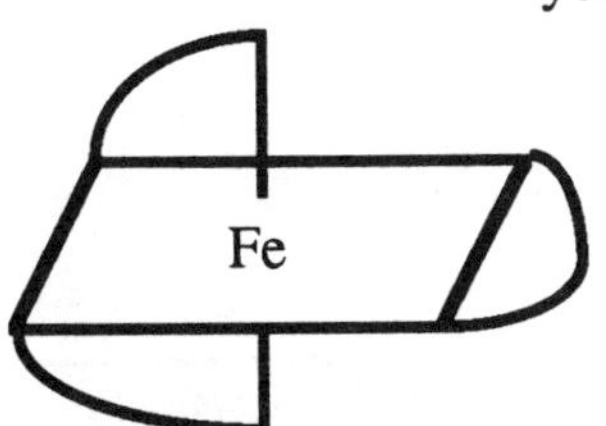

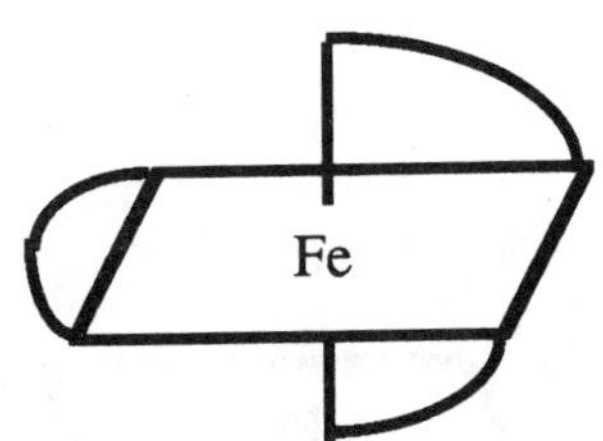

The two mirror images shown above are non superimposable and therefore possess a chiral center. (Another version is shown as Figure 23.26 of your text). The curved lines represent the ethylenediamine ligands.

b. cis-[<u>Co</u>$(en)_2Br_2$]$^+$ yes

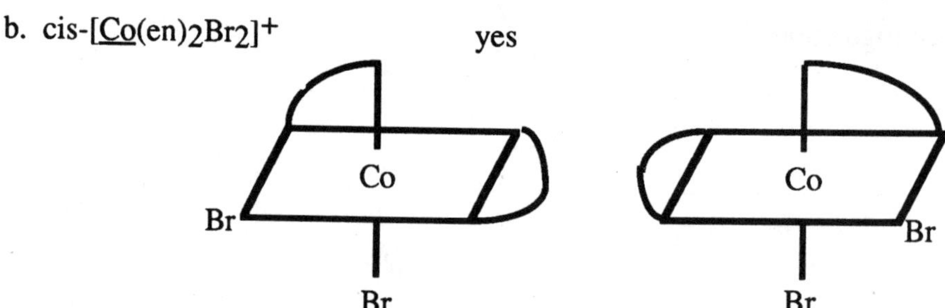

(There is more than one isomer of this ion.) The *trans* isomer doesn't have a chiral center and is shown below.

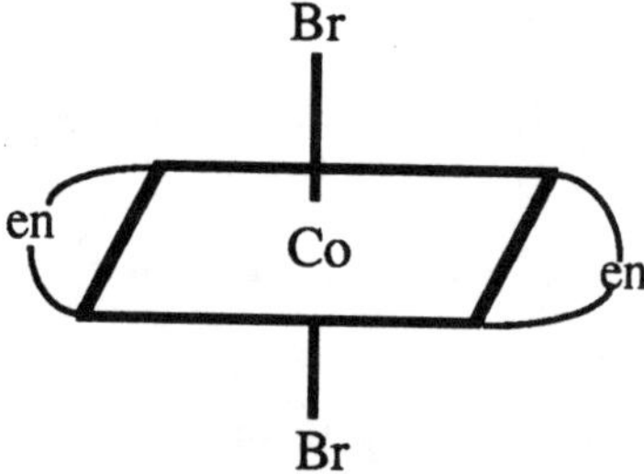

c. fac-[<u>Co</u>$(en)(H_2O)Cl_3$]$^+$ no mer-[<u>Co</u>$(en)(H_2O)Cl_3$]$^+$

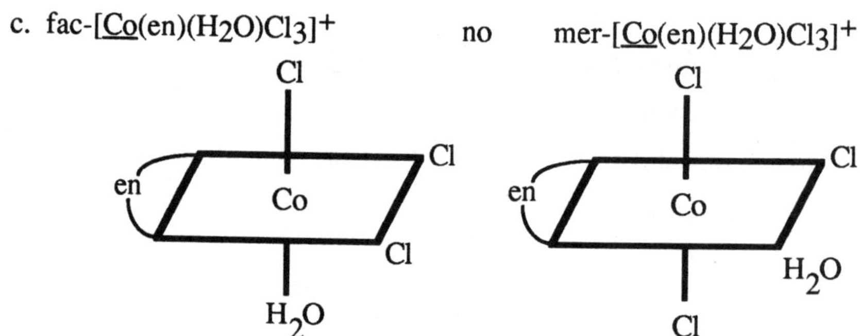

d. $Pt(NH_3)(H_2O)Cl(NO_2)$ no

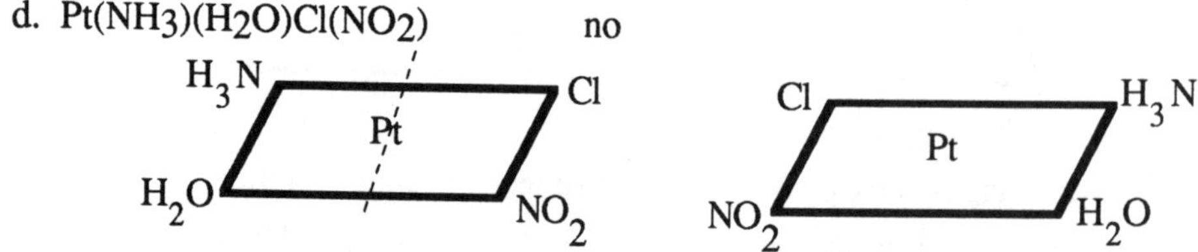

Above are two mirror images of the complex. Rotation of the first complex along the dotted axis by 180° results in the second complex. There are no nonsuperimposable isomers.

54. Ligands arranged in increasing crystal field splitting energy:

Example 23.5 illustrates two cobalt complexes with NH_3 and F^-. Ammonia has a larger Δ_o than does fluoride, thus is a stronger field ligand than fluoride. ($F^- < NH_3$)
In the text immediately preceding the cited example, there are diagrams for the hexaaquo- and hexacyano- complexes of iron(II). The cyano- complex is a low-spin complex while the aquo- complex is high spin. This tells us that cyano is a stronger field ligand than aquo. ($H_2O < CN^-$). Continued analysis of a series of common complexes leads us to the **spectrochemical series.** This series is a large number of ligands for which the crystal field splitting energies have been arranged. This series tells us that:

$$(F^- < H_2O < NH_3 < CN^-).$$

Conceptual Questions

56. Aqueous cobalt(III) sulfate is diamagnetic. H_2O provides a large enough Δ_o to force $[Co(H_2O)_6]^{3+}$ to have no unpaired electrons. An excess of F^- results in the conversion of the hexaaquo complex to the $[CoF_6]^{3-}$ ion. The weaker F^- ligands don't separate the d orbital energies as much as H_2O, resulting in four unpaired electrons—a paramagnetic complex.

58. Describe an experiment to determine if:
   a. the $[Fe(H_2O)_6]Cl_2$ complex is a low spin or high spin complex.

   One can determine the paramagnetism of the compound. Low spin iron(II) has no unpaired electrons while high spin iron(II) would have 4 unpaired electrons. The magnitude of the magnetic moment would discriminate between the two. (See study question 40a)

b. nickel in $K_2[NiCl_4]$ is in a square planar or tetrahedral environment.

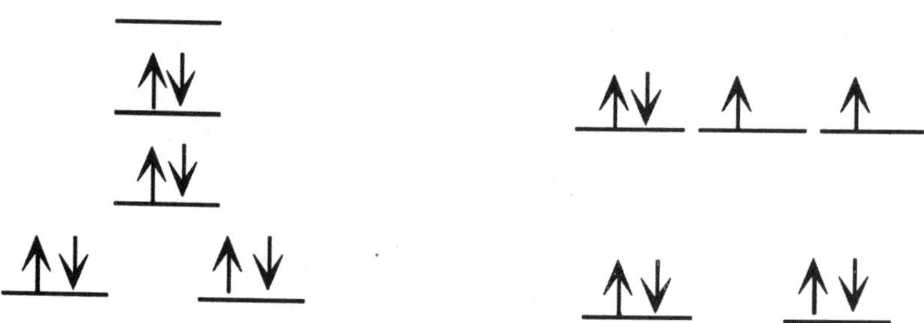

d- orbitals for square-planar complex d- orbitals for tetrahedral complex

Square-planar nickel(II) is diamagnetic while tetrahedral nickel(II) is paramagnetic. Measuring the magnetic moment would discriminate between the two.

60. a. Structures for the fac- and mer- isomers of $Cr(dien)Cl_3$. In these diagrams the curved lines represent the $H_2N-CH_2CH_2-NH-CH_2CH_2-NH_2$ ligand—with attachments to the metal ion through the electron pair on the N atoms.

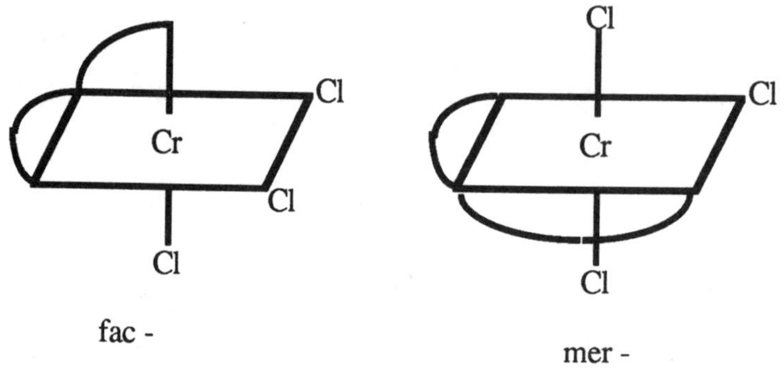

b. Two different isomers of mer-$Cr(dien)Cl_3$:

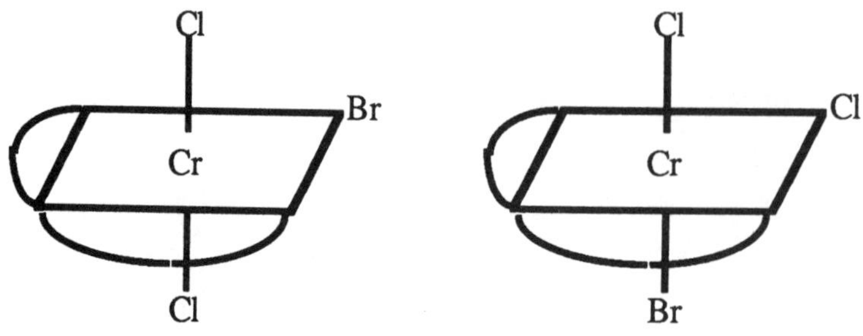

c. The geometric isomers for isomers of $[Cr(dien)_2]^{3+}$

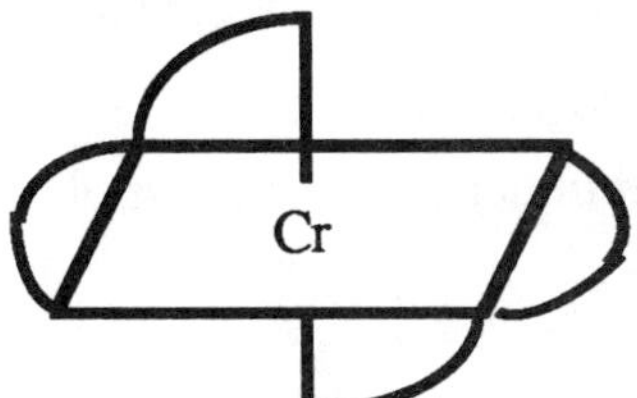

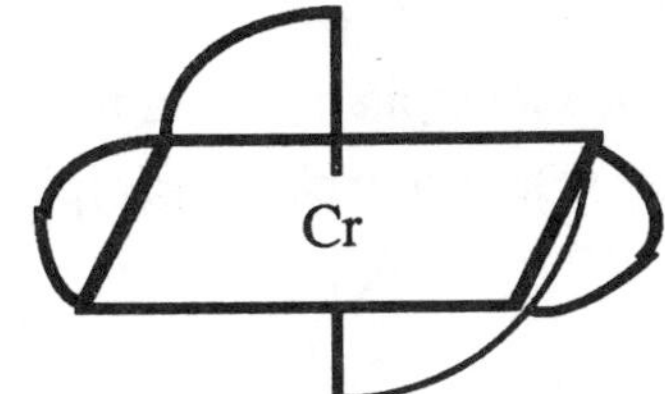

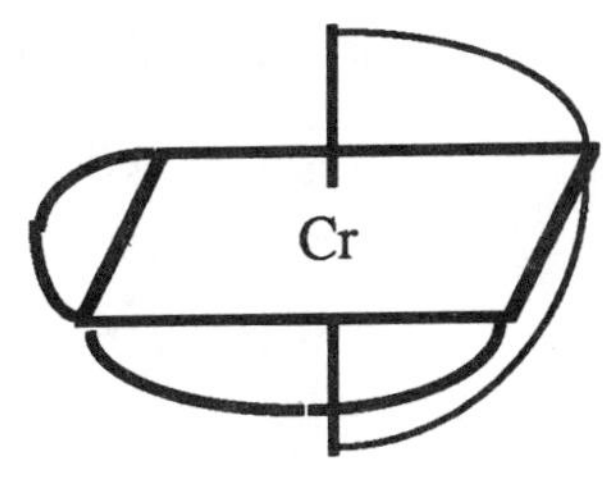

fac- mer -

62. For the tris(carbonato)cobaltate(III) complex:

a. The predicted color:

The spectrum shown in Section 23.6 indicates that the 640 nm wavelength is in the "orange" part of the visible spectrum. Hence this complex should be blue—or more specifically cyan.

b. Carbonate ion placed in the spectrochemical series

Examination of Table 23.4 indicates that an absorption of 640 nm is between fluoride and oxalate (in that table)—that is Δ_o is relatively small. Hence the carbonate ligand would belong a bit higher than fluoride but lower than oxalate.

c. Magnetism of the tris(carbonato)cobaltate(III) complex:

Since Δ_o is small, the complex will behave much like CoF_6^{3-}. With 4 unpaired electrons, the complex is **paramagnetic.**

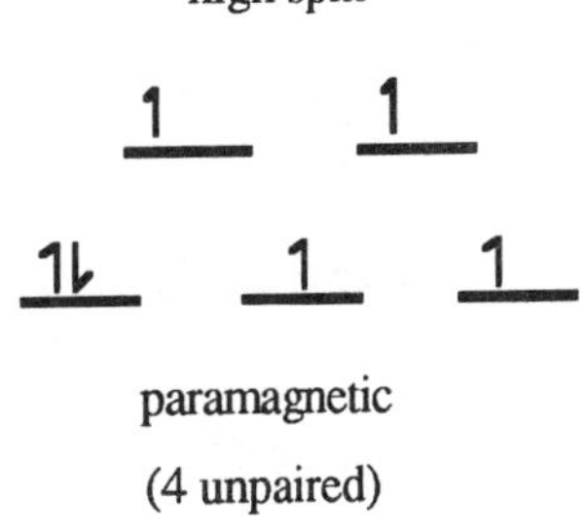

Challenging Questions

64. A + $BaCl_2$ ppt ($BaSO_4$) ⇒ A = $[Co(NH_3)_5Br]SO_4$

B + $AgNO_3$ ppt (AgBr) ⇒ B = $[Co(NH_3)_5SO_4]Br$

The complex has the structure $[Co(NH_3)_5Br]SO_4$ or $[Co(NH_3)_5(SO_4)]Br$. What is true is that the ion in the outer-sphere of the coordination complex (the anion listed outside the []) is separated from the rest of the coordination complex when the complex is dissolved (remember that ionic compounds separate into their component ions.) So one could add say $AgNO_3$ to a solution of the complex ion. Since AgBr is an insoluble salt, while Ag_2SO_4 is a soluble salt, the appearance of a precipitate would indicate that the structure of the complex was: $[Co(NH_3)_5(SO_4)]Br$. The lack of precipitate would indicate that the complex was the sulfate salt.

66. $Pt(NO_2)Cl(NH_3)_2$ diamminechloronitritoplatinum(II)

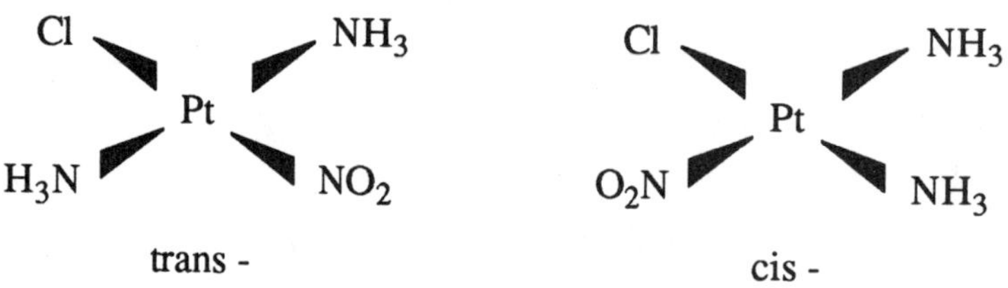

These are geometric isomers.

68. The use of permanganate ion as an oxidizing agent rests on the change that occurs with Mn in acid solution. ($MnO_4^- \rightarrow Mn^{2+}$ or a 5 - electron change).
Calculate the # of moles of permanganate—and the # of mol of electrons:

$$\frac{0.0174 \text{ mol } MnO_4^-}{1 L} \bullet \frac{0.01247 \text{ L}}{1} \bullet \frac{5 \text{ mol } e^-}{1 \text{ mol } MnO_4^-} = 1.08 \times 10^{-3} \text{ mol } e^-$$

Calculate the # of moles of the Uranium ion present:

$$0.213 \text{ g } UO_2(NO_3)_2 \bullet \frac{1 \text{ mol } UO_2(NO_3)_2}{394.0 \text{ g } UO_2(NO_3)_2} = 5.41 \times 10^{-4} \text{ mol } UO_2(NO_3)_2$$ and the stoichiometry indicates that there is **one** U atom per molecule of $UO_2(NO_3)_2$, there are 5.41×10^{-4} mol U. Now we can calculate the # of mol of electron/mol of U.

$$\frac{1.08 \times 10^{-3} \text{ mol } e^-}{5.41 \times 10^{-4} \text{ mol } UO_2(NO_3)_2} = 2 \text{ mol } e^-/\text{ mol U.}$$

Since the **ending** oxidation state of U was U(VI), the reduced form of U **must have been the +4 oxidation state.**

The balanced net ionic equation for the oxidation is then:

$$MnO_4^- + U^{4+} \rightarrow Mn^{2+} + UO_2^{2+}$$

Balancing the equation as shown earlier (in Chapter 21):

$MnO_4^- \rightarrow Mn^{2+}$	$U^{4+} \rightarrow UO_2^{2+}$
$MnO_4^- \rightarrow Mn^{2+} + 4\ H_2O$	$2\ H_2O + U^{4+} \rightarrow UO_2^{2+}$
$8\ H^+ + MnO_4^- \rightarrow Mn^{2+} + 4\ H_2O$	$2\ H_2O + U^{4+} \rightarrow UO_2^{2+} + 4\ H^+$
$8\ H^+ + MnO_4^- + 5\ e^- \rightarrow Mn^{2+} + 4\ H_2O$	$2\ H_2O + U^{4+} \rightarrow UO_2^{2+} + 4\ H^+ + 2e^-$
$16\ H^+ + 2MnO_4^- + 10\ e^- \rightarrow 2Mn^{2+} + 8\ H_2O$	$10\ H_2O + 5U^{4+} \rightarrow 5UO_2^{2+} + 20\ H^+ + 10e^-$

Adding and canceling any duplications:

$$16\ H^+ + 2\ MnO_4^- + 10\ e^- \rightarrow 2\ Mn^{2+} + 8\ H_2O$$

$$10\ H_2O + 5\ U^{4+} \rightarrow 5UO_2^{2+} + 20\ H^+ + 10e^-$$

$$2\ H_2O(\ell) + 5\ U^{4+}(aq) + 2\ MnO_4^-(aq) \rightarrow 5\ UO_2^{2+}(aq) + 4\ H^+(aq) + 2\ Mn^{2+}(aq)$$

70. A 2.1309 g sample of alloy contains Cu and Al. The unbalanced reactions can be written as follows:

1. alloy + HCl + HNO_3 $\rightarrow$ $Cu^{2+}(aq)$ + $Al^{3+}(aq)$
2. $Cu^{2+}(aq) + Al^{3+}(aq) + NH_3(aq) \rightarrow [Cu(NH_3)_4]^{2+} + Al(OH)_3(s)$
3. $Al(OH)_3(s)$ + heat $\rightarrow$ $Al_2O_3(s)$

To identify the percentage composition, note that **all** the aluminum in the Al_2O_3 originated in the alloy. Determine the mass of aluminum present in the 3.8249 g Al_2O_3.

$$\text{mass of Al} = 3.8249\ \text{g}\ Al_2O_3 \bullet \frac{53.963\ \text{g Al}}{101.963\ \text{g}\ Al_2O_3} = 2.0243\ \text{g Al}$$

$$\%\ \text{Al} = \frac{2.0243\ \text{g Al}}{2.1309\ \text{g alloy}} \bullet 100 = 95.000\ \%\ \text{Al}$$

Knowing that the sample contains **only** Cu and Al, the Cu percentage is: 5.000 % Cu. One can verify this percentage by the calculation:

$$\%\ \text{Cu} = \frac{(2.1309 - 2.0243)\ \text{g Cu}}{2.1309\ \text{g alloy}} \bullet 100 = 5.000\ \%\ \text{Cu}$$

Chapter 24: Nuclear Chemistry

Nuclear Reactions

12. Balance the following nuclear equations, supplying the missing particle. [The missing particle is emboldened.]

a. $^{54}_{26}Fe + ^{4}_{2}He \rightarrow 2\,^{1}_{1}H + \mathbf{^{56}_{26}Fe}$

b. $^{27}_{13}Al + ^{4}_{2}He \rightarrow ^{30}_{15}P + \mathbf{^{1}_{0}n}$

c. $^{32}_{16}S + ^{1}_{0}n \rightarrow ^{1}_{1}H + \mathbf{^{32}_{15}P}$

d. $^{96}_{42}Mo + ^{2}_{1}H \rightarrow ^{1}_{0}n + \mathbf{^{97}_{43}Tc}$

e. $^{98}_{42}Mo + ^{1}_{0}n \rightarrow ^{99}_{43}Tc + \mathbf{^{0}_{-1}e}$

14. Balance the following nuclear equations, supplying the missing particle. [The missing particle is emboldened.]

a. $^{111}_{47}Ag \rightarrow ^{111}_{48}Cd + \mathbf{^{0}_{-1}e}$

b. $^{87}_{36}Kr \rightarrow ^{0}_{-1}e + \mathbf{^{87}_{37}Rb}$

c. $^{231}_{91}Pa \rightarrow ^{227}_{89}Ac + \mathbf{^{4}_{2}He}$

d. $^{230}_{90}Th \rightarrow ^{4}_{2}He + \mathbf{^{226}_{88}Ra}$

e. $^{82}_{35}Br \rightarrow ^{82}_{36}Kr + \mathbf{^{0}_{-1}e}$

f. $\mathbf{^{24}_{11}Na} \rightarrow ^{24}_{12}Mg + ^{0}_{-1}e$

16. $^{235}_{92}U \rightarrow {}^{231}_{90}Th \rightarrow {}^{231}_{91}Pa \rightarrow {}^{227}_{89}Ac \rightarrow {}^{227}_{90}Th \rightarrow {}^{223}_{88}Ra \rightarrow {}^{219}_{86}Rn \rightarrow {}^{215}_{84}Po$

$+ {}^{4}_{2}He \quad + {}^{0}_{-1}e \quad + {}^{4}_{2}He \quad + {}^{0}_{-1}e \quad + {}^{4}_{2}He \quad + {}^{4}_{2}He \quad + {}^{4}_{2}He$

and continuing (from Po-215)we have:

$^{215}_{84}Po \rightarrow {}^{211}_{82}Pb \rightarrow {}^{211}_{83}Bi \rightarrow {}^{211}_{84}Po \rightarrow {}^{207}_{82}Pb$

$+ {}^{4}_{2}He \quad + {}^{0}_{-1}e \quad + {}^{0}_{-1}e \quad + {}^{4}_{2}He$

18. The particle emitted in the following reactions:

[The missing particle is emboldened.]

a. $^{198}_{79}Au \rightarrow {}^{198}_{80}Hg + \mathbf{{}^{0}_{-1}e}$

b. $^{222}_{86}Rn \rightarrow {}^{218}_{84}Po + \mathbf{{}^{4}_{2}He}$

c. $^{137}_{55}Cs \rightarrow {}^{137}_{56}Ba + \mathbf{{}^{0}_{-1}e}$

d. $^{110}_{49}In \rightarrow {}^{110}_{48}Cd + \mathbf{{}^{0}_{1}n}$

20. The change in mass (Δ m) for ^{10}B is:

$$\Delta m = 10.01294 - [5(1.00783) + 5(1.00867)]$$
$$= 10.01294 - 10.0825 = -0.06956 \text{ g/mol}$$

while that for ^{11}B is:

$$\Delta m = 11.00931 - [5(1.00783) + 6(1.00867)]$$
$$= 11.00931 - 11.09117 = -0.08186 \text{ g/mol}$$

The binding energies for the two isotopes are:

$$^{10}B: \ \Delta E = (-6.956 \times 10^{-5} \text{ kg/mol})(3.00 \times 10^{8} \text{ m/s})^2 \left(\frac{1 \text{ J}}{1 \text{ kg} \cdot \text{m}^2 \cdot \text{s}^{-2}}\right)$$
$$= -6.260 \times 10^{12} \text{ J/mol}$$

^{11}B : $\Delta E = (-8.186 \times 10^{-5}\ kg/mol)(3.00 \times 10^{8}\ m/s)^2 \left(\frac{1\ J}{1\ kg \cdot m^2 \cdot s^{-2}}\right)$

$= -7.367 \times 10^{12}\ J/mol$

The **binding energy per nucleon** is:

For ^{10}B : $\frac{6.260 \times 10^{9}\ kJ}{10\ mol\ nucleons} = 6.26 \times 10^{8} \frac{kJ}{mol\ nucleon}$ and

For ^{11}B : $\frac{7.367 \times 10^{9}\ kJ}{11\ mol\ nucleon} = 6.70 \times 10^{8} \frac{kJ}{mol\ nucleon}$

22. The binding energy per nucleon for calcium-40:
The change in mass (Δ m) for ^{40}Ca is:

$$\Delta m = 39.96259 - [20(1.00783) + 20(1.00867)]$$
$$= 39.96259 - 40.3300 = -0.3674\ g/mol$$

The energy change is then:

$$\Delta E = (-3.674 \times 10^{-4}\ kg/mol)(3.00 \times 10^{8}\ m/s)^2 \left(\frac{1\ J}{1\ kg \cdot m^2 \cdot s^{-2}}\right)$$
$$= -3.307 \times 10^{13}\ J/mol \text{ or } -3.307 \times 10^{10}\ kJ/mol$$

This energy can be converted into the **binding energy per mole of nucleons:**

$$\frac{3.307 \times 10^{10}\ kJ/mol}{40\ mol\ nucleons} = 8.267 \times 10^{8} \frac{kJ}{mol\ of\ nucleons}$$

[Comparing this value to those of Figure 24.5, we see that they compare quite well!!]

24. Binding energy per nucleon for O-15, O-16, and O-17
The change in mass (Δ m) for ^{15}O is:

$$\Delta m = 15.003065 - [8(1.00783) + 7(1.00867)]$$
$$= -0.12027\ g/mol$$

The energy change is then:

$$\Delta E = (-1.2027 \times 10^{-4}\ kg/mol)(3.00 \times 10^{8}\ m/s)^2 \left(\frac{1\ J}{1\ kg \cdot m^2 \cdot s^{-2}}\right)$$
$$= -1.082 \times 10^{13}\ J/mol \text{ or } -1.082 \times 10^{10}\ kJ/mol$$

The energy can also be expressed as the **binding energy per nucleon**, by dividing the energy just calculated by the number of nucleons (15 for this isotope):

$$\frac{1.082 \times 10^{10} \text{ kJ/mol}}{15 \text{ mol nucleons}} = 7.216 \times 10^{8} \frac{\text{kJ}}{\text{mol of nucleons}}$$

Using the same series of calculations for O-16 and O-17 we obtain:

Isotope	# neutrons	Δ m(g/mol)	Δ E (kJ/mol)	BE per nucleon:
O-15	7	-0.12027	-1.082×10^{10}	7.216×10^{8}
O-16	8	-0.13708	-1.234×10^{10}	7.711×10^{8}
O-17	9	-0.14154	-1.274×10^{10}	7.493×10^{8}

Examining the Binding Energies per nucleon (right column) we can see that O-16 (with magic numbers for both protons and neutrons) has the largest BE of the three isotopes—an observation that is consistent with our expectations!

Rates of Disintegration Reactions:

26. For ^{64}Cu, $t_{1/2}$ =12.8 hr

The fraction remaining as ^{64}Cu following n half-lives is equal to $(\frac{1}{2})^n$. Note that 64 hrs corresponds to exactly **five** half-lives.

The <u>fraction</u> remaining as ^{64}Cu is $(\frac{1}{2})^5$ or $\frac{1}{32}$ or 0.03125.

The mass remaining is:

(0.03125)(15.0 mg) = 0.469 mg.

28. a. The equation for β–decay of ^{131}I is:

$$^{131}_{53}\text{I} \rightarrow {}^{0}_{-1}\text{e} + {}^{131}_{54}\text{Xe}$$

b. The amount of ^{131}I remaining after 32.2 days:

For ^{131}I, $t_{1/2}$ is 8.05 days--so 32.2 days is exactly **four** half-lives:
The fraction of ^{131}I remaining is $(\frac{1}{2})^4$ or $\frac{1}{16}$ or 0.0625.

The amount of the original 25.0 mg remaining will be :

(0.0625)(25.0 mg) = 1.56 mg

30. To determine the mass of Gallium-67 left after 13 days:, determine the rate constant for the decay:

$$k = \frac{0.693}{t_{1/2}} = \frac{0.693}{78.25\ \text{hrs}} = 8.856 \times 10^{-3}\ \text{hrs}^{-1}$$

and $\ln(x) = -k \cdot t = -\ 8.856 \times 10^{-3}\ \text{hrs}^{-1} \cdot \frac{13\ \text{days}}{1} \cdot \frac{24\ \text{hrs}}{1\ \text{day}} = -\ 2.763.$

where x represents the fraction of Gallium-67 remaining. Solving for x gives:

$x = 0.06309$ (or 63.09%)

The amount of Gallium-67 remaining is then (0.06309)(0.15 mg) = 9.5×10^{-3} mg

32. For the decomposition of Radon-222:

a. The balanced equation for the decomposition of Rn-222 with α particle emission.

$${}^{222}_{86}\text{Rn} \rightarrow {}^{4}_{2}\text{He} + {}^{218}_{84}\text{Po}$$

b. Time required for the sample to decrease to 10.0 % of its original activity:

Since this decay follows 1st order kinetics, we can calculate a rate constant:

$$k = \frac{0.693}{t_{1/2}} = \frac{0.693}{3.82\ \text{days}} = 0.181\ \text{days}^{-1}$$

With this rate constant , using the 1st order integrated rate equation, we can calculate the time required:

$$\ln\left(\frac{10.0}{100}\right) = -\ (0.181\ \text{days}^{-1}) \cdot t$$

$$\frac{-\ 2.302}{-\ 0.181\ \text{days}^{-1}} = t = 12.7\ \text{days}$$

34. The age of the fragment can be determined if:

(a) we calculate a rate constant and,

(b) we use the 1st order integrated rate equation (much as we did in question 32b above)

(a) $k = \frac{0.693}{t_{1/2}} = \frac{0.693}{5730\ \text{yr}} = 1.21 \times 10^{-4}\ \text{yr}^{-1}$

(b) Now we can calculate the time required for the Carbon-14 : Carbon-12 to decay to 72 % of that ratio in living organisms.

$$\ln\left(\frac{72}{100}\right) = -\ (1.21 \times 10^{-4}\ \text{yr}^{-1}) \cdot t$$

$$\frac{-0.3285}{-1.21 \times 10^{-4}\ yr^{-1}} = t = 2700 \text{ years} \qquad (2 \text{ sf})$$

36. For the decay of Cobalt-60, $t_{1/2}$ is 5.27 yrs:

a. Time for Co-60 to decrease to 1/8 of its original activity:

Following the methodology of questions 32 and 34 to determine the rate constant:

$$k = \frac{0.693}{t_{1/2}} = \frac{0.693}{5.27\ yr} = 0.131\ yr^{-1}$$

Substituting into the equation:

$\ln(\frac{1}{8}) = -\ 0.131 yr^{-1} \bullet t$ and

$\ln(0.125) = -0.131\ yr^{-1} \bullet t$ and solving for t =15.8 yrs

A " short-cut" is available here if you notice that 1/8 corresponds to $(\frac{1}{2})^3$. Said another way, one-eighth of the Co-60 will remain after **three half-lives** have passed, so $3 \bullet 5.27$ yrs = 15.8 years !!

b. Fraction of Co-60 remaining as Co-60 after 1.0 years:

Now we can solve for the fraction on the "left-hand side" of the rate equation:

ln (fraction remaining) = $-\ k \bullet t$

ln (fraction remaining) = $-\ 0.131\ yr^{-1} \bullet 1.0$ yr

ln (fraction remaining) = -0.131 and $e^{-0.131}$ = fraction remaining

fraction remaining = 0.877 , so $\frac{87.7}{100}$ (or 87.7 %) remains after 1.0 years.

38. If the ratio of $\frac{\text{Pb-206}}{\text{U-238}}$ is 0.33, what we know is that 1/4 of the uranium has undergone decay. This would imply 3/4 of the uranium remains as U-238 [The ratio of Pb/U would be 0.25/0.75 or 1/3. We can then calculate a rate constant and use the 1st order rate equation to calculate the time required for this to occur:

$$k = \frac{0.693}{4.5 \times 10^9\ yr} = 1.54 \times 10^{-10}\ yr^{-1}$$

and $\ln(\frac{75}{100}) = -\ 1.54 \times 10^{-10}\ yr^{-1} \bullet t$

$-0.2877 = -\ 1.54 \times 10^{-10}\ yr^{-1} \bullet t$ and $t = \frac{-0.2877}{-\ 1.54 \times 10^{-10}\ yr^{-1}} = 1.9 \times 10^9$ yr

40. a. The equation for the neutron bombardment of Sodium-23:

$${}^{23}_{11}Na + {}^{1}_{0}n \rightarrow {}^{24}_{11}Na$$

and the decay of Sodium-24 by β emission :

$$^{24}_{11}Na \rightarrow \ ^{0}_{-1}e + \ ^{24}_{12}Mg$$

b. From the data provided, calculate the half-life:

Since the activity (measured in dpm) is proportional to the number of atoms:

we can write $\frac{A}{A_0} = \frac{N}{N_0}$ and using Equation 24.3 we can write:

$$\ln\left(\frac{A}{A_0}\right) = \ln\left(\frac{N}{N_0}\right) = -k \bullet t$$

So we can substitute the Activities at various times to solve for k:

Let's pick the first and last data points:

$$\ln\left(\frac{1.01 \times 10^4 \text{ dpm}}{2.54 \times 10^4 \text{ dpm}}\right) = -k \bullet (20 \text{ hours})$$

$$\frac{\ln(0.398)}{20 \text{ hours}} = -0.0461 \quad = -k$$

So $k = 0.0461 \text{ hr}^{-1}$

Now we can calculate the half-life, since $t_{1/2} = \frac{0.693}{k}$

$$t_{1/2} = \frac{0.693}{0.0461 \text{ hr}^{-1}} = 15.1 \text{ hrs}$$

Additional info:

Using the activity at 10 hours (1.60×10^4 dpm), compared to the original activity (2.54×10^4 dpm), you'll get a $k = 0.0462 \text{ hr}^{-1}$, and from that a $t_{1/2} = 15.0$ hrs.

Nuclear Transmutations

42. The formation of ^{241}Am from ^{239}Pu may be written:

$$^{239}_{94}Pu + 2\ ^{1}_{0}n \rightarrow \ ^{241}_{95}Am + \ ^{0}_{-1}e$$

44. A proposed method for producing ^{246}Cf:

$$^{238}_{92}U + \ ^{12}_{6}C \rightarrow \ ^{246}_{98}Cf + 4\ ^{1}_{0}n$$

46. Complete the following equations using deuterium bombardment:

[The missing particle is emboldened.]

a. ${}^{114}_{48}\text{Cd} + {}^{2}_{1}\text{D} \rightarrow \mathbf{{}^{115}_{48}Cd} + {}^{1}_{1}\text{H}$

c. ${}^{40}_{20}\text{Ca} + {}^{2}_{1}\text{D} \rightarrow {}^{38}_{19}\text{K} + \mathbf{{}^{4}_{2}He}$

b. ${}^{6}_{3}\text{Cd} + {}^{2}_{1}\text{D} \rightarrow \mathbf{{}^{7}_{4}Be} + {}^{1}_{0}\text{n}$

d. $\mathbf{{}^{63}_{29}Cu} + {}^{2}_{1}\text{D} \rightarrow {}^{65}_{30}\text{Zn} + \gamma$

48. The equation for the bombardment of Boron-10 with a neutron, and the subsequent release of an alpha particle:

$${}^{10}_{5}\text{B} + {}^{1}_{0}\text{n} \rightarrow {}^{7}_{3}\text{Li} + {}^{4}_{2}\text{He}$$

General Questions

50. Balance the following nuclear reactions, supplying the missing particle.

[The missing particle is emboldened.]

a. ${}^{13}_{6}\text{C} + \mathbf{{}^{1}_{0}n} \rightarrow {}^{14}_{6}\text{C}$

b. ${}^{40}_{18}\text{Ar} + \mathbf{{}^{4}_{2}He} \rightarrow {}^{43}_{19}\text{K} + {}^{1}_{1}\text{H}$

c. ${}^{250}_{98}\text{Cf} + {}^{11}_{5}\text{B} \rightarrow 4\ {}^{1}_{0}\text{n} + \mathbf{{}^{257}_{103}Lr}$

d. ${}^{53}_{24}\text{Cr} + {}^{4}_{2}\text{He} \rightarrow \mathbf{{}^{1}_{0}n} + {}^{56}_{26}\text{Fe}$

e. ${}^{212}_{84}\text{Po} \rightarrow {}^{208}_{82}\text{Pb} + \mathbf{{}^{4}_{2}He}$

f. ${}^{122}_{53}\text{I} \rightarrow \mathbf{{}^{122}_{52}Te} + {}^{0}_{1}\text{e}$

g. $\mathbf{{}^{23}_{10}Ne} \rightarrow {}^{23}_{11}\text{Na} + {}^{0}_{-1}\text{e}$

h. ${}^{137}_{53}\text{I} \rightarrow {}^{1}_{0}\text{n} + \mathbf{{}^{136}_{53}I}$

52. The fossil cells age can be calculated using the first order kinetics expressions:

Since $t_{1/2} = 4.9 \times 10^{10}$ yr, $k = \dfrac{0.693}{4.9 \times 10^{10}\ \text{yr}}$ and $\ln\left(\dfrac{C_t}{C_o}\right) = -kt$

then $\ln(0.951) = -\dfrac{0.693}{4.9 \times 10^{10}\ \text{yr}} \cdot t$ and

$\dfrac{4.9 \times 10^{10}\ \text{yr}}{0.693} \cdot \ln(0.951) = t$ and $3.6 \times 10^{9}\ \text{yr} = t$

The fossil cells are about 3.6 billion years old.

54. The time when natural uranium contained 3.0% ^{235}U is:

Since $t_{1/2} = 7.04 \times 10^8$ yr , $k = \dfrac{0.693}{7.04 \times 10^8 \text{ yr}}$ and $\ln\left(\dfrac{C_t}{C_o}\right) = -kt$

then $\ln\left(\dfrac{0.72}{3.0}\right) = -\dfrac{0.693}{7.04 \times 10^8 \text{ yr}} \cdot t$ and

$\dfrac{7.04 \times 10^8 \text{ yr}}{0.693} \cdot \ln\left(\dfrac{0.72}{3.0}\right) = t$ and $t = 1.5 \times 10^9$ years

56. The energy liberated by one pound of ^{235}U:

$$\frac{2.1 \times 10^{10} \text{ kJ}}{1 \text{ mol } ^{235}U} \cdot \frac{1 \text{ mol } ^{235}U}{235 \text{ g U}} \cdot \frac{454 \text{ g U}}{1 \text{ pound U}} = \frac{4.1 \times 10^{10} \text{ kJ}}{1 \text{ pound U}}$$

comparing this amount of energy to the energy per ton of coal yields:

$$\frac{4.1 \times 10^{10} \text{ kJ}}{1 \text{ pound U}} \cdot \frac{1 \text{ ton coal}}{2.6 \times 10^7 \text{ kJ}} = \frac{1600 \text{ tons coal}}{1 \text{ pound U}}$$

58. The equation for the collison may be written: ${}^{0}_{1}e$ (a positron) + ${}^{0}_{-1}e$ (a beta particle) $\rightarrow 2\gamma$

The masses of the positron and electron are equal, and the accepted mass of the electron is 9.109389×10^{-28} g.

Hence we are converting $(2 \cdot 9.109 \times 10^{-28})$ g into energy!

a. We can calculate the Energy with Einstein's famous equation (equation 24.1):

$$\Delta E = (\Delta m)c^2$$

$$\Delta E = (2 \cdot 9.109 \times 10^{-28} \text{ g}) \cdot \frac{1 \text{ kg}}{1000 \text{ g}} \cdot (3.00 \times 10^8 \tfrac{\text{m}}{\text{s}})^2$$

$$\Delta E = 1.64 \times 10^{-13} \text{ J} \qquad [\text{Recall that } 1 \frac{\text{kg} \cdot \text{m}^2}{\text{s}^2} = 1 \text{ J}]$$

b. The frequency of the two γ rays emitted would be:

$E = h\nu$ (Planck's equation) where $h = 6.626 \times 10^{-34}$ J • s

Recall that 2 photons were emitted, hence the energy of **each** is half the energy calculated above:

$$\nu = \frac{E}{h} = \frac{8.20 \times 10^{-14} \text{ J}}{6.626 \times 10^{-34} \text{ J} \cdot \text{s}} = 1.237 \times 10^{20} \text{ s}^{-1} \text{ or } 1.237 \times 10^{20} \text{ Hz}$$

60. This "dilution" problem may be solved using the equation which is useful for solutions:

$$M_c \times V_c = M_d \times V_d$$

where c and d represent the concentrated and diluted states, respectively. We'll use the number of disintegrations/second as our "molarity."

$$2.0 \times 10^6 \text{ dps} \bullet 1.0 \text{ mL} = 1.5 \times 10^4 \text{ dps} \bullet V_d$$

$$130 \text{ mL} = V_d$$

The approximate volume of the circulatory system is 130 mL.

Conceptual Questions

62. For the radioactive decay series of U-238:

a. Why can the masses be expressed as $m = 4n + 2$?

These masses correlate in this fashion since the **principle** mode of decay in the U-238 series is **alpha particle** emission. The mass of an alpha particle is **4** (since it is basically 2 neutrons + 2 protons). Hence the loss of an alpha particle (say from U-238 to Th-234) results in a **mass loss of 4 units.**The series results in a total loss of 8 alpha particles and 6 beta particle. Since the beta particles (electrons) have an insignificant mass (compared to a neutron or a proton), the loss of a beta particle--or for that matter 6 of them--does not significantly affect the masses of the daughter products. These masses correspond to n values :n=(51 -> 59).

b. Equations corresponding to the decay series for U-235 and Th-232 :

The U-235 series corresponds to the equation **m = 4n-1**, with n values: (52 -> 59). The Th-232 series corresponds to the equation **m = 4n-4**

The isotopic masses for these 3 series are summarized in the table below:

n	= 4n+2	= 4n-1	= 4n-4
51	206		
52	210	207	
53	214	211	208
54	218	215	212
55	222	219	216
56	226	223	220
57	230	227	224
58	234	231	228
59	238	235	232

From an empirical standpoint, the masses of the most massive isotopes in the three series differ by 3 (U-238 → U–235) and (U–235 → Th-232). So given that the algorithm for the U-238 series is 4n+2, subtracting 3 gives 4n-1, while subtracting 3 more gives 4n-4. Note that these two series could also be represented by (4n+3) and (4n) respectively—reflecting the difference of a mass of 4.

c. Identify the series to which each of the following isotopes belong:

Isotope	series
226-Ra:	U-238 (4n + 2)
215-At:	U-235 (4n-1) or (4n +3)
228-Th:	Th-232(4n-4) or (4n)
210- Bi:	U-238 (4n + 2)

d. Why is the series "4n+1" missing in the earth's crust?

To occur in the earth's crust, an element must be very stable—that is have a very long half-life or be non-radioactive! From hydrogen to lawrencium, with the exception of two isotopes of hydrogen (protium and tritium), every isotope of every element has a nucleus containing at least one neutron for every proton. With the mass of the neutron and proton being 1, the change in mass number would have to change by a factor of 2—so 4n+1 would not lead to a long-lived isotope and would not be found in the earth's crust.

64. The isotope ^{231}Pa belongs to the ^{235}U series—as seen in Question 62. The reactions producing this isotope from U-235 are (1) α-particle emission followed by (2) β- particle emission. These decays may be written:

$$^{235}_{92}U \rightarrow {}^{231}_{90}Th + {}^{4}_{2}He$$

$$^{231}_{90}Th \rightarrow {}^{231}_{91}Pa + {}^{0}_{-1}e$$